AF777495

Springer Aerospace Technology

More information about this series at http://www.springer.com/series/8613

Leonardo Mazzini

Flexible Spacecraft Dynamics, Control and Guidance

Technologies by Giovanni Campolo

Springer

Leonardo Mazzini
Thales Alenia Space Italia
Rome
Italy

ISSN 1869-1730 ISSN 1869-1749 (electronic)
Springer Aerospace Technology
ISBN 978-3-319-25538-5 ISBN 978-3-319-25540-8 (eBook)
DOI 10.1007/978-3-319-25540-8

Library of Congress Control Number: 2015952038

Springer Cham Heidelberg New York Dordrecht London
© Springer International Publishing Switzerland 2016
This work is subject to copyright. All rights are reserved by the Publisher, whether the whole or part of the material is concerned, specifically the rights of translation, reprinting, reuse of illustrations, recitation, broadcasting, reproduction on microfilms or in any other physical way, and transmission or information storage and retrieval, electronic adaptation, computer software, or by similar or dissimilar methodology now known or hereafter developed.
The use of general descriptive names, registered names, trademarks, service marks, etc. in this publication does not imply, even in the absence of a specific statement, that such names are exempt from the relevant protective laws and regulations and therefore free for general use.
The publisher, the authors and the editors are safe to assume that the advice and information in this book are believed to be true and accurate at the date of publication. Neither the publisher nor the authors or the editors give a warranty, express or implied, with respect to the material contained herein or for any errors or omissions that may have been made.

Printed on acid-free paper

Springer International Publishing AG Switzerland is part of Springer Science+Business Media
(www.springer.com)

I dedicate this work to my wife Michela, who with love and participation, has always supported me in my studies and my professional life.

Preface

Flexible Spacecraft Dynamics, Control and Guidance is a text conceived as a guide to the design of attitude and orbit control systems (AOCS) for modern space vehicles.

In addition to the various classic topics which we have reviewed from a modern perspective, this text deals with new issues of AOCS such as spacecraft flexibility, agility and guidance.

Attitude and orbit control (AOC) is a discipline where engineers can study the equations governing attitude, orbital, and deformation dynamics of a satellite, methods to measure the status of the satellite and to determine the control actions to be implemented on board the satellite in order to get the desired dynamics, and technologies that allow the production of these control actions from measurements.

The author's intention is to provide a manual that allows a satellite designer or a student to learn a theoretical and practical engineering approach to the design of attitude and orbit control systems, providing a panorama of applications of practical and theoretical interest.

The experience of the author in designing AOC systems was useful in defining the road map of the text, introducing topics that are important for current and future applications; these systems include various satellites for low orbit (Radarsat 2, Cosmo, Sentinel 1, and Cosmo Second Generation) and for geostationary orbit (Artemis, Sicral 1, Sicral 1-B, Sicral 2, and Atlantic Bird), participation to the design of new low orbit constellations (Globalstar 1 and 2) and study phases for future satellites generations (Neosat and Galileo Second Generation).

The AOC topic, due to its utility for industrial applications, has been the subject of many books. The reader will find an elementary bibliography in Chap. 1.

Most of these texts are focused on rigid body and spinning body dynamics: Kaplan [1], Thomson [2], and Hughes [3], while few others introduce examples dealing with the flexible spacecraft as Rimrott [4] and Sidi [5]. Sidi [5] provides also a good presentation of synthesis of AOC mode design in frequency domain.

Only Kane et al. [6] and Likins et al. [7] deal with the issue of a complex satellite with appendages and flexibilities, but these texts are mainly focused on providing a global approach to the dynamics, without discussing the applications to attitude control.

Among the available literature, Wertz [8] (1978) is surely the more complete and was reprinted several times up to 2002. It provides a wide treatment of orbital and attitude control including space environment interactions, attitude determination, control system architectures, frequency design techniques, and a review of the sensor and actuators technologies, but this text is now four decades old.

We have to say that today satellites are rarely based on spin stabilization. They are mostly three axes stabilized, which is why we have not treated the spinning body among the applications, preferring to highlight new topics such as the satellite's flexibility effect due to large appendages, the ability of fast repointings, the so-called agility and the optimal guidance, which, for various reasons, look more interesting for modern applications.

Today, the design techniques have evolved into more complex methods, like, synthesis of multi-input multi-output (MIMO) systems with use of optimal robust control theory. This is why, even if we have provided a description of the satellite AOCS's operating modes in terms of SISO (single-input single-output) models, discussing them in terms of frequency methods, in order to have an immediate and simple understanding, we have treated the control synthesis subject focusing on optimal control theory and MIMO systems.

In more recent years, many textbooks have reviewed the classic table of content of the old textbooks, introducing new subjects.

We mention a few of them: Tewari [9] who dealt with satellite attitude control and orbit control and guidance, starting from the perspective of the optimal control theory, Noton [10] who dealt with optimal orbital guidance with low thrust, and Bryson [11] who dealt with some classic AOC problems using the optimal linear control theory.

A recent book by Landis and Crassidis [12] presents the AOC topic focusing on the many new algorithms for attitude determination based on modern sensor technology and optimal estimation theory.

My primary motivation to write a new book is the fact that the classic texts are quite old today, AOC architectures have considerably changed, and even though rigid body motion and stability of spinning bodies are still very important, some new issues of dynamics and control need to be treated in a form suitable for immediate application.

For instance we propose in this text the following subjects: AOC modes design methods, flexibility theory, optimal control theory, optimal navigation and guidance, a modern description of sensors and actuators, attitude agility, and plasmic low thrust propulsion; all of these are important new topics for AOCS.

I have tried to be as general as possible on the theory that these notions can be used not only to convey an understanding of the problems but also for practical use.

My approach was to present the study of attitude control along with that of orbital control. This type of presentation can be useful to AOC designers as satellites are becoming more and more autonomous and implement orbital control tasks directly on board.

Two important chapters (Chap. 4 on attitude control and Chap. 9 on sensor and actuator technologies) were written by my colleague and friend G. Campolo. Giovanni has provided an invaluable specific contribution to make the text more usable for real engineering practice.

Ostuni, Porta Nova
August 2015

Leonardo Mazzini

References

1. M.H. Kaplan, *Modern Spacecraft Dynamics and Control* (Wiley, New York, 1976)
2. W.T. Thomson, *Introduction to Space Dynamics* (Dover Publications, New York, 1986)
3. P. Hughes, *Spacecraft Attitude Dynamics* (Dover Publications, New York, 2004)
4. F.P.J. Rimrott, *Introductory Attitude Dynamics* (Springer, New York, 1989)
5. M.J. Sidi, *Spacecraft Dynamics and Control* (Cambridge University Press, New York, 1997)
6. T.R. Kane, P.W. Likins, D.A. Levinson, *Spacecraft Dynamics* (McGraw-Hill, New York, 1983)
7. P.W. Likins, R.E. Roberson, J. Wittenburg, *Dynamics of Flexible Spacecraft, Department of General Mechanics* (Springer, Dubrovnik, 1972)
8. J.R. Wertz, *Spacecraft Attitude Determination and Control* (D. Reidel Publishing Company, Dordrecht, 2002)
9. A. Tewari, *Advanced Control of Aircraft, Spacecraft and Rockets* (Wiley, New York, 2011)
10. M. Noton, *Spacecraft Navigation and Guidance* (Springer, London, 1998)
11. A.E. Bryson, *Control of Spacecraft and Aircraft* (Princeton University Press, London, 1994)
12. F. L. Markley, J.L. Crassidis, *Fundamentals of Spacecraft Attitude Determination and Control* (Springer, New York, 2015)

Acknowledgments

I was introduced to spacecraft dynamics by my mentor Prof. Paolo Santini and his work on space structure dynamics. I am indebted to him for his spirit of investigation, his curiosity, and all those characteristics that allow one to develop a good pathological passion for research.

The theoretical part of this book is derived from the lessons I devised for the master in Space Science and Technology at the University of Roma Tor Vergata, invited by Prof.sa Alessandra Celletti. I thank Alessandra for her competence and the energy spent in supporting the growth of space culture in our country and for the important support she has given to this book.

Giovanni and I have worked on AOCS in Alenia Spazio, now Thales Alenia Space, and we are indebted to many collaborators and managers of this company for their support and trust. I personally thank them all and in particular Francesca Perrella for her interesting contributions in reviewing and discussing many applications and the orbital mechanics.

Contents

Acronyms

AKE	Attitude Knowledge Error
AOC	Attitude and Orbit Control
AOCS	Attitude and Orbit Control System
APE	Attitude Performance Error
APS	Active Pixel Sensor
ASIC	Application Specific Integrated Circuit
BRF	Body Reference Frame
CCD	Charged-Coupled Device
CMG	Control Moment Gyro
CMOS	Complementary Metal-Oxide Semiconductors
COD	Coarse Orbit Determination
CoM	Center of Mass
CoM-BRF	BRF with origin in CoM
COTS	Commercial Off The Shelf
CP	Center of Pressure
ECEF	Earth Centered Earth Fixed
ECI	Earth Centered Inertial frame
EKF	Extended Kalman Filter
FDIR	Fault Detection Isolation and Recovery
FOG	Fiber Optic Gyro
FOV	Field of View
FPGA	Field Programmable Gate Array
FSS	Fine Sun Sensor
GD	Gregorian Date
GEO	Geosynchronous Orbit
GIT	Gridded Ion Thrusters
GNSS	Global Navigation System
GPS	Global Positioning System
GTO	Geosynchronous Transfer Orbit
H/W	Hardware in the sense of Satellite Equipments
HET	Hall Effect Thrusters

IGRF	International Geomagnetic Reference Field
IGSO	Inclined Geosynchronous Orbit
JD	Julian Date
l.h.s.	left-hand side
LAE	Liquid Apogee Engine
LEO	Low Earth Orbit
LEOP	Launch and Early Orbit Phase
LQR	Linear Quadratic Regulator
LTI	Linear Time Invariant
LV	Launch Vehicle
MEO	Medium altitude Earth Orbit
MGM	Magnetometer
MGT	Magneto-torquer
MIMO	Multi-Input Multi-Output systems
MJD	Modified Julian Date
MW	Momentum Wheel
MWIT	Microwave Ion Thrusters
NPM	Normal Pointing Mode
OBC	Onboard Computer
OBS	Onboard Software
OCM	Orbit Control Mode
ORF	Orbital Reference Frame
P/L	Payload
PAD	Precise Attitude Determination
PDE	Partial Differential Equation
POD	Precise Orbit Determination
PPS	Plasmic Propulsion System
PSD	Power Spectral Density
r.h.s.	right-hand side
RCT	Reaction Control Thrusters
RFIT	Radio Frequency Ion Thrusters
RG	Rate Gyro
RIG	Rate Integrating Gyro
RLG	Ring Laser Gyro
RW	Reaction Wheels
S/C	Spacecraft
S/W	Software, specifically the satellite onboard computer AOCS Software
SAM-EM	Sun Acquisition Emergency Mode
SAR	Synthetic Aperture Radar
SAW	Solar Array Wing
SHM	Safe Hold Mode
SISO	Single-Input Single-Output systems
SSO	Sun Synchronous Orbit
STR	Star Tracker sensor
SVD	Singular Value Decomposition

TBVP	Two Boundary Value Problem
TC	Telecommand
TM	Telemetry
TOD	True of Date
UTC	Universal Time Coordinated
XIPS	Xenon Ion Propulsion System

Chapter 1
Introduction

Abstract The complexity of artificial satellites has grown very fast in the last 20 years. This complexity applies to any platform and payload technological domain; for what concerns the subject of this book we observe a clear trend in dynamic, functional and computational requirements, from quite static systems of the past century to increased capabilities in satellites' orbital and attitude agility and autonomy, which have become important features of current AOCS systems. The subjects presented in this text have been selected to provide theoretical tools to design modern satellite attitude and orbit control systems. In particular we present AOC modes and functions, the dynamics of flexible satellites with examples, attitude control synthesis methods, the optimization of orbital transfers for finite and infinite thrust and most commonly used sensors and actuators with a specific focus on plasmic propulsion.

The complexity of artificial satellites has grown very fast in the last 20 years. This complexity applies to any platform and payload technological domain; for what concerns the subject of this book we observe a clear trend in dynamic, functional and computational requirements, from quite static systems of past century to new satellites with increased capabilities in orbital and attitude agility and autonomy, which have become important features of current AOCS systems. This transformation took place thanks to a new generation of more accurate and intelligent sensors and also thanks to the enormous growth of on-board processing capability.

Moreover, as a general trend, payload functional complexity has produced an increase of satellites' installed solar array power and a need for large antennas and telescopes. These evolutions led to the installation of large flexible appendages on relatively light platforms, making the subject of treating low frequency flexibilities with a relatively high modal mass an important problem in AOC design.

The need for lighter platforms implementing an augmented capacity of orbital transfer has led to search for more efficient plasmic propulsion systems, which require new AOCS concepts, including autonomous navigation and guidance.

The subjects presented in this text have been selected to provide theoretical tools to design modern satellite attitude and orbit control systems.

© Springer International Publishing Switzerland 2016

L. Mazzini, *Flexible Spacecraft Dynamics, Control and Guidance*,
Springer Aerospace Technology, DOI 10.1007/978-3-319-25540-8_1

Before continuing with the description of the content of the book, we shall go through the main satellite applications and explain what are the latest technological advances in AOCS along with possible future trends.

We will consider in this review Earth observation applications optical and radar, telecommunication and navigation, scientific applications and microsatellites.

1.1 Radar Earth Observation

The typical Earth observation satellite flies in low Earth orbit (LEO), typically between 400–1000 km. The launcher injects these satellites directly into their final orbit, therefore the satellite propulsion system is typically limited. The choice of the altitude is a compromise between the need to have a good image resolution, that calls for lower altitudes, and the need to have a short revisit time over the same target. The orbits chosen are often almost polar, dawn/dusk Sun-synchronous, but also low inclination orbits can be selected when the service is provided over low latitude regions.

AOCS requirements depend on the specific instrument (synthetic aperture radar, radar altimeters, scatterometers): we refer here to high quality synthetic aperture radar satellites, as the one shown in Fig. 1.1, which usually require a significant amount of solar array power and integrate quite large antennas. This makes the issue of flexibility effects very important in AOCS design. When these satellites have a completely active antenna their attitude is normally quite stable, apart from periodic repointings and low rate attitude steering (so called yaw steering) to compensate for the Earth's rotation effects on the doppler shift of the received ground scattered electromagnetic wave. When the antenna does not implement an electronic steering

Fig. 1.1 Active array synthetic aperture radar satellite

capability, the AOCS design must provide an important attitude agility in its place. This calls for a high bandwidth rate control, fed by gyros with low noise performance figures and flexible modes' disturbance rejection filters, control moment gyro (CMG) actuators with important torque and momentum figures, and suitable control laws. When the radar has a very large antenna, thruster pulses for orbit corrections can produce intolerable flexible mode excitations. The AOCS uses star trackers (STT) to determine the attitude and GNSS receivers with a precise orbit determination filter to determine in real time the satellite's position with very high accuracy. The attitude and orbit knowledge are significant in providing a good image target geo-location.

1.2 Optical Earth Observation

The considerations made for radar satellites in terms of orbital selection apply here as well, except for the fact that the orbits chosen are almost polar, day/night Sun-synchronous, in order to have a long passage with Sun illuminating the targets at midday, which provides a good and stable image quality.

During the day passage satellite points towards desired targets using a very accurate star tracker while the on-board navigation system provides the information of the satellite's position in order to precisely geolocate the image. These satellites are typically as compact as possible with very small and rigid solar arrays, and the telescope body solidly integrated within the platform (see Fig. 1.2).

The AOCS requirements depend on the specific instrument used (hyper-spectral, panchromatic, thermal-infrared); we refer here to high quality panchromatic satellites which typically require important stability and pointing performances to support

Fig. 1.2 Optical panchromatic satellite

the high resolution capacity. The main problems related to the design of a high performance AOCS for these applications are a high frequency attitude stability during imaging and a large attitude agility in order to produce many images of the same location. The first issue calls for a high bandwidth rate control fed by gyros with low noise performance figures and flexible mode fast dumping; the agility instead calls for the use of CMG actuators. The appendages are designed to be extremely stiff in order to avoid oscillations after the agile maneuvers. The CMG shall provide very high torques in the range of many Newton-meters, but they shall be also extremely well-balanced in order not to produce micro-vibration levels such to affect the image quality. Other important issues related to AOCS design are: star trackers (STT) working under high speed conditions, multiple STTs' optical heads and gyros for *precise attitude determination* (PAD) and very *precise orbit determination* (POD) to increase the quality of image's geo-location.

1.3 Telecom Satellites

The telecom satellites typically fly in geosynchronous (GEO) orbit, about 36000 km above the Earth. Even if some launchers are able to place medium size satellites directly in such an orbit, the classic procedure is an injection into a highly eccentric orbit called GTO (geosynchronous transfer orbit) followed by a number of satellite maneuvers aimed at reaching the final GEO. These maneuvers form what is called the LEOP (launch and early operational phase) of the telecom satellite. These satellites are equipped with very important propulsion systems which allow them to perform LEOP transfers and following station keeping maneuvers to maintain their location in GEO.

The AOCS must be designed to support LEOP maneuvers which nowadays are made simpler by using a new generation of autonomous star trackers which can provide satellite attitude measurement looking at any direction of the sky. In the past, satellites' LEOP were much more complex because using Earth sensors and Sun sensors which could work only if pointed towards the Earth and the Sun. When placed on station, telecom satellites deploy an important system of solar arrays and antennas (see in Fig. 1.3 a typical Ka-band telecom antenna system). These antennas must be accurately pointed while they are excited by orbital maneuver pulses, thus Ka-band satellites may need a dedicated antenna pointing controller.

The driving force of telecom satellite business is the commercial competition among the big players of this market. This calls for a strong focus on satellite costs and efficiency which has led in two directions: increased on-station solar array power in order to implement more communication channels and search of lower cost launching systems. The AOCS design must comply with these low cost requirements still maintaining enough flexibility.

In order to reduce launch mass and cost, the trend is to replace the satellites' propulsion with fully electric low thrust systems. The AOCS will have to perform a very long LEOP, lasting a few months, in order to reach the GEO location. The

Fig. 1.3 Telecom Ka-band multispot satellite

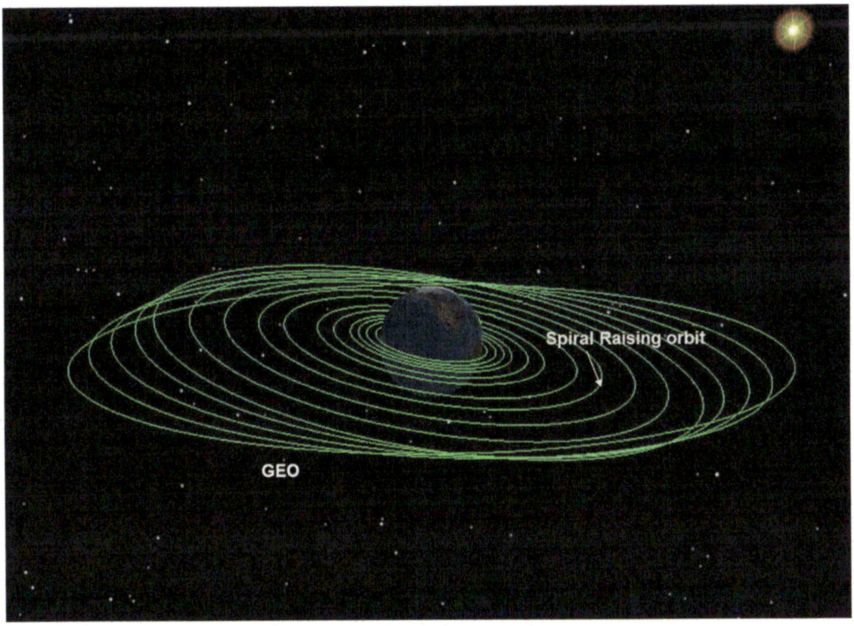

Fig. 1.4 Pictorial view of a LEO to GEO transfer by electric propulsion

launchers' injection orbits could be medium Earth orbits (MEO) or GTO but also LEO cases are under study. In Fig. 1.4 a pictorial view of a LEO to GEO spiral transfer is shown. Other trends in telecom AOCS design are relevant to gyro-less design and autonomous navigation and guidance, to reduce the costs of on-board hardware and ground station support during such a long LEOP.

1.4 The Navigation Satellites

The navigation satellites are those forming the constellations which allow to accurately measure the position of mobile ground receivers (so called GNSS receivers): the GPS constellation, the Glonass constellation and the Galileo constellation. These satellites are normally injected directly into their final MEO orbit with medium inclination, and the main requirement of AOCS design is its low cost as in the telecom case. The satellite attitude on station is pointed at the Earth and a rotation around Earth's direction is implemented in order to point the solar arrays always towards the Sun (yaw steering). One of the main issues of navigation satellites' design is the launching cost: an important novelty under study for Galileo second generation is to adopt an on-board electric propulsion system in order to give these satellites an important orbit transfer capability and use lower cost launch opportunities.

1.5 Scientific Satellites

Scientific missions give an extremely wide range of possibilities. We can have missions to Mars, to the Moon or L1, L2, L4 Lagrangian points, not to mention planetary missions to our solar system. Many scientific missions are also implemented in LEO for example to measure the physical properties of our planet. The objectives of such missions may be exploration of planets of our Solar System, observation of Sun or deep space observation (Fig. 1.5) in different bandwidths. Each scientific mission may have its peculiar problems; in general AOCS is designed to take images of stars or to point very precisely towards some spot of the Space. Issues like high precision pointing and microvibrations are common in these missions. Microvibrations produced by momentum management systems may be damped using passive or active dampers. These satellites shall be controlled with very accurate star trackers and smooth actuation systems like those using cold gas reaction control thrusters (RCTs). Large flexible appendages are quite common in these missions such as telescopes and antennas, for example, when exploring far locations of our solar system large antennas may be deployed in order to increase the communication antenna gain and ensure a good data rate.

1.6 Microsatellites

Microsatellites are a new way to conceive space missions. In this paradigm microsatellite or nanosatellites constellations or swarms (see Fig. 1.6) in low orbits should perform the same task of a single satellite at a higher orbit. An example can be provided by optical satellites: by using a smaller telescope aperture on a smaller

Fig. 1.5 Hubble space telescope exploring the deep space. *Credit* NASA

Fig. 1.6 Two nanosats from the international space station. *Credit* NASA

satellite flying at a proportionally lower orbit, we get the same resolution as if flying at a higher orbit with a larger telescope aperture on a larger satellite.

The loss of revisit time of a lower orbit microsatellite, is compensated by launching many of them (3–4). The global cost is in favor of small satellite solutions (see [1]). Microsatellite constellations can be conceived also for other applications like global internet services and local area networking. Some microsatellites in *flying formation* close to each other can be used as a single instrument by synthesizing their measurements to provide a type of information which is not reachable with a single satellite: this happens in particular in the case of optical applications, radar interferometry, bistatic radar applications and localization of electromagnetic sources. The AOCS of such systems faces important challenges: flying formation navigation and guidance (see [2]), low orbit drag compensation by small plasmic thrusters, agility, highly integrated and low cost electronics, inter-satellite link (ISL). In order to reach a high level of integration and an extremely low cost (a factor 10–100 less with respect classic satellites) microsatellites will use *commercial off the shelf* (COTS) components. The concept of reliability of a single fault tolerant satellite unit is substituted by the concept of dependability of a constellation of microsats. Such an approach open the avionic design to the use of non-Hi-Reliability parts where the quality of the system is guaranteed by specific analyses and tests performed by the satellite manufacturer.

1.7 Content of the Book

Having shortly reviewed the state of the art AOCS needs in satellites' market, we provide now a description of the contents of the book: our objective is that the presented subjects should be a useful guide in overcoming these challenges.

This book starts with a description of kinematics and reference frames used to describe satellite dynamics (Chap. 2), flight axes and inertial axes. We define here the variables that will be used to express satellite dynamics.

In the following Chap. 3 we develop the equations of the dynamics of a flexible body in space, our original approach follows the format of Lagrangian mechanics and it is suitable for control applications of large space flexible structures. The reader who is not cultivated in analytical mechanics can consult the Appendix and also the relevant bibliography. The satellite state equations are developed as a generalization of Euler's equations of motion for a rigid body which is the most used simplified model in satellite attitude control. We include also the study of *momentum management* devices and *gravitational torque* effects. The chapter ends with the description of linearized equations which are often used in controller design and a description of typical satellite's flexible modes: sloshing, solar array, antennas.

Chapter 4 is dedicated to a description of a modern architecture and the functions of an AOC system. The main AOC modes are reviewed presenting possible design solutions and examples, the subject is presented with a simple practical engineering approach and a basic control analysis background is necessary to follow the text.

In Chap. 5 the *optimal control theory* is presented using an Hamiltonian formalism and the *maximum principle*. The chapter contains a short presentation of linear optimal control using H_∞ techniques for second order systems.

Chapter 6 is dedicated to attitude control applications: we present the analysis of a flexible system, in particular how to write the non linear equations of its dynamics, how to stabilize the non linear system using Lyapunov method, how to make it linear and then use an LQR controller design technique. The last design example deals with robust control synthesis of a linearized satellite's dynamics using H_∞ synthesis in presence of flexible modes.

The following Chap. 7 deals with orbit control applications. We describe the classic station-keeping maneuvers, then using the maximum principle, we introduce the theory of orbital transfer maneuver optimization with infinite and finite thrust. We provide realistic design examples for many Earth-centered orbit transfers with low thrust guidance.

This book deals both with orbit control and attitude control. This is not a common approach, because nowadays attitude control engineers and orbital control engineers are two different professionals, with dedicated tools, methodologies and a specific background. We have noticed that, more and more often, these specialists must team together in order to design an efficient AOC system (AOCS). This text is written also for these professionals who feel that it is necessary to have a basic knowledge of the other discipline and how the two are linked together.

After these quite theoretical chapters we provide a description of the main technologies which are of interest for AOCS: propulsion systems and sensors and actuators.

In Chap. 8 we review the main satellite propulsion technologies, both chemical and electrical, and along with the general principles we develop a detailed description of plasmic propulsion systems. In the last Chap. 9 state of the art sensors and actuators are described. Modern sensors like star trackers, Sun and Earth sensors, magnetometers, gyros, GNSS receivers are described in terms of design principles, mathematical models and performances available on the market. The same applies to the actuators: wheels, control moment gyros, magneto-torquers.

References

1. A. Shao, E.A. Koltz, J.R. Wertz, Quantifying the cost reduction potential for earth observation satellites, in *AIAA Space Conference*, San Diego (2014)
2. K.T. Alfriend, S.R. Vadali, P. Gurfil, J.P. How, L.S. Breger, *Spacecraft Formation Flying, Dynamics Control and Navigation*. Elsevier Astodynamics Series (2010)

Chapter 2
The Kinematics of the Flexible Satellite

Abstract The kinematic model of a body is a system of equations that define the position in space of each particle of the body from the values of a set of variables. The time variation of these variables describes the motion of the body. In this chapter the main reference frames involved in satellite's attitude and orbit control are defined. After, we review the different ways to describe the rigid motion and how to express all the kinematic variables and their time derivatives that we need later to formulate the dynamics. Finally, the kinematics of the flexible body are described by adding a set of deformation functions on to the rigid body kinematics.

2.1 Introduction

The kinematic model of a body or many bodies, however connected, is a system of equations that define the position in space of each particle of the body from the values of a set of variables. The time variation of these variables describes the motion of the body. The kinematic model is fundamental and preliminary not only to write the dynamical equations, but also to set formally the attitude and orbit control problem, because it allows us to define mathematically what is the control objective, that is, what is our desired motion for any specific satellite.

The motion of space objects like satellites is normally described in very specific reference frames to facilitate the expression of their dynamics and to define the desired motion of the satellite in a way which is universally recognized. In this chapter the main reference frames involved in satellite's attitude and orbit control are defined. After that, we review different ways to describe rigid motion like the Euler angles and the quaternions and how it is possible to express all the kinematic variables and their time derivatives that we need to formulate the dynamics. Finally, the kinematics of the flexible body are described by adding a set of deformation functions on to the rigid body kinematics. The rigid body part of this chapter can be found in many texts like [1–4]. A good description of flexible body kinematics can be found in [5–7].

© Springer International Publishing Switzerland 2016

L. Mazzini, *Flexible Spacecraft Dynamics, Control and Guidance*,
Springer Aerospace Technology, DOI 10.1007/978-3-319-25540-8_2

2.2 Space and Time Reference Frames

The reference frames involved in attitude and orbit control for an Earth orbiting satellite are quite typical and used almost everywhere even if exceptions may exist for specific missions. We introduce the inertial reference frame where the laws of dynamics are naturally written, then a reference frame with flight axes defined by the satellite trajectory that is normally used to define the desired attitude and finally a reference frame linked to the satellite body, where we can easily express the deformations and also the sensor and actuator properties. We also need to develop all the mathematical tools to go from one reference to the other.

2.2.1 Inertial Reference Frame J2000 ECI

ECI is the acronym of Earth centered inertial. This is a quasi-inertial coordinate system and its main non-inertial acceleration is due to the gravitational attraction between the Sun and the Earth.

ECI can be assumed as inertial for most of the applications concerning Earth orbiting satellites. On the contrary, the **ECI** frame is not well-suited as an inertial system when the effect of other celestial bodies is considered (interplanetary trajectories, etc.).

At any date a specific **ECI** can be defined and all these frames differ for very small rotations. The **ECI** coordinate system J2000 is a specific **ECI** frame, defined as follows (see Fig. 2.1) on the 1st January, 2000:

- the origin O_{ECI} is placed at the center of the Earth;
- X_{ECI} lies in the equatorial plane along the intersection with the ecliptic plane towards the direction of the Sun's apparent motion ascending node; this direction, fixed with respect to the stars, is also called *first point of Aries* or *Vernal Equinox*s;
- Z_{ECI} is aligned with the Earth's rotation axis, pointing north;
- Y_{ECI} axis lies in the equatorial plane to form a right-handed frame.

As previously mentioned, at any date and time, it is possible to define different **ECI** reference frames based on the Earth's true rotation pole at that time, and an equinox (X_{ECI}) based on the intersection of the Earth's equatorial plane with the ecliptic at that time. These frames are called *true of date* or *TOD*. In the **ECI** reference frame the state of a point mass is represented either by the Cartesian position and velocity vectors or in terms of the osculating Keplerian orbit parameters (Fig. 2.2).

2.2.2 Orbit Reference Frame ORF

The orbit reference frame **ORF** is useful in defining the desired attitude of the spacecraft. Its axes $X_{ORF}, Y_{ORF}, Z_{ORF}$ are also called *flight axes*, and are defined in the following way:

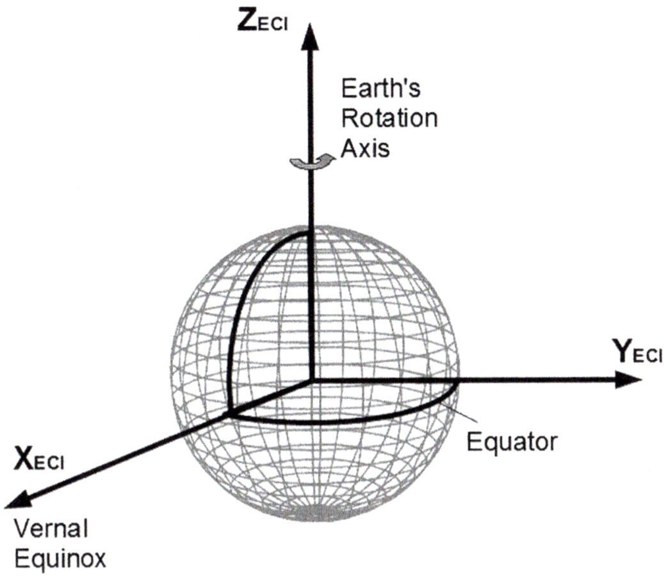

Fig. 2.1 **ECI** linertial reference frame

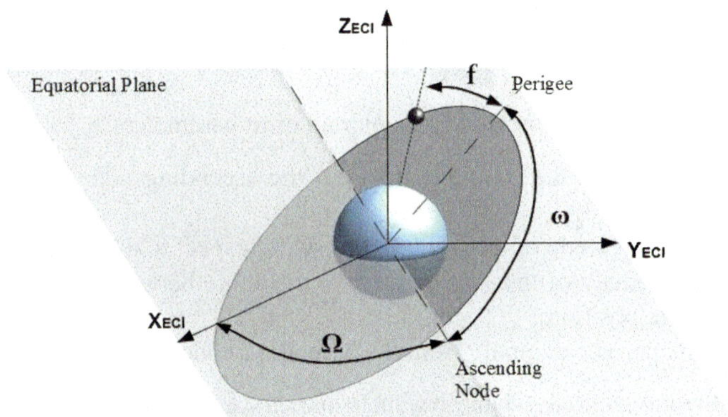

Fig. 2.2 Keplerian orbit elements

- the origin is the spacecraft's center of mass (CoM);
- the triad axes depend on the spacecraft's position and velocity (see Fig. 2.3);
- the orbital plane coincides with the X_{ORF}, Z_{ORF} one;
- positive Z_{ORF} axis points toward the Earth's center;
- X_{ORF} axis is perpendicular to the Z_{ORF} in the nominal direction of flight (i.e. X_{ORF} is closely aligned with the velocity vector);
- Y_{ORF} axis is chosen so as to form a right-handed set of coordinate axes.

Typically X_{ORF}, Y_{ORF}, Z_{ORF} are also called the *roll, pitch* and *yaw* axes respectively.

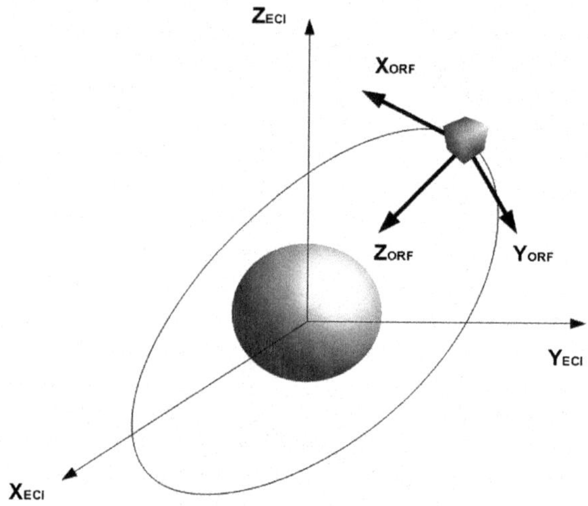

Fig. 2.3 **ORF** orbital reference frame

This reference frame is shown in Fig. 2.3.

2.2.3 *Transformation from* **ECI** *to* **ORF**

We consider the following osculating Keplerian orbit parameters:

- Ω is the RAAN (i.e. the right ascension of the ascending node of the satellite orbit);
- I is the inclination angle of the orbital plane with respect to the equator;
- f is the true anomaly of the satellite, that is the angle between the orbit perigee and the satellite position;
- g (or ω) is the perigee anomaly from the ascending node.

In the following sections, when we want to mark a specific transformation between the vector components of two triads, we will subscribe the symbol of matrix transformation M with the initials of the triads involved. So, M from the **ORF** to **ECI**, will be specifically denoted as $M_{E \hookleftarrow O}$; its transpose will perform the inverse transformation $M_{O \hookleftarrow E}$.

The rotation matrix $M_{O \hookleftarrow E}$ that transforms the components of a vector from **ECI** reference frame to **ORF** is:

$$M_{O \hookleftarrow E} = \begin{pmatrix} 0 & 1 & 0 \\ 0 & 0 & -1 \\ -1 & 0 & 0 \end{pmatrix} \cdot M_{fI\Omega}$$

with

$$M_{fI\Omega} = \begin{pmatrix} \cos(f+g) & \sin(f+g) & 0 \\ -\sin(f+g) & \cos(f+g) & 0 \\ 0 & 0 & 1 \end{pmatrix} \cdot \begin{pmatrix} 1 & 0 & 0 \\ 0 & \cos I & \sin I \\ 0 & -\sin I & \cos I \end{pmatrix} \begin{pmatrix} \cos\Omega & \sin\Omega & 0 \\ -\sin\Omega & \cos\Omega & 0 \\ 0 & 0 & 1 \end{pmatrix}$$

The matrix $M_{fI\Omega}$ is obtained making the following sequence of rotations 3-1-3:

1. counter-clockwise Ω rotation around the Z_{ECI} axis;
2. counter-clockwise I rotation around the transformed X_{ECI} axis;
3. counter-clockwise $(f+g)$ rotation around the (double) transformed Z_{ECI} axis.

Finally, to get $M_{O\leftrightarrow E}$, a counter-clockwise 90° rotation about the current (transformed) Z_{ECI} axis and a clockwise 90° rotation about the current (transformed) X_{ECI} axis are needed to meet the verse/names of the axes.

2.2.4 Body Axes

We define the body reference frame **BRF** as a frame joined to the body, with an origin attached to a point of the body.

The **BRF** coordinate system is a right-handed Cartesian system designated by the standard subscript *BRF* (e.g. X_{BRF}). See Fig. 2.4.

The definition of the **BRF** axes is usually given using three orthogonal facets of a rigid mirror cube glued to the satellite structure. The naming of the axes is arbitrary, typically the naming is given such that, when the **BRF** is aligned to the **ORF** during some operational phase, we have the same axes names in **BRF** and **ORF**. In this case X_{BRF}, Y_{BRF}, Z_{BRF} are also called the *roll, pitch* and *yaw* axes respectively.

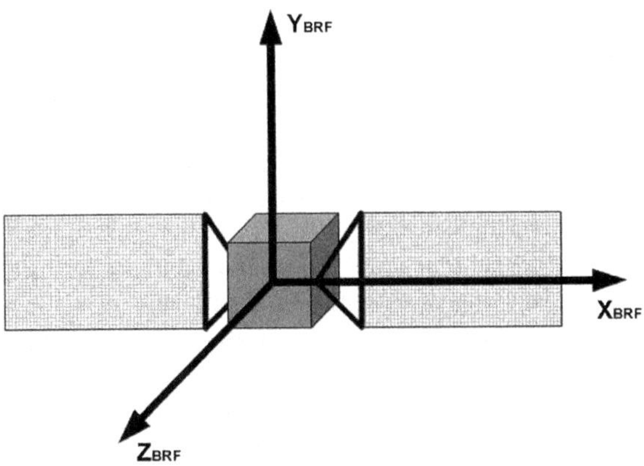

Fig. 2.4 **BRF** body reference frame

We define $M_{B\leftrightarrow O}$ the matrix that transforms the vector components from **ORF** to **BRF**. The transformation of the vector components from **ECI** to **BRF** is done by multiplying the respective orthogonal transformations, that is:

$$M_{B\leftrightarrow E} = M_{B\leftrightarrow O} M_{O\leftrightarrow E},$$

by transposing the two members we get also

$$M_{B\leftrightarrow E}^T = M_{E\leftrightarrow B} = M_{E\leftrightarrow O} M_{O\leftrightarrow B} = M_{O\leftrightarrow E}^T M_{O\leftrightarrow B}.$$

2.2.5 *Time Reference Frames*

The measurement of the time in satellite applications is generally expressed in the reference system UTC (Universal Time Coordinated), that is the standard reference for civil timekeeping and is based on the Earth's motion. The time is divided into days, hours, minutes and seconds. Each day is made up of 24 h and each hour of 60 min, but the number of seconds in a minute can be 60, or sometimes 61 or 59 to compensate for the Earth's slowing rotation rate.

The identification of a time point in the UTC reference frame is often expressed through the *Gregorian date*, written in the format DD/MM/YY hh:mm:ss.sss.

The time of an event can also be identified with its *Julian date* (JD). This is defined as the number of days elapsed since the 1st January 4713 BC at noon. The number of days has to be computed using the rules of the Julian calendar, in which the years divisible by 100 are leap years unless they are divisible by 400. For convenience of calculation a *modified Julian date* (MJD) is in use, which is defined as the cumulative sum of the days elapsed since 0h UTC of 17 November 1858. For example, January 1st 2000 at 0 h is JD 2451544.5 and MJD 51544.

2.3 Rigid Body Kinematics

The rigid body configuration is defined by a fixed distribution of mass in its body reference frame, the relative distances between the different points of the body remain constant during the motion.

We will use the subscript notation and we will imply summation over repeated subscripts, using the so called Einstein rule.

We introduce the following definitions:

- x_i is the generic point position vector in the inertial frame **ECI**;
- x_i^0 is the position vector in **ECI** of the origin dependent on the time;
- $M_{ij}(\theta)$ is the rotation matrix transforming a vector from the **BRF** to **ECI**;
- r_j is the vector that identifies the generic point of the rigid body in the body reference frame.

The rotation matrix $M = M_{E \leftrightarrow B}$ represents a point in the $SO(3)$ space[1] and depends on time. The $SO(3)$ space is a 3D variety that can be parametrized with three coordinates that we will call $\theta = (\theta_1, \theta_2, \theta_3)$. The most popular parametrizations of $SO(3)$ in the context of attitude control will be introduced later in this chapter. With the previous definitions, the kinematic model of the rigid body is:

$$x_i = x_i^o + M_{ij}(\theta) r_j \quad (2.1)$$

By time derivation we get the velocity of each generic point,

$$\begin{aligned} \dot{x}_i &= \dot{x}_i^o + \dot{M}_{ij} r_j \\ \dot{x}_i &= \dot{x}_i^o + M_{ik} e_{ksj} \omega_s r_j \end{aligned} \quad (2.2)$$

where e_{ksj} is the Levi Civita symbol, used to express 3D vector products.

For any two vectors $a_i, b_j, i = 1, 2, 3$ we have that $\sum_{i,j} e_{kij} a_i b_j = [a, b]_k = (a \times b)_k$, where $()_k$ is the kth component of the vector product.

We will use often an implicit notation for the sums where repeated indexes must be summed also when the explicit symbol of sum is missing, so that:

$$\sum_{i,j} e_{kij} a_i b_j = e_{kij} a_i b_j.$$

The pseudo-vector ω_s is the body reference frame angular speed. It is a pseudo-vector because it transforms as a vector only in transformations between right-hand triads (see [8], ω_s transforms like a vector for the transformation in $SO(3)$ but not for those in $O(3)$). It can be introduced easily from the properties of the orthogonal matrices.

By deriving the identity $M^T M = E_3$, with E_3 the 3D identity matrix, we get $\dot{M}^T M + M^T \dot{M} = 0$, so $M^T \dot{M} = -\dot{M}^T M$ is an antisymmetric matrix. We express this matrix as:

$$M^T \dot{M} = skew(\omega) \quad (2.3)$$

with

$$skew(\omega) = \begin{bmatrix} 0 & -\omega_3 & \omega_2 \\ \omega_3 & 0 & -\omega_1 \\ -\omega_2 & \omega_1 & 0 \end{bmatrix}$$

with the subscript notation we write Eq. 2.3 as:

$$\dot{M}_{ij} = M_{ie} e_{esj} \omega_s = M_{ie} skew(\omega)_{ej}.$$

---

[1] The space $SO(3)$ is the 3D rotations group of all the linear transformations of $\Re^3$ preserving distances, origin and handedness. The matrices M in this group are unitary, such that $M^T M$ is the identity matrix, with $det(M) = +1$. The space $SO(3) \subset O(3)$, the space of all the unitary matrices M.

This equation defines the angular speed of the **BRF** with respect to **ECI** in matrix form and justifies Eq. 2.2 that we have introduced to write the velocity of a point in the **BRF** triad.

In terms of transformation rules from one right hand triad to another, the three components of ω can be considered as a vector with components in the **BRF** frame, this pseudo-vector is normally called the angular velocity.

For the angular speed between orthogonal frames, we will use a specific notation, using subscripts to clarify the orthogonal frames involved and the frame where the components of ω are written. That is: $\omega = \omega_{BE,B}$ means the angular speed of the **BRF** with respect to the **ECI** in the **BRF** components.

In matrix notation the derivative of the transformation matrix from the triad **BRF** to triad **ECI** is written as:

$$\dot{M}_{E\leftrightarrow B} = M_{E\leftrightarrow B} skew(\omega_{BE,B}) \tag{2.4}$$

When we deal with the inverse matrix $M_{B\leftrightarrow E}$, by transposing the previous identity and using the properties of the antisymmetric matrices, we have:

$$\begin{aligned} \dot{M}_{B\leftrightarrow E} &= -skew(\omega_{BE,B})M_{B\leftrightarrow E} = M_{B\leftrightarrow E}\omega_{EB,E} \\ skew(\omega_{BE,B}) &= -skew(\omega_{EB,B}) = -M_{B\leftrightarrow E}\omega_{EB,E}M^T_{B\leftrightarrow E} \end{aligned} \tag{2.5}$$

A quite useful property of the angular speed is the summation rule. If we have a double orthogonal transformation: $M_{E\leftrightarrow B} = M_{E\leftrightarrow O}M_{O\leftrightarrow B}$, the angular velocity of the global transformation is the sum of the angular velocities of each transformation (the angular velocities must be expressed in the same triad before summing). We indicate with $\omega_{BE,B}$ the angular velocity of **BRF** with respect to **ECI** in **BRF** components; the summation rule can be stated as:

$$\omega_{BE,B} = \omega_{BO,B} + \omega_{OE,B} \tag{2.6}$$

This rule is proved by the following sequence of equations:

$$\begin{aligned} \dot{M}_{E\leftrightarrow B} &= M_{E\leftrightarrow B} skew(\omega_{BE,B}) \\ &= M_{E\leftrightarrow O} skew(\omega_{OE,O})M_{O\leftrightarrow B} + M_{E\leftrightarrow O}M_{O\leftrightarrow B} skew(\omega_{BO,B}) \\ &= M_{E\leftrightarrow B}M^T_{O\leftrightarrow B} skew(\omega_{OE,O})M_{O\leftrightarrow B} + M_{E\leftrightarrow B} skew(\omega_{BO,B}) \\ &= M_{E\leftrightarrow B} skew(\omega_{OE,B}) + M_{E\leftrightarrow B} skew(\omega_{BO,B}) \\ &= M_{E\leftrightarrow B} skew(\omega_{BE,B}) \\ &= M_{E\leftrightarrow B}(skew(\omega_{BO,B}) + skew(\omega_{OE,B})) \end{aligned} \tag{2.7}$$

The matrix transformation $M^T_{O\leftrightarrow B} skew(\omega_{OE,O})M_{O\leftrightarrow B} = skew(\omega_{OE,B})$ can be easily derived by Eq. 2.5 applied to **ORF** and **BRF**.

Many useful parametrizations of $SO(3)$ are provided in the literature, an elegant introduction to the subject can be found in [1].

The angular speed is a function of the $\theta = (\theta_1, \theta_2, \theta_3)$ and linearly depends on their time derivatives so that there exists a matrix μ such that:

$$\omega_e = \mu_{ek}(\theta)\dot{\theta}_k$$

Any orthogonal matrix in the $SO(3)$ space can be represented locally by three coordinates θ_i, $i = 1, 2, 3$ by which we represent the body attitude with respect to a given reference frame. We refer to *attitude* as the orientation of the body reference frame (**BRF**, whose unit vectors are here denoted by $\hat{b}_1, \hat{b}_2, \hat{b}_3$) with respect to an **ECI** (or **ORF**), of unit vectors $\hat{u}_1, \hat{u}_2, \hat{u}_3$. At least 3 variables are needed to describe the attitude.

A vector in $\{\hat{u}_i\}$ can be expressed in $\left\{\hat{b}_i\right\}$ by means of the $SO(3)$ transformation defined by the orthonormal matrix $M_{B \hookleftarrow E}$ of direction cosines:

$$\begin{pmatrix} v|_{b_1} \\ v|_{b_2} \\ v|_{b_3} \end{pmatrix} = M_{B \hookleftarrow E} \begin{pmatrix} v|_{u_1} \\ v|_{u_2} \\ v|_{u_3} \end{pmatrix}$$

$$\begin{pmatrix} v|_{u_1} \\ v|_{u_2} \\ v|_{u_3} \end{pmatrix} = M_{E \hookleftarrow B} \begin{pmatrix} v|_{b_1} \\ v|_{b_2} \\ v|_{b_3} \end{pmatrix}$$

$$M_{B \hookleftarrow E} = \begin{bmatrix} \hat{b}_1\hat{u}_1 & \hat{b}_1\hat{u}_2 & \hat{b}_1\hat{u}_3 \\ \hat{b}_2\hat{u}_1 & \hat{b}_2\hat{u}_2 & \hat{b}_2\hat{u}_3 \\ \hat{b}_3\hat{u}_1 & \hat{b}_3\hat{u}_2 & \hat{b}_3\hat{u}_3 \end{bmatrix} \tag{2.8}$$

Clearly, both matrix $M_{B \hookleftarrow E}$ or $M_{E \hookleftarrow B}$ can be used to represent the kinematics, being one the transpose of the other; therefore it is important in any calculation to establish the convention being used.

The direction cosines are a highly redundant set for the description of the attitude, in fact during the evolution of the reference frames, they must satisfy 6 equations expressing the ortho-normality condition of the triad unit vectors.

The same transformation of $SO(3)$ can be represented with a minimal (or quasi-minimal) set of parameters by considering the rigid rotation that brings the triad of axes $\left\{\hat{b}_i\right\}$ to be coincident with $\{\hat{u}_i\}$.

2.3.1 Quaternions

From Euler's *Theorem of the Principal Rotation* (see [8]) it is known that a triad **A** can be brought to coincidence with a triad **B** by a unique *principal rotation* of an angle ϕ about a *principal axis* unit vector $\hat{e} = (e_1, e_2, e_3)$, whose components e_i are the same in both triads.

The positive direction of the rotation around the unit vector $\hat{e}$ is given by the right-hand rule.

The principal rotation can also be represented by Euler's parameters or *quaternions*, which are the following set of scalar parameters:

$$q_0 = \cos \frac{\phi}{2}$$
$$q_1 = e_1 \sin \frac{\phi}{2}$$
$$q_2 = e_2 \sin \frac{\phi}{2}$$
$$q_3 = e_3 \sin \frac{\phi}{2}$$

plus the constraint equation:

$$q_0^2 + q_1^2 + q_2^2 + q_3^2 = 1$$

The components q_1, q_2, q_3 can be assembled in a column vector denoted as q and the quaternion in vector form is indicated $q = (q_0, \mathsf{q})$. We will use this format for vectors only when they are embedded as components of quaternions. Using the quaternion the transformation matrix from the triad **A** to the triad **B** is written as:

$$M_{B \leftarrow A} = \begin{bmatrix} 1 - 2(q_2^2 + q_3^2) & 2(q_1 q_2 + q_0 q_3) & 2(q_1 q_3 - q_0 q_2) \\ 2(q_2 q_1 - q_0 q_3) & 1 - 2(q_1^2 + q_3^2) & 2(q_2 q_3 + q_0 q_1) \\ 2(q_3 q_1 + q_0 q_2) & 2(q_3 q_2 - q_0 q_1) & 1 - 2(q_1^2 + q_2^2) \end{bmatrix} \tag{2.9}$$

The quaternion can be computed from the rotation matrix $M_{B \leftarrow A}$ cosine directors M_{ij} in 4 different ways.

For example:

$$\begin{aligned} q_0 &= \pm\tfrac{1}{2}(1 + M_{11} + M_{22} + M_{33})^{\frac{1}{2}} \\ q_1 &= \tfrac{1}{4q_0}(M_{23} - M_{32}) \\ q_2 &= \tfrac{1}{4q_0}(M_{31} - M_{13}) \\ q_3 &= \tfrac{1}{4q_0}(M_{12} - M_{21}) \end{aligned}$$

but also solving q_1 from the diagonal of $M_{B \leftarrow A}$:

$$\begin{aligned} q_0 &= \tfrac{1}{4q_1}(M_{32} - M_{23}) \\ q_1 &= \pm\tfrac{1}{2}(1 + M_{11} - M_{22} - M_{33})^{\frac{1}{2}} \\ q_2 &= \tfrac{1}{4q_1}(M_{12} + M_{21}) \\ q_3 &= \tfrac{1}{4q_1}(M_{31} + M_{13}) \end{aligned}$$

Similar inversion formulas exist solving q_2 or q_3 from the $M_{B\leftrightarrow A}$ diagonal.

There is an ambiguity in the sign of the first equation. In fact any quaternion $(q_0, \mathfrak{q})$ represents the same attitude matrix of $(-q_0, -\mathfrak{q})$. In practice this ambiguity can be removed by fixing a rule to choose one of the two branches, for example $q_0 > 0$, that is $-\pi < \phi < \pi$.

When the denominator in the quaternion inversion formula approaches 0 it is convenient to switch to another set of quaternion inversion formulas, so as to avoid singularities and numerical inaccuracies.

The set of the quaternions forms an algebra like the orthonormal matrices. This means that we can introduce a multiplication operator for which the identity and the inverse quaternion exist. These operations lead to the same result as if we were using the corresponding transformation matrix.

We define the identity quaternion $q = (1, 0, 0, 0)$ as the one corresponding to the identity matrix; the multiplication operation between two quaternions $q = (q_0, \mathfrak{q})$ and $p = (p_0, \mathfrak{p})$ is denoted as $q \otimes p$ and is defined as:

$$q \otimes p = (q_0 p_0 - \mathfrak{q}^T \mathfrak{p}, p_0 \mathfrak{q} + q_0 \mathfrak{p} - skew(\mathfrak{q})\mathfrak{p}) \tag{2.10}$$

It is then easy to verify that $q^{-1} = (q_0, -\mathfrak{q})$ and $(q \otimes p)^{-1} = p^{-1} \otimes q^{-1}$.

We will not develop here the complete quaternion theory, we give just the few relationships that are normally used in AOC developments.

When the quaternion represents the $M_{B\leftrightarrow E}$ matrix, being $\omega = \omega_{BE,B}$ the **BRF** angular speed with respect to **ECI** in **BRF** components, the quaternion kinematic equations are found to be (for the following relationships detailed references are: [3] Appendix A and Sect. 3.2 and [9]):

$$\begin{aligned} \dot{q} &= \tfrac{1}{2}(0, \omega) \otimes q \\ (0, \omega) &= 2\dot{q} \otimes q^{-1} \end{aligned} \tag{2.11}$$

Using Eq. 2.10, these relationships are expanded as:

$$\dot{q_0} = -\frac{1}{2}\mathfrak{q} \cdot \omega$$

$$\dot{\mathfrak{q}} = \frac{1}{2}(q_0 \omega + \mathfrak{q} \times \omega)$$

$$\omega = 2 \begin{bmatrix} -q_1 & q_0 & q_3 & -q_2 \\ -q_2 & -q_3 & q_0 & q_1 \\ -q_3 & q_2 & -q_1 & q_0 \end{bmatrix} \cdot \begin{pmatrix} \dot{q_0} \\ \dot{q_1} \\ \dot{q_2} \\ \dot{q_3} \end{pmatrix}$$

A similar expression holds for the angular acceleration:

$$\dot{\omega} = 2 \begin{bmatrix} -q_1 & q_0 & q_3 & -q_2 \\ -q_2 & -q_3 & q_0 & q_1 \\ -q_3 & q_2 & -q_1 & q_0 \end{bmatrix} \cdot \begin{pmatrix} \ddot{q}_0 \\ \ddot{q}_1 \\ \ddot{q}_2 \\ \ddot{q}_3 \end{pmatrix}$$

In conclusion the quaternion representation is normally preferred in AOC software implementations because it does not have singularities, allows to propagate the attitude state using only simple algebraic operations, it is based on the propagation of only 4 parameters while the matrix algebra uses 9 parameters, allows simple algebraic calculations. The only drawback of this representation is that it is not always immediate for the AOC engineer to understand the dynamics expressed in quaternions. This is the main reason why, when evaluating an attitude dynamics from simulations or from the satellite telemetry, the AOC engineer prefers to see directly rotation angles such as those used in the Euler angle representation.

2.3.2 *Euler Angles*

It is possible to choose as rotation axis of an *elementary rotation* one of the coordinated axes of **BRF**. The orientation of **BRF** w.r.t. to another reference system can be completely described by three subsequent elementary rotations, around the progressively dislocated axes, provided that we do not select the same axis in two subsequent rotations. The angles of these rotations are the *Euler angles*.

There exist 12 sets of Euler angles, 6 using all the permutations of 3 different axes rotations, i.e. 312, 123, 213.., and 6 using all the permutations of 2 different axes with the third rotation being of the same axis of the first one i.e. 212, 323, 313...

2.3.2.1 Euler Angles 213

The spacecraft attitude, expressed in terms of the Euler angles *roll, pitch, and yaw*, is here defined as a rotation of the **BRF** with respect to the **ORF** using the $M_{B\leftarrow O}$ matrix. When the **ORF** is aligned with the **ORF** the roll, pitch, and yaw angles are zero. We denote by:

$$\begin{aligned} \varphi &= \text{roll angle} \\ \vartheta &= \text{pitch angle} \\ \psi &= \text{yaw angle} \end{aligned}$$

When the body reference frame is not aligned with the orbital reference frame, the yaw, roll, and pitch angles are defined as the sequence of three rotations (Euler 2-1-3 sequence) required to align the **ORF** axes to the **BRF** axes. The rotation sequence is the following:

- a rotation about the Z_{ORF} axis equal in magnitude to the yaw angle,
- a rotation about the rotated X_{ORF} axis equal in magnitude to the roll angle,
- a rotation about the twice rotated Y_{ORF} axis equal in magnitude to the pitch angle.

The transformation matrix from **ORF** to **BRF** defined by the sequence of these rotations can be easily computed as the product of three consecutive single axis rotation matrices $M_{213} = M_2(\vartheta)M_1(\varphi)M_3(\psi)$, (see [9] Sect. 8.6 where M_{213} is called R_{213}, φ and ϑ are inverted and [4] Appendix E) and results to be:

$$M_{B \leftarrow O} = M_{213}(\varphi, \vartheta, \psi) = \tag{2.12}$$

$$\begin{bmatrix} \cos\vartheta\cos\psi - \sin\varphi\sin\vartheta\sin\psi, & \cos\vartheta\sin\psi + \sin\varphi\sin\vartheta\cos\psi, & -\sin\vartheta\cos\varphi \\ -\cos\varphi\sin\psi, & \cos\varphi\cos\psi, & \sin\varphi \\ \cos\psi\sin\vartheta + \sin\varphi\sin\psi\cos\vartheta, & \sin\vartheta\sin\psi - \cos\vartheta\sin\varphi\cos\psi, & \cos\varphi\cos\vartheta \end{bmatrix}$$

We define:

- $\omega = \omega_{BE,B}$ the angular rate vector of **BRF** w.r.t. **ECI** (in **BRF** components);
- $\omega_0 = \dot{f} + \dot{g}$ the scalar orbital angular rate;
- $Y_{O,B} = Y_{ORF}$ in **BRF** components, is the unit vector normal to the orbit plane in the direction opposite to the angular momentum, see Fig. 2.3;
- $-\omega_0\, Y_{O,B} = \omega_{OE,B}$ the angular rate vector of the **ORF** w.r.t. **ECI** (in **BRF** components);
- $\delta\omega = \omega_{BO,B}$ the angular rate vector of **BRF** w.r.t. **ORF** (in **BRF** components);

Following the rule of the angular speed composition (Eq. 2.6) the total angular velocity is given by:

$$\omega = -\omega_0\, Y_{O,B} + \delta\omega \tag{2.13}$$

We know (see Eq. 2.5) that

$$\dot{M}_{213}(\varphi, \vartheta, \psi) = -skew(\delta\omega)\, M_{213}(\varphi, \vartheta, \psi)$$

therefore the derivative of M_{213} can be written as:

$$\dot{M}_{213} = \begin{bmatrix} \dot{M}_{11} & \dot{M}_{12} & \dot{M}_{13} \\ \dot{M}_{21} & \dot{M}_{22} & \dot{M}_{23} \\ \dot{M}_{31} & \dot{M}_{32} & \dot{M}_{33} \end{bmatrix} = \begin{bmatrix} 0 & \delta\omega_3 & -\delta\omega_2 \\ -\delta\omega_3 & 0 & \delta\omega_1 \\ \delta\omega_2 & -\delta\omega_1 & 0 \end{bmatrix} \begin{bmatrix} M_{11} & M_{12} & M_{13} \\ M_{21} & M_{22} & M_{23} \\ M_{31} & M_{11} & M_{11} \end{bmatrix}$$

leading to:

$$\begin{pmatrix} \delta\omega_1 \\ \delta\omega_2 \\ \delta\omega_3 \end{pmatrix} = \begin{pmatrix} \dot{\varphi}\cos(\vartheta) - \dot{\psi}\cos(\varphi)\sin(\vartheta) \\ \dot{\vartheta} + \dot{\psi}\sin(\varphi) \\ \dot{\varphi}\sin(\vartheta) + \dot{\psi}\cos(\vartheta)\cos(\varphi) \end{pmatrix} = \mu(\theta) \begin{pmatrix} \dot{\varphi} \\ \dot{\vartheta} \\ \dot{\psi} \end{pmatrix}$$

with $\theta = (\varphi, \vartheta, \psi)$, and with:

$$\mu(\theta) = \begin{bmatrix} \cos\vartheta & 0 & -\sin\vartheta\cos\varphi \\ 0 & 1 & \sin\varphi \\ \sin\vartheta & 0 & \cos\vartheta\cos\varphi \end{bmatrix}$$

So, the total angular velocity can be rewritten in the form

$$\begin{pmatrix} \omega_1 \\ \omega_2 \\ \omega_3 \end{pmatrix} = \underbrace{\underbrace{M_{213}(\varphi, \vartheta, \psi) \cdot \underbrace{\begin{pmatrix} 0 \\ -\omega_0 \\ 0 \end{pmatrix}}_{ORF\_ang\_vel\_wrt\_ECI\_in\_ORF}}_{ORF\_ang\_vel\_wrt\_ECI\_in\_BRF} + \underbrace{\begin{pmatrix} \dot{\varphi}\cos(\vartheta) - \dot{\psi}\cos(\varphi)\, sin\,(\vartheta) \\ \dot{\vartheta} + \dot{\psi}\sin(\varphi) \\ \dot{\varphi} sin\,(\vartheta) + \dot{\psi}\cos(\vartheta)\cos(\varphi) \end{pmatrix}}_{BRF\_ang\_vel\_wrt\_ORF\_in\_BRF}}_{BRF\_ang\_vel\_wrt\_ECI\_in\_BRF}$$

The last equation shows an application of the well known theorem of the composition of angular speed. The derivatives of the Euler angles $(\dot{\varphi}, \dot{\vartheta}, \dot{\psi})$ are then related to the components of the angular rate $(\omega_1, \omega_2, \omega_3)$ vector by the relationship:

$$\begin{pmatrix} \dot{\varphi} \\ \dot{\vartheta} \\ \dot{\psi} \end{pmatrix} = \begin{pmatrix} \omega_1\cos(\vartheta) + \omega_0\sin(\psi) + \omega_3\sin(\vartheta) \\ \omega_1\sin(\vartheta) + \omega_0\frac{\cos(\psi)}{\sin(\varphi)} + \omega_2\frac{\cos(\varphi)}{\sin(\varphi)} - \omega_3\cos(\vartheta) \\ -\omega_1\frac{\sin(\vartheta)}{\sin(\varphi)} - \omega_0\cos(\psi) + \omega_3\frac{\cos(\vartheta)}{\sin(\varphi)} \end{pmatrix}$$

The angular rate derivatives are computed as follows (for a constant orbital rate ω_0, the generalization to a variable ω_0 is straightforward):

$$\begin{pmatrix} \dot{\omega}_1 \\ \dot{\omega}_2 \\ \dot{\omega}_3 \end{pmatrix} = \dot{M}_{213}(\varphi, \vartheta, \psi)\begin{pmatrix} 0 \\ -\omega_0 \\ 0 \end{pmatrix}$$
$$+ \begin{pmatrix} \ddot{\varphi}\cos(\vartheta) - \dot{\varphi}\dot{\vartheta}\sin(\vartheta) - \ddot{\psi}\cos(\varphi)\sin(\vartheta) + \dot{\psi}\dot{\varphi}\sin(\varphi)\sin(\vartheta) - \dot{\psi}\dot{\vartheta}\cos(\varphi)\cos(\vartheta) \\ \ddot{\vartheta} + \ddot{\psi}\sin(\varphi) + \dot{\psi}\dot{\varphi}\cos(\varphi) \\ \ddot{\varphi}\sin(\vartheta) + \dot{\varphi}\dot{\vartheta}\cos(\vartheta) + \ddot{\psi}\cos(\varphi)\cos(\vartheta) - \dot{\psi}\dot{\varphi}\cos(\varphi)\sin(\vartheta) - \dot{\psi}\dot{\vartheta}\sin(\varphi)\cos(\vartheta) \end{pmatrix}$$

The linearization for small roll and pitch angles leads to the following expressions for the rotation matrix, and the angular rates and accelerations:

$$M_{213}(\varphi, \vartheta, \psi) = \begin{bmatrix} 1 & \psi & -\vartheta \\ -\psi & 1 & \varphi \\ \vartheta & -\varphi & 1 \end{bmatrix}$$

$$\begin{pmatrix} \omega_1 \\ \omega_2 \\ \omega_3 \end{pmatrix} = \begin{pmatrix} -\omega_0\psi + \dot{\varphi} \\ -\omega_0 + \dot{\vartheta} \\ \omega_0\varphi + \dot{\psi} \end{pmatrix}$$

$$\begin{pmatrix} \dot{\omega}_1 \\ \dot{\omega}_2 \\ \dot{\omega}_3 \end{pmatrix} = \begin{pmatrix} -\omega_0\dot{\psi} + \ddot{\varphi} \\ \ddot{\vartheta} \\ \omega_0\dot{\varphi} + \ddot{\psi} \end{pmatrix}$$

This parametrization has singularities at roll values of $\varphi = \frac{\pi}{2} + n\pi$ for $n \in \mathbf{Z}$; for such values the determinant of the matrix $\mu(\theta)$ becomes null.

2.3.2.2 Euler Angles 313

The Euler angles 313 are often used to characterize the kinematics of a gyroscopic motion:

$$\begin{aligned} \varphi &= \text{spin angle} \\ \vartheta &= \text{nutation angle} \\ \psi &= \text{precession angle} \end{aligned}$$

The rotation matrix 313 is obtained by applying, in this order, the following transformations:

- a rotation about the Z_{ORF} axis equal in magnitude to the spin angle;
- a rotation about the rotated X_{ORF} axis equal in magnitude to the nutation angle;
- a rotation about the twice rotated Z_{ORF} axis equal in magnitude to the precession angle.

The transformation matrix is (see [9] Sect. 8.6 where M_{313} is called R_{313}, φ and ψ are inverted):

$$M_{B \leftrightarrow O} = M_{313}$$
$$= \begin{bmatrix} \cos(\psi)\cos(\varphi) - \sin(\psi)\sin(\varphi)\cos(\vartheta), & \sin(\psi)\cos(\varphi)\cos(\vartheta) + \cos(\psi)\sin(\varphi), & \sin(\psi)\sin(\vartheta) \\ -\sin(\psi)\cos(\varphi) - \cos(\psi)\sin(\varphi)\cos(\vartheta), & \cos(\psi)\cos(\varphi)\cos(\vartheta) - \sin(\psi)\sin(\varphi), & \cos(\psi)\sin(\vartheta) \\ \sin(\varphi)\sin(\vartheta), & -\cos(\varphi)\sin(\vartheta), & \cos(\vartheta) \end{bmatrix}$$

The expression of the angular speed is (see Appendix A.2):

$$\omega = \underbrace{\begin{bmatrix} \sin\vartheta\sin\psi & \cos\psi & 0 \\ \sin\vartheta\cos\psi & -\sin\psi & 0 \\ \cos\vartheta & 0 & 1 \end{bmatrix}}_{\mu} \cdot \begin{bmatrix} \dot{\varphi} \\ \dot{\vartheta} \\ \dot{\psi} \end{bmatrix}$$

This parametrization has singularities at nutation values of $\vartheta = n\pi$ for $n \in \mathbf{Z}$. It can be shown that every parametrization of $SO(3)$ with a minimal set of three parameters brings to the problem of singularities for some values of the parameters. In order to avoid singularities during he integration of the kinematics it is necessary to introduce a non-minimal parametrization, such as the quaternions, or to define an atlas so that each state of $SO(3)$ is mapped to two different Euler angle sets, and the singularities are avoided by defining transition rules from one set to the other.

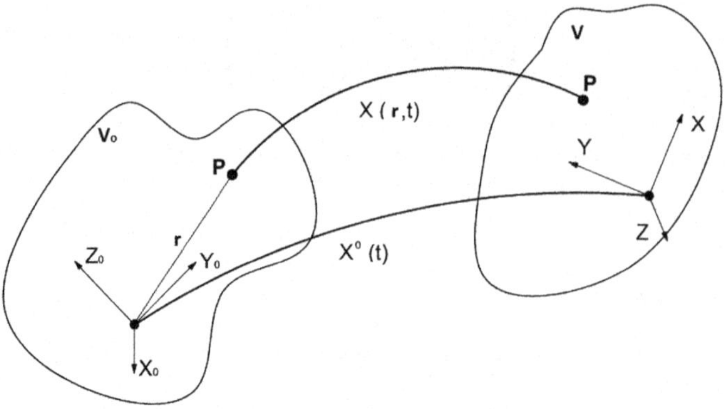

Fig. 2.5 Kinematics of the fexible body

2.4 Kinematics of the Flexible Body

The flexible body has a natural undeformed configuration defined by the mass distribution in its reference frame. During the motion of the flexible body the distances between the various points can change. The **BRF** was defined, in Sect. 2.3, with an origin attached to a special point of the body; we assume now, that this specific point is a hard point[2]—a point with a rigid small neigh-borough—of the satellite, so that we can attach to it a rigid frame and define a finite space of N_f deformation functions in this rigid frame. The deformation functions and their space derivatives are null at the origin in order to provide the continuity of the global displacement field and its derivatives.

We assume that at time t_0 the body is in the reference natural configuration where it occupies the volume V_0.

The vector r provides the coordinates of any point of the body $P \in V_0$ in the triad X_0, Y_0, Z_0. Looking at Fig. 2.5, we have that the initial triad X_0, Y_0, Z_0 moves to the triad X, Y, Z at time t. The volume of the body changes from V_0 to V.

Each generic point $P \in V_0$ of coordinates r in the triad X_0, Y_0, Z_0 moves to a new position $x(r, t) \in V$ with components always in X_0, Y_0, Z_0 . The displacement of the origin is represented by the vector $x^0(t)$.

The general mapping between V_0 and V is the kinematics of continuum mechanics as described in [10, 11]. Without loosing generality, we assume that the triad X_0, Y_0, Z_0 corresponds to **ECI** and the triad X, Y, Z corresponds to **BRF**.

We assume in addition, the displacement $x(r, t)$ as the sum of the displacement due to the rigid motion of the triads plus a small displacement due to the deformation of the body that we describe by vector functions having components in the triad X, Y, Z.

---

[2]The existence of an hard point to introduce a body triad is not necessary and can be relaxed. We will not enter int his detail.

Thus, in addition to the variables defined for the case of rigid body (see Eqs. 2.1, 2.2), we introduce the following definitions:

- $\phi_j^k(r)$ are a set of deformation functions, the lower index j spans in $1, 2, 3$ the higher index k spans in $1, 2 \ldots N_f$;
- ε_k are the amplitudes of the deformation functions and are time dependent. The index k spans in $1, 2 \ldots N_f$;

We denote the components of the transfer matrix $M_{E \leftrightarrow B}$ as M_{ij} and expand the kinematic model expressed for the rigid case by the Eqs. 2.1, 2.2; the kinematic model of the flexible body is the map $(r \in V_0,\ x^0,\ \theta,\ \varepsilon) \longrightarrow x \in V$:

$$x_i = x_i^o + \sum_{j=1,2,3} M_{ij}(\theta)(r_j + \sum_{k=1..N_f} \phi_j^k(r)\varepsilon_k),\ r \in V_0$$

Thanks to our convention on summations this equation can be written as:

$$x_i = x_i^o + M_{ij}(\theta)(r_j + \phi_j^k(r)\varepsilon_k) \tag{2.14}$$

By time derivation we get:

$$\begin{aligned} \dot{x}_i &= \dot{x}_i^o + \dot{M}_{ij}(r_j + \phi_j^k \varepsilon_k) + M_{ij}\phi_j^k \dot{\varepsilon}_k \\ \dot{x}_i &= \dot{x}_i^o + M_{ie}e_{esj}\omega_s(r_j + \phi_j^k \varepsilon_k) + M_{ij}\phi_j^k \dot{\varepsilon}_k \end{aligned} \tag{2.15}$$

where $dim[\varepsilon_k] = N_f$ is the number of shape functions describing the body deformation.

In the Eqs. 2.14, 2.15, if the index appears as superscript and subscript it is summed from 1 to N_f, so that $\phi_j^k \varepsilon_k = \sum_{k=1..N_f} \phi_j^k \varepsilon_k$; while the index is summed in $1, 2, 3$ when it appears always as subscript, so that $M_{ik}r_k = \sum_{k=1,2,3} M_{ik}r_k$.

This kinematic description includes and generalizes any description used in the *finite element method* where the deformation functions are piecewise polynomials or also continuum models using a large but finite number of eigen-funtions. This last approach is often used to produce simplified models when our body is composed of rods, beams or plates where the local eigen-modes are known to describe efficiently the kinematics.

References

1. R.H. Battin, *An Introduction to the Mathematics and Methods of Astrodynamics*, Revised Edition, AJAA Education Series (1999)
2. R. Featherstone, *Rigid Body Dynamics Algorithms* (Springer, 2007)
3. F.L. Markley, J.L. Crassidis, *Fundamentals of Spacecraft Attitude Determination and Control* (Springer, New York, 2015)
4. J.R. Wertz, *Spacecraft Attitude Determination and Control* (D. Reidel Publishing Company, Dordrecht, 2002)

5. T.R. Kane, P.W. Likins, D.A. Levinson, *Spacecraft dynamics* (McGraw-Hill, 1983)
6. P. Santini, Stability of flexible spacecraft. Acta Astronaut. **3** (1976)
7. A.A. Shabana, *Dynamics of Multibody Systems* (Cambridge University Press, 2005)
8. H. Goldstein, *Classical Mechanics* (Zanichelli, 1980)
9. J. Diebel, *Representing attitude: euler angles, quaternions and rotation vectors*, Technical Report (Standford University, Palo Alto, 2006)
10. T. Manacorda, *Introduzione alla termomeccanica dei continui* (Quaderni UMI, Pitagora Editrice, 1979)
11. C. Truesdell, W. Noll, *The Non-Linear Field Theories of Mechanics*, 3rd edn, ed. by S.S. Antman (Springer, Heidelberg, 2003)

Chapter 3
The Dynamics of the Flexible Satellite

Abstract We derive the dynamic equations of attitude, orbit and flexible modes of a generic satellite. We will follow the Lagrangian approach, and the equations will be set in a form specifically suitable for attitude and orbit control. After the derivation of the flexible body Lagrangian equations we treat some specific complements that are useful for space applications: momentum management systems, generalized forces due a central gravitational field, first invariants of motion in a central gravitational field, transformations to uncouple orbital and attitude equations and methods to linearize the equations around a reference orbital and attitude trajectory. We close the chapter by presenting a typical linear model of a flexible satellite as used in practical linear control engineering design.

3.1 Introduction

Although most of the literature on AOC systems deals with rigid body dynamics, the dynamics of the flexible modes are very important for practical engineering design. There are various reasons for this.

The satellites must be light and must embark appendages like solar arrays, antennas, radars, telescopes with a low free oscillation frequency, high modal mass and low damping. In modern satellites the trend is to have always lighter platforms with increased demands for power and performances which gradually improve the size of flexible appendages.

The flexible modes in a closed loop controlled spacecraft may affect the stability of the control systems or may produce a strong excitation of the control actuation and attitude dynamics.

Most of the books treating AOCS introduce the flexibility on simple models which cannot be exported to other design problems (see [1, 2]). The approach of [3] is effective in illustrating some of the problems in linear control design, however it does not provide a global theory to treat the flexibilities.

The flexible satellite dynamics have already been presented and derived in some monographs like [4–7]. There are two approaches to write the dynamic equations of a flexible structure: the use of the Ritz-Galerkin method or the Lagrangian method.

© Springer International Publishing Switzerland 2016

L. Mazzini, *Flexible Spacecraft Dynamics, Control and Guidance*,
Springer Aerospace Technology, DOI 10.1007/978-3-319-25540-8_3

The book [5] generalizes the Lagrangian method with an approach that encompasses articulated systems. In the works of [4, 6] the equilibrium between active and inertial forces is established using the Ritz-Galerkin approximation—an integral projection on suitable test functions—under the hypothesis of small displacements. This is the standard procedure for the finite element approach.

We develop in this chapter, using the Lagrangian method, an extended rigid body model which is of immediate use for the attitude and orbit control applications in the presence of flexibility. The rigid body part—the so-called Euler equations—is immediately recognized together with the terms (linear and non-linear) which couple the rigid body and the flexible modes dynamics.

Similar to what presented by [8], which generalizes the Lagrangian equations using the *quasi-coordinates*, our method allows for the introduction of the angular velocity in the description of the Lagrangian, giving the possibility to describe the kinematics through the standard rigid body representations studied in Chap. 2.

The designers' effort to write the full non-linear model will be simply to evaluate an extended set of inertia parameters—in addition to the mass and the inertia matrix—and to introduce a suitable elastic potential energy. Examples of the derivation of these parameters are given in Chap. 6 and in Appendix B.2.

After the derivation of the flexible body Lagrangian equations we treat some specific complements that are useful for space applications: momentum management systems, generalized forces due a central gravitational field, first invariants of motion in a central gravitational field, transformations to uncouple the orbital and the attitude equations and how to linearize these equations around a given orbital and attitude trajectory. We close the chapter by presenting a typical linear model of a flexible satellite as used in practical linear control engineering design.

We assume a basic knowledge of Lagrangian mechanics which dates back to Lagrange work [9]. In the Appendix B.1 a simple derivation of Lagrangian mechanics is presented following the classic presentation provided by [10].

Wide and modern expositions of the Lagrangian approach for multibody systems that can be found in other authors like [11, 12], where connections with *continuum mechanics* are also presented.

3.2 Lagrangian Equations

The *Lagrangian* of the satellite mechanical system is composed by the difference between its *kinetic energy* $\mathscr{T}$ and *potential energy* $\mathscr{V}$ (refer to Appendix B.1).

The potential energy can be produced by the internal interaction among satellite's particles, like the case of the elastic potential energy, or by the interaction of satellite's particles with the satellite's external world like in the case of the gravitational potential energy. Interactions that are not derived by a potential energy must be treated in terms of generalized Lagrangian forces.

Denoting by:

- q_i, $i = 1...N_f + 6$ the components of the generalized Lagrangian coordinates vector $q = (x_1^o, x_2^o, x_3^o, \theta_1, \theta_2, \theta_3, \varepsilon_1, \varepsilon_2...\varepsilon_{N_f})$. These components are the kinematic variables that describe the position of any satellite particle in **ECI** through Eq. 2.14 defined in Sect. 2.4;
- Q_i, $i = 1...N_f + 6$ the generalized Lagrangian forces. They take into account those interactions that are not included in the potential energy. They are calculated from the physical forces using $Q_i \delta q_i = \delta L$, where δL is the virtual work performed by the physical forces on the virtual displacement field generated by a virtual variation δq of the Lagrangian coordinates (see Appendix B.1);
- The Lagrangian $\mathscr{L} = \mathscr{T} - \mathscr{V}$, which is a scalar function of $(q, \dot{q})$.

From Appendix B.1, the Lagrangian equations are:

$$\frac{d}{dt}\frac{\partial \mathscr{L}}{\partial \dot{q}_i} - \frac{\partial \mathscr{L}}{\partial q_i} = Q_i \quad , \; i = 1...N_f + 6 \tag{3.1}$$

In order to specialize these equations to the flexible satellite, the first task is to derive the kinetic energy as a function of the kinematic model variables and their first time derivatives introduced in Sect. 2.4.

3.3 Kinetic Energy

The *kinetic energy* general expression is:

$$\mathscr{T} = \frac{1}{2}\int_V \rho(x) \sum_{i=1,2,3} \dot{x}_i \dot{x}_i \, dV = \frac{1}{2}\int_{V_0} \rho_o(r) \sum_{i=1,2,3} \dot{x}_i \dot{x}_i \, dV_o$$

being V_0 the volume in the reference configuration, $\rho_o(r)$ the mass density in the reference configuration, V the volume in the deformed configuration, $\rho(x)$ the mass density in the deformed configuration.

Due to the equivalence between the two integrals (see Appendix A.1) and introducing the kinematic model defined in Sect. 2.4, we can integrate the kinetic energy in the reference configuration.

Applying our convention about the sum of repeated indexes appearing in any product of arrays, we can also write:

$$\mathscr{T} = \frac{1}{2}\int_{V_0} \rho_o(r) \dot{x}_i \dot{x}_i \, dV_o \tag{3.2}$$

We remember that when the same index appears twice or more times in the same group of array factors a summation on that index is assumed. We introduce the following rule: if the index appears both as superscript and subscript it is summed from

1 to N_f, so that $\phi_j^k \varepsilon_k = \sum_{k=1\ldots N_f} \phi_j^k \varepsilon_k$; while the index is summed in 1, 2, 3 when it appears always as subscript, so that $M_{ie} e_{esj} \omega_s r_j = \sum_{e=1,2,3} M_{ie} \sum_{s,j=1,2,3} (e_{esj} \omega_s r_j)$.

We now introduce in Eq. 3.2 the expression of the body velocity field expressed by Eq. 2.15, obtaining:

$$
\begin{aligned}
\mathscr{T} = \frac{1}{2} \int_{V_0} \big[& \dot{x}_i^o \dot{x}_i^o + 2\dot{x}_i^o M_{ie} e_{esj} \omega_s r_j + 2\dot{x}_i^o M_{ie} e_{esj} \omega_s \phi_j^k \varepsilon_k \\
& + 2\dot{x}_i^o M_{ij} \phi_j^k \dot{\varepsilon}_k + M_{ie} e_{esj} \omega_s (r_j + \phi_j^k \varepsilon_k) M_{ip} e_{ptq} \omega_t (r_q + \phi_q^n \varepsilon_n) \\
& + 2 M_{ie} e_{esj} \omega_s (r_j + \phi_j^k \varepsilon_k) M_{ip} \phi_p^q \dot{\varepsilon}_q + M_{ij} \phi_j^k M_{it} \dot{\varepsilon}_k \phi_t^p \dot{\varepsilon}_p \big] \rho_o \, dV_o
\end{aligned}
$$

To carry out this computation we have used the following identities:

$$
M_{ie} M_{it} = \delta_{et} \quad \text{and} \quad e_{isj} e_{itq} = \delta_{st} \delta_{jq} - \delta_{qs} \delta_{jt}
$$

As in Sect. 2.4, M_{ij} are the components of the transfer matrix $M_{E \leftrightarrow B}$.

In order to calculate the kinetic energy we first define the following mass parameters' symbols:

$$
\begin{aligned}
m &= \int_{V_0} \rho_o \, dV_o \\
p_k &= \int_{V_0} r_k \, \rho_o \, dV_o \\
L_j^k &= \int_{V_0} \phi_j^k \rho_o \, dV_o
\end{aligned}
$$

$$
S_s^q = \int_{V_0} e_{sjp} r_j \phi_p^q \rho_o \, dV_o
$$

$$
I_{st} = \int_{V_0} (\delta_{st} r_j r_j - r_s r_t) \rho_o \, dV_o; \quad I_{st} = I_{ts}
$$

$$
T^{kp} = \int_{V_0} \phi_j^k \phi_j^p \rho_o \, dV_o; \quad T^{kp} = T^{pk}
$$

$$
J_{st}^k = \int_{V_0} (2 r_j \phi_j^k \delta_{st} - r_s \phi_t^k - r_t \phi_s^k) \rho_o \, dV_o; \quad J_{st}^k = J_{ts}^k
$$

$$
K_{st}^{ke} = \int_{V_0} (\delta_{st} \phi_j^k \phi_j^e - \phi_s^k \phi_t^e) \rho_o \, dV_o; \quad K_{st}^{ke} = K_{ts}^{ke} = K_{st}^{ek}
$$

$$
P_s^{kq} = \int_{V_0} e_{sjp} \phi_j^k \phi_p^q \rho_o \, dV_o; \quad P_s^{kq} = -P_s^{qk} \tag{3.3}
$$

The mass parameters superscript indexes range in $1...N_f$, while the subscript indexes range in $1, 2, 3$. In Appendix B.2 these parameters are calculated for a simple system of a rigid body connected with springs to one or two concentrated masses.

We get for the kinetic energy:

$$
\begin{aligned}
\mathscr{T} &= \frac{1}{2} m \dot{x}_i^o x_i^o + \dot{x}_i^o M_{ie} e_{esj} \omega_s p_j + \dot{x}_i^o M_{ie} e_{esj} \omega_s \varepsilon_k L_j^k \\
&+ \dot{x}_i^o M_{ij} \dot{\varepsilon}_k L_j^k + \frac{1}{2} (I_{st} + J_{st}^n \varepsilon_n + K_{st}^{kn} \varepsilon_k \varepsilon_n) \omega_s \omega_n \\
&+ \omega_s \dot{\varepsilon_q} (S_s^q + P_s^{kq} \varepsilon_k) + \frac{1}{2} \dot{\varepsilon}_k \dot{\varepsilon}_p T^{kp}
\end{aligned}
$$

3.3.1 Compact Notation

We introduce now a *compact notation* that allows to present the kinetic energy and then the flexible body equations in a simpler format with respect to the indicial notation. This notation uses the standard column vector and matrix notation for the operations in $\Re^3$ to replace the subscripts indexes spanning $1, 2, 3$.

The classic operations of scalar product, transpose $(\cdot)^T$, vector product $[\cdot, \cdot]$ or $(\cdot \times \cdot)$, trace $tr(\cdot)$ and matrix product are applicable and operate only in $\Re^3$ substituting the use subscripts.

The one index array is considered as a column vector; the following equalities apply: $\omega_s \omega_s = \omega^T \omega$, $\{I_{ts} \omega_s\}_{t=1,2,3} = I\omega$, $I_{ss} = tr(I)$.

In order to have vectors and matrices in $\Re^3$ from the mass parameters introduced, we first need to sum the superscript indexes. For example $I_{st} + J_{st}^n \varepsilon_n + K_{st}^{kn} \varepsilon_k \varepsilon_n$ is a 3×3 matrix when we have implemented the sums for $k, n = 1...N_f$.

To represent superscript indexes' sums $1...N_f$ we introduce the operator $\langle\ \rangle$.

When we have superscript indexes in a mass parameter A^{i_1, i_2, i_3}, summing with flexible amplitudes $\varepsilon_{i_1}, \varepsilon_{i_2}, \varepsilon_{i_3}$ or their derivatives $\dot{\varepsilon}_{i_k}$, we eliminate the indices by surrounding the sequence of symbols with a $\langle\ \rangle$ parenthesis and separate the kinematic variables with comas: $A^{i_1, i_2, i_3} \varepsilon_{i_1} \dot{\varepsilon}_{i_2} \varepsilon_{i_3} = \langle A\varepsilon, \dot{\varepsilon}, \varepsilon \rangle$, the kinematic variables must be put in the same order of the respective A index.

In most cases, the order of the sequence of the kinematic variables is not relevant because almost all mass properties are symmetric in the superscript indexes; the order is relevant only for the mass parameter P which is antisymmetric in the superscripts. We have the following identities:

$$
\langle L\varepsilon \rangle = L^k \varepsilon_k, \ \langle S\dot{\varepsilon} \rangle = S^k \dot{\varepsilon}_k \quad \text{and} \quad \langle T\dot{\varepsilon}, \dot{\varepsilon} \rangle = T^{ij} \dot{\varepsilon}_i \dot{\varepsilon}_j.
$$

The $\langle\ \rangle$ operator sums superscript indexes, the symbols encircled by these brackets have subscript or superscript indexes left free which can operate externally to the parenthesis, therefore: $\langle J\varepsilon \rangle$ or $\langle K\varepsilon, \varepsilon \rangle$ are matrices in $\Re^3$ and $\langle L\varepsilon \rangle$ is a column vector in $\Re^3$, $\omega^T\langle K\varepsilon\rangle\omega = \left\{K_{st}^{kn}\varepsilon_k\omega_s\omega_t\right\}_{n=1...N_f}$ is a N_f column vector or $\langle J\varepsilon\rangle\omega = \left\{J_{ij}^k\varepsilon_k\omega_j\right\}_{i=1,2,3}$ is a $\Re^3$ column vector.

We advice the reader that at the end of this chapter and in the following chapters when there is no ambiguity we will omit the use of $\langle\,\rangle$ parentheses, so that the $\Re^3$ column vector $\langle S\varepsilon\rangle$ can also be represented as $S\varepsilon$ indifferently, in this case S operates as a $3\times N_f$ matrix.

Calling $M_{ie}\dot{x}_i^o = v_e$, and using the compact notation $M^T\dot{x}^o = v$, we can express the *kinetic energy* as:

$$\begin{aligned}\mathcal{T} &= \frac{1}{2}mv^Tv + (p+\langle L\varepsilon\rangle)^T[v,\omega] + v^T\langle L\dot{\varepsilon}\rangle + \frac{1}{2}\omega^T(I+\langle J\varepsilon\rangle+\langle K\varepsilon,\varepsilon\rangle)\omega\\ &\quad+\omega^T\langle(S+\langle P\varepsilon\rangle)\dot{\varepsilon}\rangle + \frac{1}{2}\langle T\dot{\varepsilon},\dot{\varepsilon}\rangle\end{aligned} \tag{3.4}$$

$\mathcal{T}$ is a function of v, ω, $\dot{\varepsilon}$, ε. If θ is a generic parametrization of $SO(3)$, then

$$v = M^T(\theta)\dot{x}^o \quad , \quad \omega = \mu(\theta)\dot{\theta}$$

therefore, implicitly $\mathcal{T}$ depends on $\dot{x}^o, \theta, \dot{\theta}, \dot{\varepsilon}, \varepsilon$.

3.4 Implicit Form of the Lagrangian Equations

Let us recall the Lagrangian $\mathcal{L} = \mathcal{T} - \mathcal{V}$ and the equations of the dynamics of flexible body:

$$\frac{d}{dt}\frac{\partial\mathcal{L}}{\partial\dot{q}_i} - \frac{\partial\mathcal{L}}{\partial q_i} = Q_i, i = 1...N_f+6$$

where $q_i,\ i = 1...N_f+6$ is the vector containing all configuration variables:

$$q = (x^o_{i=1,3}, \theta_{i=1,3}, \varepsilon_{i=1,N_f})^T,$$

and $Q_i,\ i = 1...N_f+6$ is the vector that contains all components of Lagrangian forces *which are not accounted for in the potential energy*:

$$Q = (F_{i=1,3}, C_{i=1,3}, Q^{\varepsilon}_{i=1,N_f})^T.$$

The generalized Lagrangian forces Q are calculated from physical forces interacting with the satellite, not included in the potential energy, using:

$$\delta L = Q^T\delta q = F^T\delta x^o + C^T\delta\theta + Q^{\varepsilon T}\delta\varepsilon, \tag{3.5}$$

where δL is the virtual work performed by the physical forces on the virtual displacement field generated by a virtual variation δq of Lagrangian coordinates (see Appendix B.1).

Their explicit calculation in terms of force field integral will be done later in this section.

To define what interactions will be modeled as Q and what will be modeled with the potential energy $\mathscr{V}$, we divide all interactions in four sets: *conservative internal body interactions* (i.e. the conservative interactions among the particles of the body), *conservative external interactions*, *non-conservative internal body interactions*, *non-conservative external interactions*.

The forces exerted by mass ejection systems—for which the previous definition is ambiguous—will be considered as *non-conservative external interactions* and the *rocket equation* discussed in Chap. 8 will be used to calculate its effects.

Any conservative interaction, which is derived from a potential can be considered either inside the Lagrangian or as a Lagrangian force.

In order to create a flexible satellite model suitable to many different environmental conditions, we choose to consider inside the Lagrangian the *conservative internal body interactions* (the interactions among the particles of the body) and the *conservative external interactions* as generalized Lagrangian forces.

Thus, we split the potential energy in two terms $\mathscr{V} = \mathscr{V}_I + \mathscr{V}_E$. The first term $\mathscr{V}_I$ is a function only of internal variables $\varepsilon_{i=1,N_f}$, we call it the *internal potential energy* and we will include it in the Lagrangian.

The second term is the *external potential energy* due to all conservative external interactions (like gravity). We do not add the external potential energy in the Lagrangian and we consider minus the gradient of the *external potential energy* as a generalized Lagrangian force.

External non-conservative forces denoted by ($F_i^{(n.c.)}$, $C_i^{(n.c.)}$, $Q_i^{\varepsilon(n.c.)}$) will be considered as generalized Lagrangian forces.

Non-conservative internal forces like the internal damping will be treated using the so-called *dissipation function* $\mathscr{F}(\varepsilon, \dot{\varepsilon})$ (see [10]) which is a sort of potential of the internal dissipative forces, it is a function of the internal variables and their time derivatives. The dissipation function, by definition, provides the generalized Lagrangian forces due to the *non-conservative internal forces* by deriving it with respect to the $\dot{q}$ and changing the sign of the result.

With this split it follows that the Lagrangian equations, Eq. 3.1, become:

$$\frac{d}{dt}\frac{\partial \mathscr{T}}{\partial \dot{q}_i} - \frac{\partial \mathscr{T}}{\partial q_i} + \frac{\partial \mathscr{F}}{\partial \dot{q}_i} + \frac{\partial \mathscr{V}_I}{\partial q_i} = Q_i \quad , \; i = 1 \ldots N_f + 6 \tag{3.6}$$

with:

$$
\begin{aligned}
Q &= [F_{i=1\ldots3}, C_{i=1\ldots3}, Q^{\varepsilon}_{i=1\ldots N_f}]^T \\
F_i &= -\tfrac{\partial \mathscr{V}_E}{\partial x^o_i} + F_i^{(n.c.)} \\
C_i &= -\tfrac{\partial \mathscr{V}_E}{\partial \theta_i} + C_i^{(n.c.)} \\
Q^{\varepsilon}_i &= -\tfrac{\partial \mathscr{V}_E}{\partial \varepsilon_i} + Q_i^{\varepsilon(n.c.)} \\
\mathscr{L} &= \mathscr{T} - \mathscr{V}_I
\end{aligned}
\tag{3.7}
$$

$\mathscr{F}$ and $\mathscr{V}_I$ are invariant to translations and rotations being functions of only $\dot{\varepsilon}$ and ε respectively.

In the next section we derive the equations of the dynamics considering the kinetic energy an implicit function of v, ω, $\dot{\varepsilon}$, ε (following the model defined by Eq. 3.4); the resulting equations will be called in *implicit form* and are of very general application.

The expression of the kinetic energy obtained in Sect. 3.3 Eq. 3.4 will be later introduced in the *implicit form* equations to get the final flexible body equations.

3.4.1 The Equations of Translational Dynamics

For the Lagrangian equations relevant to $\dot{x}^o$, being $\mathscr{V}_I$ and $\mathscr{F}$ independent of x^o, $\dot{x}^o$ and $\mathscr{T}$ dependent on $\dot{x}^o$ through v, we get from the first three of Eq. 3.6:

$$
\frac{d}{dt}\frac{\partial \mathscr{T}}{\partial \dot{x}^o} = F \quad \Rightarrow \quad \frac{d}{dt}\left(\frac{\partial \mathscr{T}}{\partial v_s}\frac{\partial v_s}{\partial \dot{x}^o_i}\right) = \frac{d}{dt}\left(\frac{\partial \mathscr{T}}{\partial v_s} M_{is}\right) = F_i
$$

that is

$$
M_{is}\frac{d}{dt}\frac{\partial \mathscr{T}}{\partial v_s} + \frac{\partial \mathscr{T}}{\partial v_s} M_{ie} e_{ems}\omega_m = F_i
$$

Here F_i are the components of the generalized Lagrangian force corresponding to translational kinematic variables in **ECI**. We will see later that this is the sum of all external forces.

If we introduce the force components in **BRF** $f_k = M_{ik}F_i$ (or $f = M^T F$), the equations of translational dynamics can be written, multiplying by M_{ik} and summing with the usual rules, as

$$
\frac{d}{dt}\frac{\partial \mathscr{T}}{\partial v_k} + e_{kms}\omega_m \frac{\partial \mathscr{T}}{\partial v_s} = M_{ik}F_i = f_k \tag{3.8}
$$

In *compact notation* Eq. 3.8 become:

$$
\frac{d}{dt}\frac{\partial \mathscr{T}}{\partial v} + [\omega, \frac{\partial \mathscr{T}}{\partial v}] = f \tag{3.9}
$$

3.4.2 The Equations of Rotational Dynamics

To derive the equations of the rotational dynamics, let us introduce a parametrization of the $SO(3)$ space $(\theta_1, \theta_2, \theta_3)$; as shown in Sect. 2.3 the angular speed depends linearly on the derivatives of the three parameters (e.g. the Euler angles):

$$\omega_s = \mu_{sk}\dot{\theta}_k$$

The Lagrange equations for θ are obtained from the fourth, fifth and sixth equation of Eq. 3.6, being $\mathscr{V}_I$, $\mathscr{F}$ independent of θ and $\mathscr{T}$ dependent on θ through ω and v, these equations can be written as follows:

$$\frac{d}{dt}\left(\frac{\partial \mathscr{T}}{\partial \omega_s}\frac{\partial \omega_s}{\partial \dot{\theta}_k}\right) - \frac{\partial \mathscr{T}}{\partial \omega_s}\frac{\partial \omega_s}{\partial \theta_k} - \frac{\partial \mathscr{T}}{\partial v_s}\frac{\partial v_s}{\partial \theta_k} = C_k,$$

here C_k are the Lagrangian force components corresponding to θ variables. The following passages can be easily implemented:

$$\frac{d}{dt}\left(\frac{\partial \mathscr{T}}{\partial \omega_s}\mu_{sk}\right) - \frac{\partial \mathscr{T}}{\partial \omega_s}\frac{\partial \mu_{se}}{\partial \theta_k}\dot{\theta}_e = C_k + \frac{\partial \mathscr{T}}{\partial v_s}\frac{\partial v_s}{\partial \theta_k}$$

$$\left(\frac{d}{dt}\frac{\partial \mathscr{T}}{\partial \omega_s}\right)\mu_{sk} + \frac{\partial \mathscr{T}}{\partial \omega_s}\left(\frac{\partial \mu_{sk}}{\partial \theta_e} - \frac{\partial \mu_{se}}{\partial \theta_k}\right)\dot{\theta}_e = C_k + \frac{\partial \mathscr{T}}{\partial v_s}\frac{\partial v_s}{\partial \theta_k}$$

$$\frac{d}{dt}\frac{\partial \mathscr{T}}{\partial \omega_q} + \frac{\partial \mathscr{T}}{\partial \omega_s}\mu^{\#}_{qk}\left(\frac{\partial \mu_{sk}}{\partial \theta_e} - \frac{\partial \mu_{se}}{\partial \theta_k}\right)\dot{\theta}_e = \mu^{\#}_{qk}\left(C_k + \frac{\partial \mathscr{T}}{\partial v_s}\frac{\partial v_s}{\partial \theta_k}\right) \tag{3.10}$$

where $\mu^{\#}_{qk}\mu_{sk} = \delta_{qs}$ and by definition the symbol $^{\#}$ is equivalent to the transpose inverse:

$$\mu^{\#} = \mu^{-1^T} \tag{3.11}$$

The virtual work can be also expressed in terms of a **BRF** vector $c = \mu^{\#}C$; this can be verified by developing the identity $c^T\delta\omega = C^T\delta\dot{\theta}$ using $\delta\omega = \mu\delta\dot{\theta}$.

It follows $c^T\mu\delta\dot{\theta} = C^T\delta\dot{\theta}$ for any $\delta\dot{\theta}$, and transposing $\mu^T c = C$ or $c = \mu^{-1^T}C$. We will demonstrate soon that c is the total external torque (with respect to x^o) projected in **BRF**.

We now develop the two terms of Eq. 3.10 that contain the derivatives in θ_k.

First, the following identity A2.2 (Appendix A.2) can be demonstrated:

$$\mu^{\#}_{qk}\left(\frac{\partial \mu_{sk}}{\partial \theta_e} - \frac{\partial \mu_{se}}{\partial \theta_k}\right)\dot{\theta}_e = e_{qns}\omega_n$$

Therefore we have:

$$\frac{\partial \mathcal{T}}{\partial \omega_s}\mu^{\#}_{qk}\left(\frac{\partial \mu_{sk}}{\partial \theta_e}-\frac{\partial \mu_{se}}{\partial \theta_k}\right)\dot{\theta}_e = e_{qns}\omega_n\frac{\partial \mathcal{T}}{\partial \omega_s} = \left[\omega, \frac{\partial \mathcal{T}}{\partial \omega}\right]_q \tag{3.12}$$

Second, we work on the term

$$\mu^{\#}_{qk}\frac{\partial \mathcal{T}}{\partial v_s}\frac{\partial v_s}{\partial \theta_k}, \tag{3.13}$$

from

$$\frac{\partial v_s}{\partial \theta_k} = \frac{\partial M_{es}}{\partial \theta_k}\dot{x}^o_e \tag{3.14}$$

and using Eq. A2.1 (Appendix A.2) we have:

$$\mu^{\#}_{qk}\frac{\partial \mathcal{T}}{\partial v_s}\frac{\partial v_s}{\partial \theta_k} = \frac{\partial \mathcal{T}}{\partial v_s}\mu^{\#}_{qk}\frac{\partial M_{es}}{\partial \theta_k}\dot{x}^o_e = e_{qsn}\frac{\partial \mathcal{T}}{\partial v_s}v_n \tag{3.15}$$

Finally, using Eqs. 3.12, 3.15 into Eq. 3.10 we get the Lagrangian equations of the rotational dynamics in implicit form:

$$\frac{d}{dt}\frac{\partial \mathcal{T}}{\partial \omega_q} + e_{qns}\omega_n\frac{\partial \mathcal{T}}{\partial \omega_s} + e_{qns}v_n\frac{\partial \mathcal{T}}{\partial v_s} = \mu^{\#}_{qk}C_k = c_q \tag{3.16}$$

which in *compact notation* are written as:

$$\frac{d}{dt}\frac{\partial \mathcal{T}}{\partial \omega} + [\omega, \frac{\partial \mathcal{T}}{\partial \omega}] + [v, \frac{\partial \mathcal{T}}{\partial v}] = c \tag{3.17}$$

3.4.3 The Equations of Flexible Dynamics

To obtain the Lagrangian equations of the ε, we must introduce the *internal potential energy* and the *dissipation function*.

From the last N_f equations of Eq. 3.6 it follows:

$$\frac{d}{dt}\frac{\partial \mathcal{T}}{\partial \dot{\varepsilon}_s} + \frac{\partial \mathcal{V}_I}{\partial \varepsilon_s} + \frac{\partial \mathcal{F}}{\partial \dot{\varepsilon}_s} - \frac{\partial \mathcal{T}}{\partial \varepsilon_s} = Q^{\varepsilon}_s \quad , \; s = 1...N_f \tag{3.18}$$

Taking a classic linear elastic and friction model (see [10]) we have:

$$\mathcal{V}_I = \frac{1}{2}\langle \mathcal{K}\varepsilon, \varepsilon\rangle \quad \mathcal{F} = \frac{1}{2}\langle \mathcal{C}\dot{\varepsilon}, \dot{\varepsilon}\rangle \quad \mathcal{K} \geq 0 \;\; ; \;\; \mathcal{C} \geq 0$$

where $\mathscr{K}$, $\mathscr{C}$ are two $N_f \times N_f$ symmetric matrices. With this model the last N_f Eq. 3.6 can be written as:

$$\frac{d}{dt}\frac{\partial \mathscr{T}}{\partial \dot{\varepsilon}_s} + \mathscr{C}^{sk}\dot{\varepsilon}_k + \mathscr{K}^{sk}\varepsilon_k - \frac{\partial \mathscr{T}}{\partial \varepsilon_s} = Q_s^{\varepsilon} \quad , \; s = 1...N_f$$

Equation 3.18 in *compact notation* are written:

$$\frac{d}{dt}\frac{\partial \mathscr{T}}{\partial \dot{\varepsilon}} + \frac{\partial \mathscr{V}_I}{\partial \varepsilon} + \frac{\partial \mathscr{F}}{\partial \dot{\varepsilon}} - \frac{\partial \mathscr{T}}{\partial \varepsilon} = Q^{\varepsilon} \tag{3.19}$$

3.4.4 The Generalized Lagrangian Forces

We can now evaluate the generalized Lagrangian force components. Following the standard Lagrangian method, the Q_i are defined by $Q_i\delta q_i = \delta L$ where δL is the virtual work performed by the forces on the virtual displacement field generated by a virtual variation δq of the Lagrangian coordinates.

We describe the external forces applied to the particles of the body with the field of the external forces per unit mass in **ECI** components: $B_i(r)$; the virtual displacement field, can be derived by the kinematic model introduced:

$$\delta x_i = \delta x_i^o + M_{ie}e_{esj}\mu_{sk}\delta\theta_k(r_j + \phi_j^k\varepsilon_k) + M_{ij}\phi_j^k\delta\varepsilon_k$$

The virtual work is calculated by integrating the scalar product of the two fields $B_i(r)\delta x_i(r)$ in the reference configuration:

$$\delta L = \int_{V_0} B_i\, \delta x_i\, \rho_o\, dV_o = F^T\delta x^o + C^T\delta\theta + Q^{\varepsilon T}\delta\varepsilon$$

Expanding the virtual displacement we get:

$$\delta L = \left(\int_{V_o} B_i\rho_o\, dV_o\right)\delta x_i^o + \left(\int_{V_o} M_{ie}B_i(r_j + \phi_j^k\varepsilon_k)e_{sje}\rho_o\, dV_o\mu_{sk}\right)\delta\theta_k$$
$$+ \left(\int_{V_o} M_{ij}B_i\phi_j^k\rho_o\, dV_o\right)\delta\varepsilon_k$$

Therefore, having defined $b = M_{B\leftarrow E}B$ which in components is written $b_j = M_{ij}B_i$, we obtain:

$$F_i = \int_{V_o} M_{ij} b_i \rho_o \, dV_o$$

$$C_i = \mu_{si} \left(\int_{V_o} e_{sjm} \left(r_j + \phi_j^k \varepsilon_k \right) b_m \rho_o \, dV_o \right)$$

$$Q_i^\varepsilon = \int_{V_o} b_j \phi_j^i \rho_o \, dV_o$$

In the equation instead of F_i and C_i we will use more often the components in **BRF**:

$$f_i = M_{ji} F_j = \int_{V_o} b_i \rho_o \, dV_o \tag{3.20}$$

$$c_i = \mu_{iq}^{\#} C_q = \left(\int_{V_o} e_{ijm} \left(r_j + \phi_j^k \varepsilon_k \right) b_m \rho_o \, dV_o \right) \tag{3.21}$$

$$Q_i^\varepsilon = \int_{V_o} b_j \phi_j^i \rho_o \, dV_o \tag{3.22}$$

From Eqs. 3.9, 3.17, 3.19, using Eqs. 3.20–3.22, we obtain the Lagrangian equations in *implicit form*:

$$\frac{d}{dt}\frac{\partial \mathscr{T}}{\partial v} + \left[\omega, \frac{\partial \mathscr{T}}{\partial v} \right] = \int_{V_0} b \rho_o \, dV_o \triangleq f \tag{3.23}$$

$$\frac{d}{dt}\frac{\partial \mathscr{T}}{\partial \omega} + \left[\omega, \frac{\partial \mathscr{T}}{\partial \omega} \right] + \left[v, \frac{\partial \mathscr{T}}{\partial v} \right] = \int_{V} \left[(r + \phi\varepsilon), b \right] \rho_o \, dV_o \triangleq c \tag{3.24}$$

$$\frac{d}{dt}\frac{\partial \mathscr{T}}{\partial \dot{\varepsilon}} + \frac{\partial \mathscr{V}_I}{\partial \varepsilon} + \frac{\partial \mathscr{F}}{\partial \dot{\varepsilon}} - \frac{\partial \mathscr{T}}{\partial \varepsilon} = \int_{V} \phi^T b \rho_o \, dV_o \triangleq Q^\varepsilon \tag{3.25}$$

These equations are applicable to a wide class of mechanical systems; they can be applied to a specific mechanical system by introducing the specific *kinetic energy*, *internal potential energy* and *dissipation function* together with the interacting external physical force fields. The specific kinetic energy requires in particular the definition of all mass parameters.

3.5 Lagrangian Equations

In order to specify the Lagrangian Eqs. 3.23–3.25, to our particular problem we need to calculate:

$$\frac{\partial \mathscr{T}}{\partial v}, \; \frac{\partial \mathscr{T}}{\partial \omega}, \; \frac{\partial \mathscr{T}}{\partial \varepsilon}, \; \frac{\partial \mathscr{T}}{\partial \dot{\varepsilon}} \tag{3.26}$$

In compact notation, from the expression of the *kinetic energy* Eq. 3.4, we have:

$$
\begin{aligned}
\frac{\partial \mathscr{T}}{\partial v} &= mv + [\omega, p + \langle L\varepsilon\rangle] + \langle L\dot{\varepsilon}\rangle \\
\frac{\partial \mathscr{T}}{\partial \omega} &= [p + \langle L\varepsilon\rangle, v] + (I + \langle J\varepsilon\rangle + \langle K\varepsilon, \varepsilon\rangle)\omega + \langle S\dot{\varepsilon}\rangle + \langle P\varepsilon, \dot{\varepsilon}\rangle \\
\frac{\partial \mathscr{T}}{\partial \dot{\varepsilon}} &= L^T v + (S + \langle P\varepsilon\rangle)^T \omega + \langle T\dot{\varepsilon}\rangle \\
\frac{\partial \mathscr{T}}{\partial \varepsilon} &= L^T [v, \omega] + \frac{1}{2}\omega^T J\omega + \omega^T \langle K\varepsilon\rangle\omega - \omega^T \langle P\dot{\varepsilon}\rangle
\end{aligned}
$$

Note that the minus sign in the last term depends on P being antisymmetric. Using these last identities in the Eqs. 3.23–3.25, we get:

$$
\begin{aligned}
m\dot{v} + [\dot{\omega}, p + \langle L\varepsilon\rangle] + [\omega, \langle L\dot{\varepsilon}\rangle] + \langle L\ddot{\varepsilon}\rangle +\\
m[\omega, v] + [\omega, [\omega, p + \langle L\varepsilon\rangle]] + [\omega, \langle L\dot{\varepsilon}\rangle] = f
\end{aligned}
$$

$$
\begin{aligned}
&[p + \langle L\varepsilon\rangle, \dot{v}] + (I + \langle J\varepsilon\rangle + \langle K\varepsilon, \varepsilon\rangle)\dot{\omega} + (\langle J\dot{\varepsilon}\rangle + 2\langle K\varepsilon, \dot{\varepsilon}\rangle)\omega + \langle S\ddot{\varepsilon}\rangle + \langle P\varepsilon, \ddot{\varepsilon}\rangle \\
&\quad + [\omega, [p + \langle L\varepsilon\rangle, v]] + [\omega, (I + \langle J\varepsilon\rangle + \langle K\varepsilon, \varepsilon\rangle)\omega] + [\omega, \langle S\dot{\varepsilon}\rangle] \\
&\quad + [\omega, \langle P\varepsilon, \dot{\varepsilon}\rangle] + [v, [\omega, p + \langle L\varepsilon\rangle]] = c
\end{aligned}
$$

$$
\begin{aligned}
L^T \dot{v} + (S + \langle P\varepsilon\rangle)^T \dot{\omega} + \langle T\ddot{\varepsilon}\rangle + 2\omega^T \langle P\dot{\varepsilon}\rangle + \frac{\partial \mathscr{V}_I}{\partial \varepsilon} + \frac{\partial \mathscr{F}}{\partial \dot{\varepsilon}} \\
- L^T [v, \omega] - \frac{1}{2}\omega^T J\omega - \omega^T \langle K\varepsilon\rangle\omega = Q^\varepsilon
\end{aligned}
$$

These equations can be simplified by introducing new symbols for the mass properties dependent on the deformations as follows:

$$
\begin{aligned}
\tilde{J} &= J + 2\langle K\varepsilon\rangle = \partial_\varepsilon \tilde{I} \\
\tilde{I} &= I + \langle J\varepsilon\rangle + \langle K\varepsilon, \varepsilon\rangle \\
\tilde{S} &= S + \langle P\varepsilon\rangle \\
\tilde{p} &= p + \langle L\varepsilon\rangle
\end{aligned}
$$

Furthermore, calling $a = \dot{v} + [\omega, v]$ the acceleration of the origin[1] in **BRF** components and reordering, we get the following synthetic expressions for the Lagrangian equations of flexible body dynamics: m

[1]The velocity of the origin in **BRF** components has been designated with v, in order to get the inertial acceleration in **BRF**: $a = M^T(\frac{d}{dt}Mv)$, we need to bring v in **ECI** components, derive each component and then bring them again in **BRF**. This sequence of operations brings to apply the operator $\dot{v} + [\omega, v]$ to the initial vector v, this can be easily derived from Eq. 2.3 given in Sect. 2.3. The same rule applies to any vector in **BRF** when we have to derive its inertial components.

$$ma + [\dot{\omega}, \tilde{p}] + \langle L\ddot{\varepsilon}\rangle + [\omega, [\omega, \tilde{p}]] + 2[\omega, \langle L\dot{\varepsilon}\rangle] = f \tag{3.27}$$

$$[\tilde{p}, a] + \tilde{I}\dot{\omega} + \left\langle \tilde{S}\ddot{\varepsilon} \right\rangle + [\omega, \tilde{I}\omega] + [\omega, \left\langle \tilde{S}\dot{\varepsilon} \right\rangle] + \langle \tilde{J}\dot{\varepsilon}\rangle\omega = c \tag{3.28}$$

$$L^T a + \tilde{S}^T \dot{\omega} + \langle T\ddot{\varepsilon}\rangle + \frac{\partial \mathscr{V}_I}{\partial \varepsilon} + \frac{\partial \mathscr{F}}{\partial \dot{\varepsilon}} - \frac{1}{2}\omega^T \tilde{J}\omega + 2\omega^T \langle P\dot{\varepsilon}\rangle = Q^\varepsilon \tag{3.29}$$

We remark that the structure of the first two equations recalls the Euler equations of the rigid body.

The explicit form of the r.h.s. terms of these equations can be given in terms of external potential energy $\mathscr{V}_E$ and *external non-conservative forces*:

$$f_k = -M_{sk}\frac{\partial \mathscr{V}_E}{\partial x_s^o} + M_{sk}F_k^{(n.c.)} \tag{3.30}$$

$$c_k = -\mu_{kq}^{\#}\frac{\partial \mathscr{V}_E}{\partial \theta_q} + \mu_{kq}^{\#}C_k^{(n.c.)} \tag{3.31}$$

$$Q_k^\varepsilon = -\frac{\partial \mathscr{V}_E}{\partial \varepsilon_k} + Q_k^{\varepsilon(n.c.)} \tag{3.32}$$

3.6 First Integrals of the Free Flexible Body

The flexible body dynamics have *first integrals*[2] of motion or *invariants* when the external forces are null: the conservation of the *momentum* and the conservation of the *angular momentum*; if the dissipative internal forces are absent the *energy* is also preserved.

In order to derive these first integrals we recall that, when the external forces are null Eqs. 3.23–3.25 become:

$$\frac{d}{dt}\frac{\partial \mathscr{T}}{\partial v} + \left[\omega, \frac{\partial \mathscr{T}}{\partial v}\right] = 0 \tag{3.33}$$

$$\frac{d}{dt}\frac{\partial \mathscr{T}}{\partial \omega} + \left[\omega, \frac{\partial \mathscr{T}}{\partial \omega}\right] + \left[v, \frac{\partial \mathscr{T}}{\partial v}\right] = 0 \tag{3.34}$$

$$\frac{d}{dt}\frac{\partial \mathscr{T}}{\partial \dot{\varepsilon}} + \frac{\partial \mathscr{V}_I}{\partial \varepsilon} - \frac{\partial \mathscr{T}}{\partial \varepsilon} = 0 \tag{3.35}$$

[2]We define first integrals of motion or invariants of a system of differential equations functions of the state variables that are constant on any trajectory of motion. These first integrals can help to find closed form solutions. The first integrals are also called invariants.

where:

$$\frac{\partial \mathcal{T}}{\partial v} = mv + \dot{\tilde{p}} + [\omega, \tilde{p}] \quad ; \frac{\partial \mathcal{T}}{\partial \omega} = [\tilde{p}, v] + \tilde{I}\omega + \langle \tilde{S}\dot{\varepsilon} \rangle \tag{3.36}$$

In Lagrangian mechanics *energy* is defined as:

$$\mathcal{E} = \dot{q}_i \frac{\partial \mathcal{L}}{\partial \dot{q}_i} - \mathcal{L} \tag{3.37}$$

$\mathcal{E}$ is a integral of motion when we have the following conditions satisfied: all external forces null, the *Lagrangian* is independent of time, the *dissipation function* is null:

$$Q = 0, \quad \frac{\partial \mathcal{L}}{\partial t} = 0, \quad \mathcal{F} = 0 \tag{3.38}$$

The second conditions on the time dependency of the Lagrangian can be considered verified in our specific model when there are no moving parts on-board that modify the satellite's mass parameters and when there is no mass ejection system in operation.

We translate the classic definition of *energy* to our system in the form:

$$\mathcal{E} = \omega_i \frac{\partial \mathcal{T}}{\partial \omega_i} + v_i \frac{\partial \mathcal{T}}{\partial v_i} + \dot{\varepsilon}_i \frac{\partial \mathcal{T}}{\partial \dot{\varepsilon}_i} - (\mathcal{T} - \mathcal{V}_I) \tag{3.39}$$

By direct calculation, we verify that Eqs. 3.33–3.35, guarantee that $\mathcal{E}$ is constant along any trajectory. In fact deriving Eq. 3.39, we get:

$$\begin{aligned} \frac{d\mathcal{E}}{dt} &= \dot{\omega}^T \frac{\partial \mathcal{T}}{\partial \omega} + \dot{v}^T \frac{\partial \mathcal{T}}{\partial v} + \ddot{\varepsilon}^T \frac{\partial \mathcal{T}}{\partial \dot{\varepsilon}} \\ &-v^T \left[\omega, \frac{\partial \mathcal{T}}{\partial v}\right] - \omega^T \left[v, \frac{\partial \mathcal{T}}{\partial v}\right] - \dot{\varepsilon}^T \frac{\partial \mathcal{V}_I}{\partial \varepsilon} + \dot{\varepsilon}^T \frac{\partial \mathcal{T}}{\partial \varepsilon} - \dot{\mathcal{T}} + \dot{\mathcal{V}}_I = 0 \end{aligned} \tag{3.40}$$

The same conservation law applies also if we add the *external potential energy*. Specializing Eq. 3.39 to flexible bodies, the *energy* is written:

$$\mathcal{E} = \frac{1}{2} m v^T v + \tilde{p}^T [v, \omega] + v^T \langle L\dot{\varepsilon} \rangle + \frac{1}{2} \omega^T \tilde{I} \omega + \omega^T \langle \tilde{S}\dot{\varepsilon} \rangle + \frac{1}{2} \langle T\dot{\varepsilon}, \dot{\varepsilon} \rangle + \mathcal{V}_I \tag{3.41}$$

For what concerns the other invariants, they require only that all external forces are null. In this case Eqs. 3.33, 3.34 are equivalent to:

$$\frac{d}{dt}(M_{E \leftrightarrow B} \frac{\partial \mathcal{T}}{\partial v}) = 0 \Rightarrow M_{E \leftrightarrow B} \frac{\partial \mathcal{T}}{\partial v} = \mathcal{I}_{TR} \tag{3.42}$$

$$\frac{d}{dt}(M_{E \leftrightarrow B} \frac{\partial \mathcal{T}}{\partial \omega}) + M_{E \leftrightarrow B} \left[v, \frac{\partial \mathcal{T}}{\partial v}\right] = 0 \tag{3.43}$$

We see that $\mathscr{I}_{TR}$ is defined in **ECI** and is constant during the motion. It is an invariant usually called *momentum*.

We can derive a similar integral of motion for the rotational dynamics: the *angular momentum*. From Eqs. 3.42, 3.43 and using Eq. 3.36 to derive the velocity v, it follows:

$$\frac{d}{dt}(M_{E\leftrightarrow B}\frac{\partial \mathscr{T}}{\partial \omega}) + \frac{1}{m}\left[\mathscr{I}_{TR} - \frac{d}{dt}(M_{E\leftrightarrow B}\widetilde{p}), \mathscr{I}_{TR}\right] = 0 \Rightarrow \tag{3.44}$$

$$\frac{d}{dt}(M_{E\leftrightarrow B}\left(\frac{\partial \mathscr{T}}{\partial \omega} + \left[\frac{\partial \mathscr{T}}{\partial v}, \widetilde{p}/m\right]\right)) = 0 \tag{3.45}$$

where we identify the following integral of motion:

$$M_{E\leftrightarrow B}\left(\frac{\partial \mathscr{T}}{\partial \omega} + \left[\frac{\partial \mathscr{T}}{\partial v}, \widetilde{p}/m\right]\right) = \mathscr{I}_{ROT} \tag{3.46}$$

In conclusion, when the external forces are null, the 3-D vectors: $\mathscr{I}_{TR}$ defined by Eq. 3.42, $\mathscr{I}_{ROT}$ defined by Eq. 3.46 are integrals of motion, the scalar $\mathscr{E}$ defined by Eq. 3.39, is another integral of motion provided Eq. 3.38 are satisfied.

3.7 Momentum Management Systems

Momentum management systems are devices used on-board satellites to control the attitude using rotors that are intentionally accelerated or decelerated: *reaction wheels* (RW), or where the rotor axis is steered in one or two directions: *control moment gyros* (CMG). These actuation systems are also discussed in Chap. 9 where real H/W characteristics are provided.

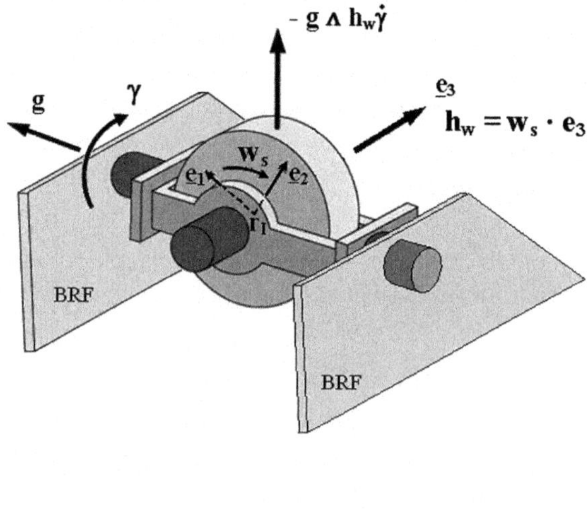

Fig. 3.1 Rotor with frames and degrees of freedom

Momentum management devices exchange their own *angular momentum* with that of the satellite under the action of specific control commands, providing the ability to control the satellite attitude until they reach a saturation condition.

When the device is saturated it is necessary to unload the stored *angular momentum* with the action of external control actuators like thrusters (RCT) or Magnetotorquers.

We will see in this section how we have to modify Eqs. 3.27–3.29 in order to take into account of such devices.

Let us consider a spinning rigid body called *rotor* represented in Fig. 3.1. The rotorspins with the scalar angular speed ω_s around the e_3 axis, in addition in the case of CMGs the spinning axis can also rotate around another axis called the gimbal axis (called g in Fig. 3.1) the unit vector g is fixed in **BRF**.

The *rotor* is integrated on the satellite body at an interface hinge point r_I in **BRF**.

We assume that the interface hinge point, the rotor geometrical center and the origin of the rotor triad **ROT** of unit vectors e_1, e_2, e_3, coincide in the same point which is fixed in **ROT** and in **BRF**.

In Fig. 3.1 a CMG is presented, the RW case is derived considering the gimbal rotation $\gamma = constant$, in this case e_3 is fixed in **BRF**.

We introduce the matrix $M_{B \leftarrow R}$ which depends on the relative kinematic between the *rotor* and the satellite body and brings from **ROT** to **BRF**; in the RW case $M_{B \leftarrow R}$ represents a single axis rotation transformation about the spin axis, in the CMG case this matrix is also a function of the angle γ.

Let us define the rotor mass properties as $m_R,\ I_R,\ p_R$ and assume for the moment $p_R = 0$.

If we want to use the Lagrangian method, we simply need to calculate the kinetic energy of this rotor, add it to the kinetic energy of the satellite and then use the Eqs. 3.23–3.25.

From the general expression of the kinetic energy Eq. 3.4 we obtain:

$$\mathscr{T}_R = \frac{1}{2} m_R v_I^T v_I + \frac{1}{2} \omega_R^T I_R \omega_R \tag{3.47}$$

The angular velocity of the rotor (in rotor frame) can be expressed in terms of $\omega_s,\ \omega,\ \dot{\gamma}$ using the summation rule of the angular velocities Eq. 2.6:

$$\omega_R = e_3 \omega_s + M_{R \leftarrow B}(\omega + g\dot{\gamma}) \tag{3.48}$$

Therefore the expression of the kinetic energy becomes:

$$\begin{aligned}\mathscr{T}_R = {} & \tfrac{1}{2} m_R v_I^T v_I + \tfrac{1}{2}\omega^T (M_{B \leftarrow R} I_R M_{R \leftarrow B})\omega \\ & + \tfrac{1}{2} g^T M_{B \leftarrow R} I_R M_{R \leftarrow B} g \dot{\gamma}^2 + g^T M_{B \leftarrow R} I_R e_3 \dot{\gamma} \omega_s + \tfrac{1}{2}\omega_s^2 e_3^T I_R e_3 \\ & + \omega^T (M_{B \leftarrow R} I_R e_3 \omega_s + M_{B \leftarrow R} I_R M_{R \leftarrow B} g \dot{\gamma})\end{aligned} \tag{3.49}$$

Considering that $v_I = v + [\omega, r_I]$[3] this expression depends quadratically on the satellite velocities ω, v and on the velocities of the interface kinematics $\dot{\gamma}$, ω_s.

The interface kinematics must be considered as exogenous time dependent variables which produce a time dependent Lagrangian.

The rotor kinetic energy must be added to the satellite's kinetic energy to derive the Lagrangian equations, applying the Lagrangian equations in implicit form Eqs. 3.23–3.25 to Eq. 3.49, we observe that:

- the first two terms, quadratic in (ω, v) can be directly summed in the expression of the satellite's kinetic energy, where they will produce an increase of the satellite inertial properties due to the contribution of the rotor inertial terms;
- The factor $M_{B\leftrightarrow R} I_R M_{R\leftrightarrow B}$ in the second term, depends on γ and on the rotor spin angle. Thus, the satellite inertia matrix depends on time, this must be taken into account in Eq. 3.24 and produces an additional term with respect to Eq. 3.28 which is the time derivative of the inertia matrix multiplied by the satellite angular speed. All these effects are normally considered negligible when the mass of CMGs is small compared to the overall satellite;
- the third, fourth and fifth terms terms have no effect on the expression of the momenta:

$$\frac{\partial \mathcal{T}_R}{\partial \omega}, \frac{\partial \mathcal{T}_R}{\partial v} \tag{3.50}$$

thus, they have no effect on the satellite dynamics.
- the last two terms have a direct effect on the rotational momentum $\frac{\partial \mathcal{T}_R}{\partial \omega}$, which will be designated shortly as h_w:

$$\frac{\partial \mathcal{T}_R}{\partial \omega} = h_w = (M_{B\leftrightarrow R} I_R e_3 \omega_s + M_{B\leftrightarrow R} I_R M_{R\leftrightarrow B} g \dot{\gamma}) \tag{3.51}$$

The effect on the satellite dynamics is a new additional term in Eq. 3.24:

$$\frac{d}{dt}\frac{\partial \mathcal{T}_R}{\partial \omega} + \left[\omega, \frac{\partial \mathcal{T}_R}{\partial \omega}\right] \tag{3.52}$$

If we assume that the rotor inertia matrix is $I_R = diag(I_t, I_t, I_r)$, we can represent the momentum as:

$$h_w = M_{B\leftrightarrow R} e_3 I_r \omega_s + I_t g \dot{\gamma} \tag{3.53}$$

and its time derivative as (where we use $\frac{d}{dt}(M_{B\leftrightarrow R} e_3) = [g, M_{B\leftrightarrow R} e_3]\dot{\gamma}$):

$$\dot{h_w} = M_{B\leftrightarrow R} e_3 I_r \dot{\omega}_s + [g, M_{B\leftrightarrow R} e_3] I_r \omega_s \dot{\gamma} + I_t g \ddot{\gamma} \tag{3.54}$$

[3]We do not treat here the effects due to the elasticity of the rotor and of the satellite.

We can decompose the kinetic energy of the rotor in three terms:

$$\begin{aligned} \mathscr{T}_R &= \mathscr{T}_{RS} + \mathscr{T}_{R0} + \omega^T h_w \\ \mathscr{T}_{RS} &= \tfrac{1}{2} m_R v_I^T v_I + \tfrac{1}{2}\omega^T (M_{B\leftrightarrow R} I_R M_{R\leftrightarrow B})\omega \\ \mathscr{T}_{R0} &= \tfrac{1}{2}\omega_s^2 I_r + \tfrac{1}{2}\dot{\gamma}^2 I_t \end{aligned} \tag{3.55}$$

The term $\mathscr{T}_{RS}$ is to be taken into account by including the rotor mass in the satellite mass matrix, $\mathscr{T}_{R0}$ is the energy contained in the rotors as if the satellite was inertial fixed and has no effect on the satellite dynamics, the term proportional to ω adds a contribution to the l.h.s of the satellite's rotational equations Eqs. 3.24 or 3.28 equal to:

$$\frac{dh_w}{dt} + [\omega, h_w] \tag{3.56}$$

where h_w and $\frac{dh_w}{dt}$ must be taken from Eqs. 3.53, 3.54. Thus the *momentum management system* effect on the satellite equations is like an *effective torque* c_e:

$$c_e = -\frac{dh_w}{dt} - [\omega, h_w] \tag{3.57}$$

Due to the linearity of Eq. 3.56 with respect to h_w, in the presence of a system of multiple N_r rotors $h_{w,i},\ i = 1, 2 \ldots N_r$ we need simply to consider the vector sum in **BRF** of each rotor momentum: $h_w = \sum h_{w,i}$.

This result is very simple and contains most of the information on *momentum management systems*.

3.7.1 The First Integrals with Momentum Management Systems

In the previous section we have seen that *momentum management systems* adds to satellite's kinetic energy the term $\mathscr{T}_R = \mathscr{T}_{RS} + \mathscr{T}_{R0} + \omega^T h_w$; consequently derivatives of kinetic energy used in Sect. 3.6 to calculate the first integrals of motion must be updated to consider the rotor momentum:

$$\frac{\partial \mathscr{T}}{\partial v} = mv + \dot{\tilde{p}} + [\omega, \tilde{p}] \quad ; \quad \frac{\partial \mathscr{T}}{\partial \omega} = [\tilde{p}, v] + \tilde{I}\omega + \langle \tilde{S}\dot{\varepsilon} \rangle + h_w \tag{3.58}$$

In these equations we assume that satellite inertia properties include also rotor properties coming from $\mathscr{T}_{RS}$.

We recall that $h_w = \sum h_{w,i}$, where the sum is extended to all rotors.

The derivation of invariants remains the same described in Sect. 3.6. The *momentum* is:

$$\mathscr{I}_{TR} = M_{E \leftrightarrow B}(mv + \dot{\tilde{p}} + [\omega, \tilde{p}]) \tag{3.59}$$

The *angular momentum* is:

$$\mathscr{I}_{ROT} = M_{E \leftrightarrow B}([\tilde{p}, v] + \tilde{I}\omega + \langle \tilde{S}\dot{\varepsilon} \rangle + h_w + [M_{B \leftrightarrow E}\mathscr{I}_{TR}, \tilde{p}/m]) \tag{3.60}$$

Thus, the effect of a *momentum management system* is to add to $\mathscr{I}_{ROT}$ the term $M_{E \leftrightarrow B}h_w$.

By varying the rotor *angular momentum* the total $\mathscr{I}_{ROT}$ will not change, consequently, an opposite change in the satellite part of it, will result.

Concerning *energy*, the total mechanical energy of the system is not conserved because the motors of the *momentum management system* transform electric energy into mechanic energy injected into the rotor.

However, if the rotor momentum h_w is constant, the energy defined by Eq. 3.39 is still made constant by Eqs. 3.33–3.35 even with the additional $\mathscr{T}_R$ term.

Using Eq. 3.39 the term $\omega^T h_w$ gives no contribution to energy and we obtain again Eq. 3.41. Therefore the form:

$$\begin{aligned}\mathscr{E} = \tfrac{1}{2}mv^T v + \tilde{p}^T[v, \omega] + v^T\langle L\dot{\varepsilon}\rangle + \tfrac{1}{2}\omega^T \tilde{I}\omega \\ +\omega^T\langle \tilde{S}\dot{\varepsilon}\rangle + \tfrac{1}{2}\langle T\dot{\varepsilon}, \dot{\varepsilon}\rangle + \mathscr{V}_I\end{aligned} \tag{3.61}$$

is conserved during the motion if the conditions Eq. 3.38 are valid and the h_w is constant. In the case of a varying rotor momentum a useful conserved scalar quantity is the modulus of $\mathscr{I}_{ROT}$, which will be indicated with $|\mathscr{I}_{ROT}|$,

3.7.2 The Multibody Method

Lagrangian mechanics do not provide explicitly the value of the interface torques and forces which are passed from the satellite body to the rotor through the constraints. This type of forces can be important to size the *momentum management system* parts and to evaluate the high frequency vibration effects on the satellite.

In order to have this additional information we will use another method directly calculating the forces that are exchanged at the interface. We have to consider the rotor and the satellite as two separate bodies, to write the dynamic equations of each of the two bodies and to impose the interface kinematic constraints in order to calculate the forces exchanged at the interface. We call this approach the *multibody method*. The amount of calculations that will be necessary for this purpose illustrate quite well how powerful the Lagrangian approach is.

We then write Eqs. 3.27–3.29 for the satellite body without the rotor in **BRF** and those of the rotor in **ROT**. There are some steps of this procedure which are independent of the specific kinematics of the hinge:

- forces and torques that the rotor exerts on satellite's body are opposite to those exerted by satellite's body on the rotor; even if we have superficial forces, we assume that the integrated effect of such superficial forces is applied at the interface hinge point. Therefore if f_I and c_I are the force and torque (calculated at the hinge point) that the satellite exerts on the rotor in **ROT** components, the generalized forces to be used in satellite equations are[4]:

$$\begin{aligned} f &= -M_{B\leftrightarrow R} f_I \\ c &= -r_I \times M_{B\leftrightarrow R} f_I - M_{B\leftrightarrow R} c_I \end{aligned} \tag{3.62}$$

- We can express the kinematic variables to be used in the rotor equations as function of the **BRF** kinematics and the kinematic variables defining the relative motion of the interface (we use here note (see footnote 1) on $v_I = v + [\omega, r_I]$, and the rule for the composition of angular velocities Eq. 2.6):

$$\begin{aligned} a_{I,R} &= M_{R\leftrightarrow B} a_{I,B} = M_{R\leftrightarrow B}(a + [\dot{\omega}, r_I] + [\omega, [\omega, r_I]] \\ \omega_{RE,R} &= \omega_{RB,R} + \omega_{BE,R} = \omega_{RB,R} + M_{R\leftrightarrow B}\omega \\ \dot{\omega}_{RE,R} &= \dot{\omega}_{RB,R} - M_{R\leftrightarrow B}(\omega_{RB,B} \times \omega) + M_{R\leftrightarrow B}\dot{\omega} \\ &= \dot{\omega}_{RB,R} - \omega_{RB,R} \times \omega_{BE,R} + \dot{\omega}_{BE,R} \\ \omega &= \omega_{BE,B} \\ \dot{\omega} &= \dot{\omega}_{BE,B} \end{aligned} \tag{3.63}$$

In these equations a, ω, $\dot{\omega}$ are the kinematic variables of the spacecraft body, while $\omega_{RB,R}$, $\dot{\omega}_{RB,R}$, $M_{R\leftrightarrow B}$ depend only on the angular rotations of the interface and their first and second derivatives.

- Introducing the kinematic positions Eq. 3.63 into Eqs. 3.27, 3.28 written for the rotor body—with inertial properties p_R, I_R, m_R—we get:

$$\begin{aligned} f_I &= m_R a_{I,R} + [\dot{\omega}_{RB,R} + \dot{\omega}_{BE,R} - [\omega_{RB,R}, \omega_{BE,R}], p_R] \\ &\quad + [\omega_{RB,R} + \omega_{BE,R}, [\omega_{RB,R} + \omega_{BE,R}, p_R]] \\ c_I &= [p_R, a_{I,R}] + I_R(\dot{\omega}_{RB,R} + \dot{\omega}_{BE,R} - [\omega_{RB,R}, \omega_{BE,R}]) \\ &\quad + [(\omega_{RB,R} + \omega_{BE,R}), I_R(\omega_{RB,R} + \omega_{BE,R})] \end{aligned} \tag{3.64}$$

where:

$$\begin{aligned} \omega_{BE,R} &= M_{R\leftrightarrow B}\omega \\ \dot{\omega}_{BE,R} &= M_{R\leftrightarrow B}\dot{\omega} \end{aligned} \tag{3.65}$$

These equations provide the forces acting on the rotor.

[4]The expression of the generalized Lagrangian forces Q_k^ε acting on the satellite are null because we have assumed that the momentum management system is mounted on a hard point. Actually at frequencies $\gg$10 Hz where the disturbances due to the rotor unbalances are generated, the satellite structure cannot be considered rigid and their effects are normally evaluated in the frame of the satellite's structural dynamics.

Using Eq. 3.64 in Eq. 3.62, to eliminate the interface forces, we can calculate the effect of the rotor on satellite dynamics as a function of the rotor interface kinematic.

- In rotor dynamics Eq. 3.64, we introduce the position $p_R = 0$ which implies that the center of mass of the rotor is fixed in **BRF**. This feature of the rotor is called the *rotor static balance*.
  The simplification $p_R = 0$ which is realized in perfectly balanced rotors can be realized in real *momentum management systems* with certain errors. From rotor dynamics Eq. 3.64, considering that usually $|\omega_{RB,R}| \geq 10000\ |\omega_{BE,R}|$, the term $[\omega_{RB,R}, [\omega_{RB,R}, p_R]] = f_I$ is typically the most important balancing error.
  For a high speed rotor with $\omega_{RB,R} = e_3\omega_s$, this term produces an interface force

$$\begin{array}{l} f_I = (e_3 \times (e_3 \times p_R))\omega_s{}^2, \ \text{or} \\ f_I = -(E_3 - e_3 e_3{}^T)p_R\omega_s{}^2 \end{array} \tag{3.66}$$

  constant in **ROT** and orthogonal to e_3, which transformed with $M_{B \leftrightarrow R}$ becomes a rotating force in **BRF**; this force may excite the satellite's high frequency flexible modes.
  This is the effect of the so-called rotor *static unbalance.*
  The level of the unbalance and its stability versus time are important quality parameters of a *momentum management system.*
- The matrix of the moment of inertia of the rotor will be considered diagonal, with equal inertias of the two axes orthogonal to the rotation one $I_R = diag(I_t, I_t, I_r)$. This property of the rotor is called the *rotor dynamic balance*.
  We now consider the effect of a small *dynamic unbalance* in the rotor inertia matrix:

$$I_R = diag(I_t, I_t, I_r) + \delta I \tag{3.67}$$

  where the small inertia matrix δI takes into account dynamic balance errors.
  From Eq. 3.64 we see that the main effect is the one proportional to $\omega_s{}^2$:

$$c_I = [\omega_{RB,R}, \delta I \omega_{RB,R}] \tag{3.68}$$

  introducing $\omega_{RB,R} = e_3\omega_s$ we have:

$$\begin{array}{l} c_I = [e_3, \delta I e_3]\omega_s{}^2, \ \text{or} \\ c_I = (-\delta I_{23}, \delta I_{13}, 0)^T \omega_s{}^2 \end{array} \tag{3.69}$$

  c_I is a constant torque in **ROT** which becomes a rotating torque in **BRF**. The *dynamic unbalance* must be minimized with an appropriate balancing procedure. In fact the parasitic effect of the dynamic unbalance produces attitude high frequency oscillations called *attitude jittering*, which may impact on payload performances, rotor ball bearings life degradation and may introduce noise in measurements.

- When we apply Eq. 3.64 to 3.62, all force and torque terms in Eq. 3.64 that depend only on the satellite rate and acceleration variables, namely:

$$
\begin{aligned}
f_{I0} &= m_R a_{I,R} \\
c_{I0} &= I_R \dot{\omega}_{BE,R} + [\omega_{BE,R}, I_R \omega_{BE,R}]
\end{aligned}
\tag{3.70}
$$

are the forces necessary to keep the rotor integrated with the satellite. When writing the satellite dynamic equations, f_{I0}, c_{I0} are not considered in the calculations because their effect is taken into account including the rotor mass in the satellite's inertia properties. We need to consider them only when we evaluate the torques exchanged at the interface.
The residual torque $c_I - c_{I0}$ realizes the *effective torque* c_e of the *momentum management system* acting on the satellite including rotor system:

$$
\begin{aligned}
&I_R \dot{\omega}_{RB,R} - I_R(\omega_{RB,R} \times \omega_{BE,R}) + \omega_{RB,R} \times (I_R \omega_{RB,R}) + \omega_{RB,R} \times (I_R \omega_{BE,R}) \\
&+ \omega_{BE,R} \times I_R \omega_{RB,R} = c_I - c_{I0} = -M_{R \leftrightarrow B} c_e
\end{aligned}
\tag{3.71}
$$

Equation 3.71 contain all terms necessary to discuss the RW and CMG control dynamics.

3.7.3 Reaction Wheels

In the ideal reaction wheel the rotor inertia matrix is diagonal, with transverse inertias equal to each other. The center of mass of the rotor is along the spin axis (so that it can be considered the hinge point r_I fixed in **BRF**).

The rotor mass distribution is invariant with respect to the wheel rotation angle so that: $M_{B \leftrightarrow R} I_R M_{R \leftrightarrow B} = const.$ and $I_R = diag(I_t, I_t, I_r)$.

The kinematics of the hinge will be represented as $\omega_{RB,R} = e_3 \omega_s$ with e_3 the third unit vector of the rotating triad **ROT**, e_3 is fixed also in **BRF**.

In these conditions, we can apply the following simplifications in Eq. 3.71:

$$
\begin{aligned}
\omega_{RB,R} \times (I_R \omega_{BE,R}) &= I_R(\omega_{RB,R} \times \omega_{BE,R}) \\
\omega_{RB,R} \times (I_R \omega_{RB,R}) &= 0
\end{aligned}
\tag{3.72}
$$

With these simplifications, we have for the torque that the satellite exerts on the rotor c_I:

$$
c_I - c_{I0} = I_R e_3 \dot{\omega}_s + [\omega_{BE,R}, I_R e_3 \omega_s] = -M_{R \leftrightarrow B} c_e \tag{3.73}
$$

The *effective torque* acting on the satellite plus rotor becomes:

$$
c_e = -\dot{h}_w - \omega \times h_w \tag{3.74}
$$

with:

$$\begin{aligned} h_w &= I_R e_3 \omega_s = e_3 I_r \omega_s \\ \dot{h}_w &= e_3 I_r \dot{\omega}_s \end{aligned} \tag{3.75}$$

The same conclusion was obtained using Eqs. 3.53, 3.54 and 3.56.

The torque that the satellite provides to the RW can be divided in two parts, one passing through the interface (the ball bearings) and another one along the spin axis, this last component is imposed by the motor (generally a DC/DC brush-less motor), naturally here we have to consider also the term c_{I0}[5]. Projecting the torque along the spin axis we find that in order to change the wheel speed, the motor must provide a torque c_w to the wheel:

$$c_w = c_I{}^T e_3 = \dot{h}_w + c_{I0}^T e_3 = I_r \dot{\omega}_s + c_{I0}^T e_3 \tag{3.76}$$

Because we have always a certain friction torque along the spin axis due to the contact forces of the ball bearings, the motor has to provide this torque on top of the friction torque. If c_f is the friction torque applied to the rotor by the ball bearings ($sign(c_f) = -sign(\omega_s)$), the real motor torque will be:

$$c_w = I_r \dot{\omega}_s + c_{I0}^T e_3 - c_f \tag{3.77}$$

The electrical power which is injected in the satellite by the motor is $P_E = c_w \omega_s$. Typically in the wheel electronics there is an internal high frequency control loop that takes care to impose the torque necessary to have a requested momentum variation whatever is the level of friction or the satellite rate.

It is usual to write the *effective torque* $c_e = -\dot{h}_w - \omega \times h_w$ directly embedded in the l.h.s of the satellite equations thereby arriving at Eq. 3.87.

Using a system of three or four wheels (see Chap. 9), by commanding a variation of the rotors' speeds it is possible to control the satellite attitude in three axes, provided that motors' torque capability is not exceeded.

3.7.4 Control Moment Gyros

In the control moment gyro (CMG), the rotor, as in the RW, is a wheel rotating at a high speed ω_s around the spin axis e_3. This time the spin axis can be rotated around an orthogonal axis of unit vector g which is called the gimbal axis and fixed in **BRF** (see Fig. 3.1). While rotating the spin axis e_3 describes a circumference in **BRF**. The CMG kinematics can be established as: $\omega_{RB,R} = e_3 \omega_s + M_{R \leftrightarrow B} g \dot{\gamma}$, where:

- e_3 is the third unit vector in **ROT**;
- g is a fixed unit vector in **BRF** called the gimbal axis;

[5]We remember that C_{I0} from Eq. 3.70 is the part of the interface torque necessary to transmit the satellite motion to the rotor.

- γ is the gimbal rotation angle;
- $e_3{}^T M_{R\leftrightarrow B} g = 0$;
- $\frac{d}{dt}(M_{R\leftrightarrow B} g) = -e_3 \times (M_{R\leftrightarrow B} g)\omega_s$;
- $\dot{\omega}_{RB,R} = e_3\dot{\omega}_s - e_3 \times (M_{R\leftrightarrow B} g)\omega_s\dot{\gamma} + (M_{R\leftrightarrow B} g)\ddot{\gamma}$.

Using these kinematics in Eq. 3.71 we get the interface torque.

We will not evaluate the quadratic terms which are of the order $|\dot{\gamma}||\omega_{BE,R}|$ and $|\dot{\gamma}|^2$ because typically they are many orders of magnitude less than the others; the matrix of the moment of inertia of the rotor will be considered $I_R = diag(I_t, I_t, I_r)$.

With these assumptions the terms of Eq. 3.71 can be developed as follows:

$$\begin{aligned}
&I_R\dot{\omega}_{RB,R} = e_3 I_r\dot{\omega}_s - [e_3, M_{R\leftrightarrow B} g] I_t\omega_s\dot{\gamma} + M_{R\leftrightarrow B} g I_t\ddot{\gamma} \\
&[\omega_{RB,R}, (I_R\omega_{BE,R})] - I_R[\omega_{RB,R}, \omega_{BE,R}] = 0 \\
&[\omega_{BE,R}, I_R\omega_{RB,R}] = [\omega_{BE,R}, e_3] I_r\omega_s \\
&[\omega_{RB,R}, (I_R\omega_{RB,R})] = [M_{R\leftrightarrow B} g, e_3] I_r\omega_s\dot{\gamma} + [e_3, M_{R\leftrightarrow B} g] I_t\omega_s\dot{\gamma}
\end{aligned}$$

Summing up all the terms in Eq. 3.71 using the simplifications introduced and bringing the vectors in **BRF** we get:

$$\begin{aligned}
&c_I - c_{I0} = M_{R\leftrightarrow B}(\dot{h}_w + [\omega, h_w]) \\
&c_e = -\dot{h}_w - [\omega, h_w] \\
&h_w = M_{B\leftrightarrow R} e_3 I_r\omega_s + I_t g\dot{\gamma} \\
&\dot{h}_w = M_{B\leftrightarrow R} e_3 I_r\dot{\omega}_s + [g, M_{B\leftrightarrow R} e_3] I_r\omega_s\dot{\gamma} + I_t g\ddot{\gamma}
\end{aligned} \tag{3.78}$$

Thus we have obtained the same expressions of Eqs. 3.53, 3.54 and 3.56.

The actuation of the control commands $\dot{\omega}_s$, $\dot{\gamma}$ is performed with motors which must exert scalar torques c_w, c_g on the CMG given by the projection of the interface torques on the spin axis and on the gimbal axis, plus the friction torques present in the transmission chains of the two axe[6]:

$$\begin{aligned}
&c_w = c_I^T e_3 = I_r\dot{\omega}_s + \omega^T[g, M_{B\leftrightarrow R} e_3] I_t\dot{\gamma} + c_{I0}^T e_3 - c_{fw} \\
&c_g = c_I^T M_{R\leftrightarrow B} g = I_t\ddot{\gamma} + \omega^T[M_{B\leftrightarrow R} e_3, g] I_r\omega_s + c_{I0}^T M_{R\leftrightarrow B} g - c_{fg}
\end{aligned} \tag{3.79}$$

The electrical power which is injected in the satellite by the motors is

$$P_E = c_w\omega_s + c_g\dot{\gamma}.$$

Typically in the CMG electronics there is an internal high frequency control loop that takes care to impose the torque necessary to have a requested momentum variation.

In Eq. 3.78 h_w is a time varying vector in **BRF** and the *effective torque*

$$c_e = -\dot{h}_w - \omega \times h_w$$

[6]We use here the product rule: $a^T[b, c] = b^T[c, a]$.

is the same we have already found dealing the RWs and discussing the *momentum management systems* with the Lagrangian method.

The more important term of c_e which is applied to the satellite by the CMG as a consequence of a gimbal rotation is $-[g, M_{B\leftrightarrow R}e_3]I_r\omega_s\dot{\gamma}$ and from this formula we can see that steering at 1 rad/s a 30 Nms momentum wheel, we can obtain on the satellite a torque of 30 Nm, if we had used RWs to obtain these level of torques, using standard available commercial motors (0,2 Nm) we should have used at least 150 RWs.

The difficulty in CMG management stands in the fact that in $[g, M_{B\leftrightarrow R}e_3]I_r\omega_s\dot{\gamma}$ the vector $M_{B\leftrightarrow R}e_3$ continuously moves in **BRF** as a consequence of the commands applied, and the direction of the instantaneous output torque moves as well.

A system of three or four CMGs can produce very high torques, but the gimbal rates to be commanded change as a function of the CMG gimbal angles.

A CMG system is also subjected to saturation (like any other actuator) and singular configurations in which it is not possible to produce a 3 axes torque (see Chap. 9). Additional information on the geometrical problems related to commanding a system of CMGs can be found in the extensive available literature (see for example [13]).

3.8 Multibody Systems

In the frame of *momentum management systems*, we have provided two methods to connect additional bodies to the satellite: the Lagrangian method and the multibody method where the interface forces are calculated explicitly using the kinematical constraints of the interface, which in the case of the previous section where the imposed relative rotations and rotation rates of the rotor.

If the rigid relative motion of these bodies is small, the kinematic representation we have introduced in Chap. 2 can be applied also to the *multibody articulated systems* (see [14]). We have to describe the rigid relative motion between the central body and the appendage using a set of deformation functions that are null in the central body and represent the congruent small rigid rotations of the appendage. This approach can be used in the quite common case of antennas with decentralized controlled pointing.

If we have additional appendages attached to satellite body, rigid or flexible, that must perform a wide relative motion controlled with motors, then the non-linear kinematics of the relative motion must be introduced. Examples may be steerable antennas or telescopes.

In all these cases, we can apply the Lagrangian method or the multibody approach (for an ample explanation of the multibody equations see [11, 12]), resolving the interface forces in terms of the interface kinematical constraints using Eqs. 3.62–3.64.

When the multibody has a *star topology* like the typical satellite, which means that there is a central body where all other bodies, called appendages, are connected, the multibody method described in Sect. 3.7.2 can be applied as we did for the *momentum management systems*.

In the more general case where the appendages are interconnected to each other, it may be difficult to provide a general kinematic model or to resolve in closed form the interface forces; in these cases the bodies can be integrated separately and the resolution of interconnection forces, is usually done numerically.

The general multibody approach will always be the same: we apply Eqs. 3.27–3.29 to each of the N bodies (each one with its **BRF**). The forces exchanged at each interface between each couple of bodies will be unknown. For each body to body interface the number of unknown forces equal the number of kinematical constraints between the two bodies (up to a maximum of 6 unknowns for each interface).

The congruence relationships between the kinematic variables of the different bodies, which express how the bodies' kinematics are linked to each other, provide the additional set of equations necessary to resolve the system. There are specific numerical procedures for solving this problem for a general multibody topology (again refer to specialized texts like [11]).

3.9 The Center of Mass

We introduce in this section a particular kinematic model which allows to simplify the expressions of the integrals of motion and the Lagrangian equations. Let us introduce the point x^g:

$$x^g = x^o + M_{E \hookleftarrow B}(p + \langle L\varepsilon \rangle)/m$$

with velocity and acceleration v^g, a^g. This point is called *center of mass* (CoM); it represents the mass averaged position vector of the body and its position in **BRF** depends also on ε.

If we use the CoM as the new origin displacing the **BRF** origin to it, we obtain a new frame CoM-**BRF** and a new kinematic model, which generates the same displacements of the previous model using different Lagrangian variables:

$$x = x^g + M_{E \hookleftarrow B}(r - p/m + \langle (\phi - L/m)\varepsilon \rangle) \tag{3.80}$$

We need to recalculate all mass properties for this new kinematic model, using the new CoM-**BRF** translated body coordinates and the new CoM-**BRF** translated deformation functions:

$$r_g = r - p/m, \quad \phi_g = \phi - L/m$$

The *internal potential energy* and the *dissipation functions* that depend only on the ε will not change because those two functions are obviously invariant to a space translation of the deformation functions.

The new mass parameters' symbols will be marked with a sub-index G which we use to indicate they refer to CoM-**BRF**.

It is easy to verify that with this new kinematic model $p_G = 0$ and $L_G = 0$; therefore $\tilde{p_G} = p_G + \langle L_G \varepsilon \rangle = 0$ and using these new coordinates the expression of the first integrals Eqs. 3.59, 3.60, 3.41 are simplified in:

$$\mathscr{I}_{TR} = M_{E \leftrightarrow B}(mv^g) \tag{3.81}$$

$$\mathscr{I}_{ROT} = M_{E \leftrightarrow B}(\tilde{I_G}\omega + \left\langle \tilde{S_G}\dot{\varepsilon} \right\rangle + h_w) \tag{3.82}$$

$$|\mathscr{I}_{ROT}| = \sqrt{(\tilde{I_G}\omega + \left\langle \tilde{S_G}\dot{\varepsilon} \right\rangle + h_w)^T (\tilde{I_G}\omega + \left\langle \tilde{S_G}\dot{\varepsilon} \right\rangle + h_w)} \tag{3.83}$$

$$\mathscr{E} = \frac{1}{2} m v^{gT} v^g + \frac{1}{2}\omega^T \tilde{I_G}\omega + \omega^T \left\langle \tilde{S_G}\dot{\varepsilon} \right\rangle + \frac{1}{2}\langle T_G \dot{\varepsilon}, \dot{\varepsilon} \rangle + \mathscr{V}_I \tag{3.84}$$

We can appreciate how the invariants are simplified in this formulation and in particular how h_w can control $\tilde{I_G}\omega$ provided that the *angular momentum* is not absorbed in the flexible modes (the term $\left\langle \tilde{S_G}\dot{\varepsilon} \right\rangle$).

Let us provide the mass properties in the new CoM-**BRF** from the mass properties of a generic **BRF**. We call $d = p/m$, E_3 the 3×3 identity matrix and we use the compact notation for the subscripts. Then we have:

$$m_G = \int_{V_0} \rho_o \, dV_o = m$$

$$p_G = \int_{V_0} (r - d) \, \rho_o \, dV_o = 0$$

$$L_G^k = \int_{V_0} (\phi^k - L^k/m) \rho_o \, dV_o = 0$$

$$S_G^q = S^q - [d, L^q]$$

$$I_G = I - m(d^T d E_3 - dd^T)$$

$$T_G^{kq} = T^{kq} - L^{kT} L^q / m$$

$$J_G^k = J^k - 2d^T L^k E_3 + dL^{kT} + L^k d^T$$

$$K_G^{kq} = K^{kq} - L^{kT} L^q E_3 / m + L^k L^{qT} / m$$

$$P_G^{kq} = P^{kq} - [L^k, L^q]$$

When the *external interactions* are not null the invariants are not constant, but the use of CoM-**BRF** provides advantages as well in the expression of the dynamic equations.

In the new kinematics also the generalized forces have a different expression; from Eqs. 3.20–3.22, introducing the new kinematics Eq. 3.80 in the virtual displacement introduced in Sect. 3.4.4, we get the new expressions:

$$\begin{aligned} f_G &= f \\ c_G &= c - [\tilde{p}, f]/m = c - [p + \langle L\varepsilon \rangle, f]/m \\ Q_G{}^{\varepsilon} &= Q^{\varepsilon} - L^T f/m \end{aligned} \tag{3.85}$$

where f, c, Q^{ε} are the generalized forces calculated in the—body fixed origin—**BRF**, where also p, L are calculated, using Eqs. 3.20–3.22. In particular c is the torque calculated from x^o and Q^{ε} is calculated from the virtual work done by the external forces on the untranslated deformation functions ϕ.

From Eq. 3.85 a constant generalized force f produces a static Q_G and c_G.

As an example during orbital maneuvers with high thrust the static acceleration produces a static deflection of the flexible modes and a torque dependent on the flexible mode deflection (static and dynamic).

The equations of motion of flexible bodies Eqs. 3.27–3.29, written in the *center of mass*, become:

$$ma^g = f \tag{3.86}$$

$$\tilde{I_G}\dot{\omega} + \left\langle \tilde{S_G}\ddot{\varepsilon} \right\rangle + \dot{h_w} + [\omega, \tilde{I_G}\omega + h_w] + [\omega, \left\langle \tilde{S_G}\dot{\varepsilon} \right\rangle] + \langle \tilde{J_G}\dot{\varepsilon} \rangle \omega = c - [\tilde{p}, f]/m \tag{3.87}$$

$$\tilde{S_G}^T \dot{\omega} + \langle T_G \ddot{\varepsilon} \rangle + \frac{\partial \mathscr{V}_I}{\partial \varepsilon} + \frac{\partial \mathscr{F}}{\partial \dot{\varepsilon}} - \frac{1}{2}\omega^T \tilde{J_G}\omega + 2\omega^T \langle P_G \dot{\varepsilon} \rangle = Q^{\varepsilon} - L^T f/m \tag{3.88}$$

The external forces f, c, Q^{ε} are those due to external interactions: control actions exerted by mass ejection devices like RCTs (see Chap. 8) or torque-rods, external environmental interactions like gravity, solar pressure, atmospheric drag, electromagnetic interactions. The *momentum management systems* action is taken into account by the terms h_w, $\dot{h_w}$ in Eq. 3.87.

It is important to note that—apart when using mass ejection systems where the mass and the inertia properties are time varying—we can write Eqs. 3.86, 3.87, in terms of the invariants introduced. Equation 3.86 is equivalent to:

$$\frac{d\mathscr{I}_{TR}}{dt} = M_{E \hookleftarrow B} f = F \tag{3.89}$$

and, Eq. 3.87 is equivalent to:

$$\frac{d\mathscr{I}_{ROT}}{dt} = M_{E \hookleftarrow B} c_G = C_G \tag{3.90}$$

The second relationship can be easily recognized using the note (see footnote 1) that provides the rule for the derivation in **ECI** components, and considering that in Eq. 3.87, the inertia properties $\tilde{I_G}$, $\tilde{S_G}$ are functions of the deformations.

3.10 Dynamics in the Gravitational Field

In this section we include some complements to the basic Lagrangian equations that are specific for space applications. Let us evaluate the interaction of a gravitational central field with the flexible body. This is an example, relevant to space applications, of how to deal with external forces produced by a conservative field.

The Earth's *gravitational potential energy* of a unitary mass is expressed as (see [15]):

$$\begin{aligned}\mathcal{V}_g = -\tfrac{\mu_e}{|x|}\Bigg\{1 - \sum_{k=2}^{\infty} J_k \left(\tfrac{R_e}{|x|}\right)^k P_k(\sin(\mathfrak{L})) \\ + \sum_{k=2}^{\infty}\sum_{j=2}^{\infty} J_{k,j} P_k^j (\sin(\mathfrak{L}))\cos\left[m\left(\mathfrak{l} - \mathfrak{l}_{k,j}\right)\right]\Bigg\}\end{aligned} \tag{3.91}$$

where $\mathfrak{L}$ and $\mathfrak{l}$ are the body particle latitude and longitude in the **ECEF**[7] frame, $|x|$ is the body particle distance from the Earth center, $\mu_e = GM_e$ where G is the Gravitational constant M_e and R_e are the Earth mass and radius.

We will consider only the principal term due to a perfectly spherical Earth but the following derivation can be extended to each term of this series.

A central gravity field of strength μ_e applied to a space vehicle produces a potential energy which is the integral of the potential energy of each mass particle:

$$\mathcal{V}_g = -\int_V \frac{\mu_e}{|x|}\rho dV = -\int_{V_0} \frac{\mu_e}{|x|}\rho_o dV_o$$

where V_0 is the volume in the reference configuration, $\rho_o(r)$ the mass density in the reference configuration, V the volume in the deformed configuration, $\rho(r)$ the mass density in the deformed configuration. Due to the equivalence between the two integrals (see Appendix A.1), we can integrate the *gravitational potential energy* in the reference configuration as we did for the kinetic energy.

We now introduce the kinematic model in order to express the potential in terms of the kinematic variables.

$$x_i = x_i^o + M_{ij}(r_j + \phi_j^k(r)\varepsilon_k)$$

Let us pose for simplicity $\xi = r + \langle\phi\varepsilon\rangle$, let $k = M(\theta)^T x^o/|x^o|$ be the unit vector of the anti-Earth direction in **BRF** and expand in Taylor series the potential up to the second order in ξ:

$$1/|x| = 1/|x^o|(1 - k^T\xi/|x^o| - 1/2\xi^T\xi/|x^o|^2 + 3/2(k^T\xi)^2/|x^o|^2). \tag{3.92}$$

[7] **ECEF** is similar to **ECI** presented in Chap. 2, but it is fixed with the Earth. The Z axis is aligned with the **ECI** Z axis, orthogonal to the equatorial plane, the X axis points to the Greenwich meridian, the Y axis completes the triad. Therefore, omitting Earth precession and nutation corrections, the transformation from **ECI** to **ECEF** is accomplished with a simple rotation in the X/Y plane.

We introduce this form in the potential energy per unit mass and integrate in V_0 using the following identities:

$$\tilde{p} = \int_{V_0} \xi \rho_o dV_o, \quad m = \int_{V_0} \rho_o dV_o,$$
$$\tilde{I} = \int_{V_0} (\xi^T \xi - \xi \xi^T) \rho_o dV_o, \quad tr(\tilde{I}) = 2 \int_{V_0} \xi^T \xi \rho_o dV_o$$

we get for the potential energy of a flexible body under a central gravitational field:

$$\mathscr{V}_g = -\frac{\mu_e m}{|x^o|} + \frac{\mu_e k^T \tilde{p}}{|x^o|^2} - \frac{\mu_e tr(\tilde{I})}{2|x^o|^3} + \frac{3\mu_e k^T \tilde{I} k}{2|x^o|^3}$$

We now derive the expression of the generalized forces; by definition (see Eq. 3.30) we have:

$$f_k = -M_{sk} \frac{\partial \mathscr{V}_g}{\partial x_s^o}, \quad c_k = -\mu_{kq}^{\#} \frac{\partial \mathscr{V}_g}{\partial \theta_q}, \quad Q_k^{\varepsilon} = -\frac{\partial \mathscr{V}_g}{\partial \varepsilon_k}$$

We use in the derivation the following identities:

$$\mu_{qk}^{\#} \frac{\partial M_{es}}{\partial \theta_k} = M_{en} e_{nqs}, \quad k_i = M_{ij} x_j^o / |x^o|, \quad \frac{\partial |x^o|^{-n}}{\partial x_i^o} = -n x_i^o / |x^o|^{n+2},$$

thus obtaining:

$$f_k = -\frac{\mu_e m k_k}{|x^o|^2} - \frac{\mu_e \tilde{p}_k}{|x^o|^3} + \frac{3\mu_e k_k k_i \tilde{p}_i}{|x^o|^3} - \frac{3\mu_e k_k \tilde{I}_{ii}}{2|x^o|^4} - \frac{3\mu_e \tilde{I}_{ki} k_i}{|x^o|^4} + \frac{15\mu_e k_k \tilde{I}_{ij} k_i k_j}{2|x^o|^4}$$

$$c_k = -\frac{\mu_e e_{kij} \tilde{p}_i k_j}{|x^o|^2} + \frac{3\mu_e e_{kij} k_i \tilde{I}_{jm} k_m}{|x^o|^3}$$

$$Q_k^{\varepsilon} = -\frac{\mu_e L_j^k k_j}{|x^o|^2} + \frac{\mu_e \tilde{J}_{ii}^k}{2|x^o|^3} - \frac{3\mu_e \tilde{J}_{ij}^k k_i k_j}{2|x^o|^3}$$

This expression of c_k is commonly called *gravity gradient torque*. These Lagrangian forces can be placed directly as right members of the flexible body Lagrangian equations: Eqs. 3.27–3.29.

3.10.1 Uncoupling Attitude and Orbit Dynamics

We have introduced in Sect. 3.9 a particular kinematic model in which we can simplify the integrals of motion and the Lagrangian equations and the expressions of the gravitational generalized Lagrangian forces defined by Eq. 3.93. In this specific kinematic description the mass properties p_G, L_G are null, and the Lagrangian forces must be modified as per Eq. 3.85:

$$
\begin{aligned}
f_G &= f \\
c_G &= c - [\tilde{p}, f]/m = c - [p + \langle L\varepsilon \rangle, f]/m \\
Q_G{}^{\varepsilon} &= Q^{\varepsilon} - L^T f/m
\end{aligned}
\tag{3.93}
$$

The equations of motion of the flexible body Eqs. 3.86–3.88 subject to a central gravity field, written in CoM-**BRF** are:

$$
\begin{aligned}
& ma^g = -\mu_e m \frac{k}{|x^g|^2} - 3\mu_e \frac{tr\left(\tilde{I_G}\right)k}{2|x^g|^4} - 3\mu_e \frac{\tilde{I_G}k}{|x^g|^4} + \frac{15\mu_e k\left(k^T \tilde{I_G} k\right)}{2|x^g|^4} + f \\
& \tilde{I_G}\dot{\omega} + \left\langle \tilde{S_G}\ddot{\varepsilon} \right\rangle + \dot{h_w} + [\omega, \tilde{I_G}\omega + h_w] + [\omega, \left\langle \tilde{S_G}\dot{\varepsilon} \right\rangle] + \langle \tilde{J_G}\dot{\varepsilon} \rangle \omega = \frac{3\mu_e [k, \tilde{I_G}k]}{|x^g|^3} + c - [\tilde{p}, f]/m \\
& \tilde{S_G}^T \dot{\omega} + \langle T_G \ddot{\varepsilon} \rangle + \frac{\partial \mathscr{V}_I}{\partial \varepsilon} + \frac{\partial \mathscr{F}}{\partial \dot{\varepsilon}} - \frac{1}{2}\omega^T \tilde{J_G}\omega + 2\omega^T \langle P_G \dot{\varepsilon} \rangle = \frac{\mu_e\, tr(\tilde{J_G})}{2|x^g|^3} - \frac{3\mu_e k^T \tilde{J_G} k}{2|x^g|^3} + Q^{\varepsilon} - L^T f/m
\end{aligned}
\tag{3.94}
$$

The external forces f, c, Q^{ε}, calculated in any—body fixed origin—**BRF** by using Eq. 3.85 in which also p, L in the r.h.s. are calculated, are due to the *external non conservative interactions* like control actions exerted by mass ejection actuators (RCTs) and environmental interactions additional to the central body gravitational force like solar pressure, atmospheric drag, electro-magnetic interactions, additional gravitational perturbations—those not considered in the *external potential energy*.

We see that the orbital dynamics are affected by forces that depend on attitude and deformation. These forces are very small: in fact, they are considered null in applications.

The motion of the *center of mass* can be studied separately with respect to the attitude and flexible modes dynamics in most engineering cases. The ratio of the norm of the most important coupling term with the norm of the central Newtonian force is equal to the ratio between the trace of the body inertia and the square of the orbital range multiplied by the mass.

For example, in case of a bar-shaped body 100 m long positioned in a 10000 km radius orbit, this ratio is about:

$$
(3/2)\frac{tr(\tilde{I_G})}{m\,|x^g|^2} \approx 10^{-10}
$$

By considering the coupling not relevant, we can integrate the orbit equations separately from attitude and deformation ones; the translational dynamic equations are then reduced to Kepler equations:

$$
a^g = M^T \ddot{x^g} = -\mu_e \frac{k}{|x^g|^2} + f/m
\tag{3.95}
$$

On the contrary, the interaction of the gravity field with attitude dynamics is often noticeable and at least for the lower orbits is to be taken into account.

The AOCS, when RCTs are used, (see Sect. 8.2) generates important resultant forces f/m to react to an attitude error; in addition the AOCS generates torques and deformation forces (see Eq. 3.85) when trying to control the satellite orbit (see in [16, 17] specific thruster modulation methods to limit the excitation of flexible modes during orbital maneuvers).

Therefore the coupling between orbital, attitude and deformation dynamics is often maintained via the AOCS active control and must be carefully analyzed by the designer.

3.10.2 First Integrals in a Central Gravitational Field

The central gravitational field forces are such that $\int_{V_0}[x, f_{grav}]\rho_o dV_o = 0$, where f_{grav} is the force per unit mass acting on the flexible body.

If we introduce in this equation the kinematic model $x = x^o + M_{E\leftrightarrow B}\xi$ where $\xi = r + \langle\phi\varepsilon\rangle$ like in Sect. 3.10, it results that the generalized Lagrangian forces f and c generated by the central gravitational field and calculated with Eqs. 3.20, 3.21 satisfy the following relationship:

$$\int_{V_0}[x, f_{grav}]\rho_o dV_o = M_{E\leftrightarrow B}(c + |x^o|[k, f]) = 0 \tag{3.96}$$

We can derive easily an invariant from this equation. Starting from:

$$\frac{d}{dt}\frac{\partial \mathcal{T}}{\partial v} + \left[\omega, \frac{\partial \mathcal{T}}{\partial v}\right] = f \qquad [\mathbf{1}]$$

$$\frac{d}{dt}\frac{\partial \mathcal{T}}{\partial \omega} + \left[\omega, \frac{\partial \mathcal{T}}{\partial \omega}\right] + \left[v, \frac{\partial \mathcal{T}}{\partial v}\right] = c \qquad [\mathbf{2}]$$

As in the previous section we can write the above expressions as:

$$\frac{d}{dt}(M_{E\leftrightarrow B}\frac{\partial \mathcal{T}}{\partial v}) = F \qquad [\mathbf{1}]$$

$$\frac{d}{dt}(M_{E\leftrightarrow B}\frac{\partial \mathcal{T}}{\partial \omega}) + M_{E\leftrightarrow B}\left[v, \frac{\partial \mathcal{T}}{\partial v}\right] = C \qquad [\mathbf{2}]$$

For central fields $[\mathbf{2}] + [x^o, [\mathbf{1}]] = 0$, namely

$$[\mathbf{2}] + [x^o, [\mathbf{1}]] = \frac{d}{dt}(M_{E\leftrightarrow B}\frac{\partial \mathcal{T}}{\partial \omega}) + M_{E\leftrightarrow B}\left[v, \frac{\partial \mathcal{T}}{\partial v}\right] + [x^o, \frac{d}{dt}(M_{E\leftrightarrow B}\frac{\partial \mathcal{T}}{\partial v})] = 0$$

$$\Rightarrow \frac{d}{dt}(M_{E\leftrightarrow B}\frac{\partial \mathcal{T}}{\partial \omega}) + M_{E\leftrightarrow B}\left[v, \frac{\partial \mathcal{T}}{\partial v}\right] + \frac{d}{dt}[x^o, M_{E\leftrightarrow B}\frac{\partial \mathcal{T}}{\partial v}] - M_{E\leftrightarrow B}[v, \frac{\partial \mathcal{T}}{\partial v}] = 0$$

$$\Rightarrow \frac{d}{dt}(M_{E\leftrightarrow B}(\frac{\partial \mathcal{T}}{\partial \omega} + |x^o|\left[k, \frac{\partial \mathcal{T}}{\partial v}\right])) = 0$$

The last passage follows from $x^o = |x^o|\, M_{E \leftrightarrow B}\, k$ and from the invariance of the vector product ($[M_{E \leftrightarrow B}a, M_{E \leftrightarrow B}b] = M_{E \leftrightarrow B}[a, b], \forall a, b \in \Re^3$). The invariant *angular momentum* within a central gravitational field becomes:

$$M_{E \leftrightarrow B}(\frac{\partial \mathscr{T}}{\partial \omega} + |x^o| \left[k, \frac{\partial \mathscr{T}}{\partial v}\right]) = \mathscr{I}_{ROT}$$

Using the CoM-**BRF** we get for the flexible body:

$$M_{E \leftrightarrow B}(\tilde{I_G}\omega + \langle \tilde{S_G}\dot{\varepsilon} \rangle + h_w + m|x^g| \left[k, v^g\right]) = \mathscr{I}_{ROT} \tag{3.97}$$

The *angular momentum* is then the standard orbital momentum plus the free body one. This invariance applies only if the external interactions are constituted by a central gravity field.

It is interesting to compare Eq. 3.97 with Eq. 3.90 and the second equations of Eq. 3.94, in the contest of a central gravity field.

We discover that the attitude part of the *angular momentum*:

$$M_{E \leftrightarrow B}(\tilde{I_G}\omega + \langle \tilde{S_G}\dot{\varepsilon} \rangle + h_w) \tag{3.98}$$

is not conserved in a central gravity field because of the coupling term with the gravity field. If the gravity gradient torque is negligible because the orbit is very high or the inertia matrix is almost spherical, then we have the conservation of both attitude and orbital components of *angular momentum*.

Regarding *energy*, we can derive a new invariant in the case of a central gravitational field by adding to Eq. 3.84 the contribution of the gravitational potential energy:

$$\begin{aligned} \mathscr{E} = \tfrac{1}{2} m {v^g}^T v^g + \tfrac{1}{2}\omega^T \tilde{I_G}\omega + \omega^T \left\langle \tilde{S_G}\dot{\varepsilon} \right\rangle + \tfrac{1}{2}\langle T_G \dot{\varepsilon}, \dot{\varepsilon} \rangle \\ + \tfrac{1}{2}\langle K\varepsilon, \varepsilon \rangle - \tfrac{\mu_e m}{|x^g|} - \tfrac{\mu_e tr(\tilde{I_G})}{2|x^g|^3} + \tfrac{3\mu_e k^T \tilde{I_G} k}{2|x^g|^3} \end{aligned} \tag{3.99}$$

As in the case of Eq. 3.84, the invariant defined by Eq. 3.99 is constant only if h_w is constant and conditions of Eq. 3.38 are satisfied.

In Eqs. 3.97 and 3.99 the orbital terms are many order of magnitude above the attitude and the *momentum management system* ones. We have already seen that the coupling terms allow the passage of *energy* and *angular momentum* from the orbital state to the attitude states at least in low orbits, the reverse passage is really negligible. This is why we can integrate the orbit equations without considering the variation of attitude rates.

3.11 Linearized Dynamics

One of the applications of space flexible body equations is the control of an artificial satellites nearby an orbital and attitude trajectory that has been chosen in order to fulfill the space vehicle mission tasks; these desired dynamics are called the *reference trajectory*.

The choice of this trajectory is defined by the specific application and is one of the main topics in system space engineering. It is common practice to linearize the dynamic equations Eq. 3.94 of the satellite around such a trajectory in order to simplify the control design task.

This section is devoted to illustrate the methodology of linearizing the dynamics' equations.

We choose to work with the CoM-**BRF** so we linearize equations Eq. 3.94 and the origin coincides with the *center of mass* $x^o = x^g$.

As first step, the kinematic variables must be written as the sum of the *reference trajectory* kinematic variables $\overline{x}^g, \overline{\theta}, \overline{\varepsilon} = 0$ which are over-lined in the following, plus a small perturbation term dependent on the perturbation variables: $\xi,\ \delta\phi,\ \varepsilon$.

The perturbation variables will be kept in the equations only to first order.

The perturbation variables are defined in the following Eqs. 3.100–3.102:

$$\begin{cases} x^g = \overline{x}^g + M(\overline{\theta})\xi \\ M(\theta) = M(\overline{\theta}) + \delta M = M(\overline{\theta}) + \frac{\partial M}{\partial \theta}\delta\theta \end{cases} \tag{3.100}$$

where ξ is the relative motion of the body with respect to the *reference trajectory* $\overline{x}^g$ projected in CoM-**BRF** and $M(\overline{\theta}) = M_{E\leftrightarrow B}(\overline{\theta})$ is the reference attitude matrix from CoM-**BRF** to **ECI**.

Concerning the angular motion, from Appendix A.2 we get[8]:

$$\frac{\partial M}{\partial \theta}\delta\theta = M(\theta)\, skew(\mu\delta\theta) \tag{3.101}$$

being μ the matrix defined by:

$$\omega = \mu(\theta)\dot{\theta}$$

We introduce a new perturbation variable which is the projection of the angular perturbation vector in body axes,

$$\overline{\mu}\delta\theta = \delta\phi \tag{3.102}$$

with

$$\overline{\mu} = \mu(\overline{\theta}) \tag{3.103}$$

[8]In Eq. A2.1 the Levi-Civita operator is here replaced by the skew matrix. In fact the application of the Levi-Civita operator on a vector A produces a matrix whose ij-th element is $skew(A)_{ij} = e_{ikj}A_k$.

From Eqs. 3.100–3.102, it follows (E_N is the $N \times N$ identity matrix) that:

$$M(\theta) = M(\overline{\theta})(E_3 + skew\,(\delta\phi))$$

Having defined the perturbation variables we pass to introduce them into the dynamic equations. The dynamic equations are written in terms of the following kinematic variables:

$$\begin{cases} Ma^g = \ddot{x}^g \\ skew(\omega) = M^T\dot{M} \\ \dot{\omega} \end{cases} \tag{3.104}$$

We can evaluate these kinematic variables in function of ξ, $\delta\phi$ and their derivatives. Substituting in these expressions the reference trajectory variables plus the perturbation variables, we get to first order in the perturbation:

$$\begin{aligned} \ddot{x}^g &= \ddot{\overline{x}}^g + \overline{M}\delta a \\ \omega &= \overline{\omega} + \delta\omega \\ \dot{\omega} &= \dot{\overline{\omega}} + \delta\dot{\omega} \end{aligned}$$

with

$$\begin{aligned} \delta a &= \ddot{\xi} + [\overline{\omega}, [\overline{\omega}, \xi]] + [\dot{\overline{\omega}}, \xi] + 2[\overline{\omega}, \dot{\xi}] \\ \delta\omega &= \delta\dot{\phi} + [\overline{\omega}, \delta\phi] \\ \delta\dot{\omega} &= \delta\ddot{\phi} + [\overline{\omega}, \delta\dot{\phi}] + [\dot{\overline{\omega}}, \delta\phi] \end{aligned}$$

The first equation can be obtained deriving twice the first of Eq. 3.100 and noting that $\dot{\overline{M}} = \overline{M}skew(\overline{\omega})$.

The second and third equations can be proved from $M = \overline{M}(E_3 + skew(\delta\phi))$. Keeping only the first order terms the following identities apply:

$$\begin{aligned} skew(\omega) = M^T\dot{M} &= (E_3 - skew(\delta\phi))\,\overline{M}^T\left(\dot{\overline{M}}\,(E_3 + skew(\delta\phi)) + \overline{M}skew(\delta\dot{\phi})\right) \\ &= (E_3 - skew(\delta\phi))\left(skew(\overline{\omega})\,(E_3 + skew(\delta\phi)) + skew(\delta\dot{\phi})\right) \\ &= skew\,(\overline{\omega}) + (skew\,(\overline{\omega})\,skew\,(\delta\phi) - skew\,(\delta\phi)\,skew\,(\overline{\omega})) + skew\,\left(\delta\dot{\phi}\right) \end{aligned} \tag{3.105}$$

Applying the property of skew matrices, being a, b, c 3D vectors:

$$skew(a)skew(b) - skew(b)skew(a) = skew(a \times b)$$

and the distributivity property of the vector cross product with respect to the sum, Eq. 3.105 becomes:

$$M^T\dot{M} = skew(\omega) = skew(\overline{\omega} + \overline{\omega} \times \delta\phi + \delta\dot{\phi})$$

which implies

$$\Rightarrow \begin{cases} \omega = \overline{\omega} + \underbrace{\overline{\omega} \times \delta\phi + \delta\dot{\phi}}_{\delta\omega} \\ \dot{\omega} = \dot{\overline{\omega}} + \underbrace{\dot{\overline{\omega}} \times \delta\phi + \overline{\omega} \times \delta\dot{\phi} + \delta\ddot{\phi}}_{\delta\dot{\omega}} \end{cases} \tag{3.106}$$

The anti-Earth direction in CoM-**BRF**:

$$k = M(\theta)^T \frac{x^g}{|x^g|} = (E_3 - skew(\delta\phi))M(\overline{\theta})^T \frac{x^g}{|x^g|} \tag{3.107}$$

can be split in a part that is derived by the desired attitude and a perturbation term, $k = \overline{k} + \delta k$, where $\overline{k} = M(\overline{\theta})^T \frac{\overline{x}^g}{|\overline{x}^g|}$ and $\delta k = [\overline{k}, \delta\phi] + (E_3 - \overline{k}\overline{k}^T)\frac{\xi}{|\overline{x}^g|}$.

Also the *momentum management system* h_w will be split in two parts $h_w = \overline{h}_w + \delta h_w$, where the over-lined part is the is the control actions necessary to produce the equilibrium of the reference dynamics. If the $\overline{h}_w$ is not sufficient to provide an equilibrium of the reference attitude dynamic we have also to introduce an additional control torque $\overline{c}_c$ produced by mass ejection systems or torque-rods and c will be split into $c = \overline{c}_c + \delta c$.

We introduce now in Eq. 3.94, the kinematic variables as the sum of an over-lined term plus a perturbation term and request the 0th order equations and the first order ones are separately satisfied. We consider $\overline{h}_w, \overline{c}_c$ 0th order and $f, \delta c, Q^\varepsilon$ first order like the kinematic perturbation variables. Only the gravity gradient will be kept among the gravitational coupling terms.

The *reference trajectory* satisfies:

$$\begin{aligned} m\ddot{\overline{x}}^g &= -\tfrac{\mu_e m \overline{x}^g}{|\overline{x}^g|^3} \\ I_G\dot{\overline{\omega}} + \dot{\overline{h}}_w + \overline{\omega} \times (I_G\overline{\omega} + \overline{h}_w) &= \overline{c}_c + \tfrac{3\mu_e[\overline{k}, I_G\overline{k}]}{|x^g|^3} \\ \overline{\varepsilon} &= 0 \end{aligned} \tag{3.108}$$

The perturbation variables solve the following first order equations:

$$\begin{aligned} m(\ddot{\xi} + [\overline{\omega}, [\overline{\omega}, \xi]] + [\dot{\overline{\omega}}, \xi] + 2[\overline{\omega}, \dot{\xi}]) &= f - \tfrac{\mu_e m x^g}{|x^g|^3} + \tfrac{\mu_e m \overline{x}^g}{|\overline{x}^g|^3} \\ &= f - \tfrac{\mu_e m (I - 3\overline{k}\overline{k}^T)\xi}{|\overline{x}^g|^3} \end{aligned} \tag{3.109}$$

$$\begin{aligned} I_G\delta\dot{\omega} + \langle S_G\ddot{\varepsilon}\rangle + \delta\dot{h}_w + [\overline{\omega}, I_G\delta\omega + \delta h_w] + [\delta\omega, I_G\overline{\omega} + \overline{h}_w] + [\overline{\omega}, \langle S_G\dot{\varepsilon}\rangle] + \langle J_G\dot{\varepsilon}\rangle\overline{\omega} \\ = \delta c - [\tilde{p}, f/m] + \tfrac{3\mu_e}{|\overline{x}^g|^3}([\delta k, I_G\overline{k}] + [\overline{k}, I_G\delta k]) \end{aligned} \tag{3.110}$$

$$S_G^T\delta\dot{\omega} + \langle T_G\ddot{\varepsilon}\rangle + \frac{\partial \mathscr{V}_I}{\partial \varepsilon} + \frac{\partial \mathscr{F}}{\partial \dot{\varepsilon}} - \frac{1}{2}\overline{\omega}^T J_G\overline{\omega} + 2\overline{\omega}^T\langle P_G\dot{\varepsilon}\rangle = Q^\varepsilon - L^T f/m \tag{3.111}$$

In these equations the generalized Lagrangian forces f, $c = \overline{c}_c + \delta c$, Q^{ε} include the control actions exerted by mass ejection systems or the control actions of other actuators like magneto-torquers, solar sails, plus the environmental forces: solar pressures, drag, magnetic forces and higher terms of the gravitational potential.

Thus we can consider f, c as the sum of two factors, the external disturbances f_d, c_d, and the control action f_c, $c_c = \overline{c}_c + \delta c_c$ managed by the AOCS. So $f = f_d + f_c$ and $c = \overline{c}_c + \delta c_c + c_d$.

Equations 3.108–3.111 must be specialized to the specific *reference trajectory* desired.

Remarkable cases applicable to the presented linearization are:

- inertially pointing satellite $\Longrightarrow \overline{\omega} = 0$;
- Earth (central body) pointing satellite $\Longrightarrow \overline{\omega} = \omega_0$;
- spinning satellite $\Longrightarrow \overline{\omega} = \omega_{Spin}$.

Let us consider the problem of a satellite having a circular orbit as reference trajectory and an Earth pointing (with CoM-**BRF**) aligned to **ORF** as reference attitude, then $\overline{\omega} = -\omega_0 e_2$ with constant $\omega_0 = \sqrt{\mu_e/|\overline{x}^g|^3}$ and $\overline{k} = -e_3$.

Equation 3.108 become:

$$\overline{c}_c + \frac{3\mu_e[e_3, I_G e_3]}{|x^g|^3} = {\omega_0}^2[e_2, I_G e_2] - \omega_0[e_2, \overline{h}_w] + \dot{\overline{h}}_w \tag{3.112}$$

In this case from Eq. 3.112 we see that the *reference trajectory* can be kept with a static control torque (i.e. $\dot{\overline{h}}_w = 0$). The torque balance along e_1 and e_3 can be obtained biasing $\overline{h}_w$ in the e_1 and e_3 plane, the torque balance along e_2 must be obtained using $\overline{c}_c$ torque-rods or RCTs. In high orbits like GEO all these effects are very small. We note that a momentum bias along the e_2 direction—orthogonal to the orbital plane—does not disturb the Earth pointing *reference trajectory*, thus the designer could decide to bias in e_2 direction the *momentum management system* to obtain advantages in the equation of the perturbation variables like a gyroscopic stiffness.

The perturbations due to additional un-modeled external interactions like solar pressure, control action un-nominal performance, orbital perturbation, dynamic transient due to the actuation system or the flexible modes excitation will be managed by the control actions: $\delta\dot{h}_w$ and δc_c and f_c which the on-board controller (or the ground station typically for the orbital control) determines from the perturbation variables error signals, produced by the exogenous disturbs c_d and f_d.

The attitude perturbation variables will be expressed as roll, pitch, yaw Euler angles: $\delta\phi = (\varphi, \vartheta, \psi)^T$.

The choice of the Euler sequence is irrelevant using a linearized kinematics. As the satellite points towards the Earth, from Eq. 2.13 the over-lined angular velocity of CoM-**BRF** is:

$$\overline{\omega} = -\omega_0 e_2 = \begin{pmatrix} 0 \\ -\omega_0 \\ 0 \end{pmatrix} \tag{3.113}$$

while $\delta\omega$, $\delta\dot{\omega}$ are expressed, from Eq. 3.106, as:

$$\delta\omega = \begin{pmatrix} -\omega_0\psi + \dot{\varphi} \\ \dot{\vartheta} \\ \omega_0\varphi + \dot{\psi} \end{pmatrix} \tag{3.114}$$

$$\delta\dot{\omega} = \begin{pmatrix} -\omega_0\dot{\psi} - \dot{\omega}_0\psi + \ddot{\varphi} \\ \ddot{\vartheta} \\ \omega_0\dot{\varphi} + \dot{\omega}_0\varphi + \ddot{\psi} \end{pmatrix} \tag{3.115}$$

Introducing Eqs. 3.113–3.115 into Eqs. 3.109–3.111 we get:

$$m\left(\ddot{\xi} + \omega_0^2\,[e_2,[e_2,\xi]] - \dot{\omega}_0\,[e_2,\xi] - 2\omega_0\left[e_2,\dot{\xi}\right]\right) = f - \frac{\mu_e m(I - 3e_3{e_3}^T)\xi}{|\overline{x}^g|^3} \tag{3.116}$$

which can be further simplified in[9]:

$$\ddot{\xi} + \omega_0^2 e_2{e_2}^T\xi - 3\omega_0^2 e_3{e_3}^T\xi - 2\omega_0 e_2 \times \dot{\xi} = f/m \tag{3.117}$$

$$\begin{aligned} & I_G\begin{pmatrix} -\omega_0\dot{\psi} - \dot{\omega}_0\psi + \ddot{\varphi} \\ \ddot{\vartheta} \\ \omega_0\dot{\varphi} + \dot{\omega}_0\varphi + \ddot{\psi} \end{pmatrix} + \langle S_G\ddot{\varepsilon}\rangle + \delta\dot{h}_w + \left[-\omega_0 e_2, I_G\begin{pmatrix} -\omega_0\psi + \dot{\varphi} \\ \dot{\vartheta} \\ \omega_0\varphi + \dot{\psi} \end{pmatrix} + \delta h_w\right] \\ & + \left[\begin{pmatrix} -\omega_0\psi + \dot{\varphi} \\ \dot{\vartheta} \\ \omega_0\varphi + \dot{\psi} \end{pmatrix}, -\omega_0 I_G e_2 + \overline{h}_w\right] - \omega_0\left[e_2, \langle S_G\dot{\varepsilon}\rangle\right] - \langle J_G\dot{\varepsilon}\rangle\omega_0 e_2 \\ & = \delta c - [\tilde{p}, f/m] - 3{\omega_0}^2([\delta k, I_G e_3] + [e_3, I_G\delta k]) \end{aligned} \tag{3.118}$$

in this last equation δk is specialized as: $\delta k = -[e_3,\delta\phi] + (E_3 - e_3e_3^T)\frac{\xi}{|\overline{x}^g|}$.

$$\begin{aligned} & S_G^T\begin{pmatrix} -\omega_0\dot{\psi} - \dot{\omega}_0\psi + \ddot{\varphi} \\ \dot{\omega}_0 + \ddot{\vartheta} \\ \omega_0\dot{\varphi} + \dot{\omega}_0\varphi + \ddot{\psi} \end{pmatrix} + \langle T_G\ddot{\varepsilon}\rangle + \frac{\partial\mathcal{V}_I}{\partial\varepsilon} + \frac{\partial\mathcal{F}}{\partial\dot{\varepsilon}} - \frac{1}{2}\omega_0^2 e_2^T J_G e_2 - 2\omega_0 e_2^T\langle P_G\dot{\varepsilon}\rangle \\ & = Q^\varepsilon - L^T f/m \end{aligned} \tag{3.119}$$

Equation 3.117 is used to study the orbital motion of a single satellite around its reference orbit and it is also the classic equation used to study *flying formations* (see [18]), i.e. the orbital motion that each satellite of the formation performs around the formation common *reference orbit*. In the case $f = 0$ these equations can be easily integrated to provide the basic types of *flying formations*:

[9]We use here the double vector product rule: $[a,[b,c]] = b(c^Ta) - c(a^Tb)$.

$$
\begin{aligned}
\xi_1(t) &= -\tfrac{2\dot{\xi}_{30}}{\omega_0}\cos(\omega_0 t) + (-6\xi_{30} + \tfrac{4\dot{\xi}_{10}}{\omega_0})\sin(\omega_0 t) + (6\omega_0\xi_{30} - 3\dot{\xi}_{10})t + \tfrac{2\dot{\xi}_{30}}{\omega_0} + \xi_{10} \\
\xi_2(t) &= \tfrac{\dot{\xi}_{20}}{\omega_0}\sin(\omega_0 t) + \xi_{20}\cos(\omega_0 t) \\
\xi_3(t) &= \tfrac{\dot{\xi}_{30}}{\omega_0}\sin(\omega_0 t) - (3\xi_{30} - \tfrac{2\dot{\xi}_{10}}{\omega_0})\cos(\omega_0 t) + 4\xi_{30} - \tfrac{2\dot{\xi}_{10}}{\omega_0}
\end{aligned}
\tag{3.120}
$$

The dynamics of rotations and flexibilities are normally used for the design of controllers in a linear context.

This model can be further simplified when used for control applications because we need mainly to represent the dynamics close to the open loop cross-over frequency (see Sect. 5.5). We distinguish low and high bandwidth dynamic models.

Low bandwidth dynamic models are those including only the low frequency eigenmodes of Eq. 3.118, for example the orbital coupling (terms due to ω_0) and the nutation modes (terms due to a large $\overline{h}_w$ in the e_2 direction). These models are used with the design of low bandwidth controllers (see e.g. [3]).

The high bandwidth dynamic linear models are those used when designing high bandwidth controllers in the presence of flexibilities, they are treated with more detail in the next section. In this case we may consider negligible the effects of the orbital coupling and nutation on the stability and performance of the attitude control.

3.12 Spacecraft Linear Model

We present here a typical linear model of a spacecraft suitable for linear controller design (see Chap. 5) in presence of flexibilities. This type of model will be useful to assess the closed loop short term behavior of the dynamics and the system stability. If we want to model the long term attitude dynamics—under the action of solar pressure for example—then we have to consider the full linear model Eqs. 3.118, 3.119.

This model will be derived from Eqs. 3.118, 3.119 with some additional hypotheses:

- In Eqs. 3.118, 3.119 we set $\omega_0 = 0$. The orbital dynamics are considered to have a negligible effect. This can be assumed when the orbital frequency is by far too low with respect to the controller bandwidth.
- The potential energy and the dissipation functions are quadratic positive semi definite functions in their arguments. They can be represented with semi-positive symmetric matrices K and D, such that:

$$
\mathscr{V}_I = \frac{1}{2}\varepsilon^T K \varepsilon \geq 0, \mathscr{F} = \frac{1}{2}\dot{\varepsilon}^T D \dot{\varepsilon} \geq 0 \tag{3.121}
$$

Moreover, we assume that a linear transformation exists $\varepsilon = W\eta$, such that $W^T T_G W = E_{N_f}$ and $W^T K W = \Omega^2$.

Ω^2 is a diagonal matrix listing all squared angular frequencies of the appendages as cantilevered to satellite $\Omega = diag(\Omega_i), i = 1...N_f$. The matrix W is the modal matrix of the cantilevered satellite.
We assume also that the same transformation diagonalizes the damping matrix D, which is truly the case when D is a linear combination of K and T_G. We assume $D = qK$ for some q to guarantee the existence of a real modal matrix. Under this last hypothesis $W^T DW = 2\Sigma\Omega = q\Omega^2$, with $\Sigma = diag(\zeta_i), i = 1...N_f$ each ζ_i being the damping factor of the corresponding eigen-mode.
If we substitute in Eqs. 3.118, 3.119 $\varepsilon = W\eta$ and pre-multiply Eq. 3.119 by W^T, we obtain the same equations in η with new system parameters: $S_G = S_G W$, $L = LW, T_G = E_{N_f}, K = \Omega^2$ and $D = 2\Sigma\Omega$.

- We can use the transformation W also to equalize the dimensions of S_G. So far we have not done any hypothesis on the dimension of ε, which could be of any dimension provided that, when multiplied by deformation functions they provide a displacement in meters (the ε could be angles or lengths or other variables used in the particular structural model that describes the appendage deformation).
We apply to the transformation coefficients of W a dimension such that the identity matrix $W^T T_G W = E_{N_f}$ results non-dimensional and $W^T KW = \Omega^2$ has dimension $(rad/sec)^2$. In such a case we will have $dim(\eta) = m\sqrt{Kg},\ dim(S_G) = m\sqrt{Kg},\ dim(L) = \sqrt{Kg},\ dim(Q^\eta) = m\sqrt{Kg}/sec^2$.
As an example, if all ε have the dimensions in meters, then $dim(W) = 1/\sqrt{Kg}$ in all coefficients. This method has the advantage to simplify the interface data between the designer of the appendage and the AOCS designer who needs to know only the table of the S_G numbers.
With the satellite inertia matrix I_G and S_G, Ω and Σ we can solve the complete dynamics when $Q^\eta = 0$, but in order to calculate the real displacement of the appendages we need also to know the transformation W.
- Typically the matrices S_A and L_A for each appendage are given in a reference frame with the origin at the root of the appendage r_A normalized with the transformation provided in the previous bullet and possibly also rotated with respect to **BRF** of a matrix $M_{B\leftrightarrow A}$. In order to bring the information from the appendage reference frame to **BRF** it is necessary to implement a transformation to the integrals already defined in Sect. 3.3, in those integrals we must simply substitute r with $M_{B\leftrightarrow A}r + r_A$ and $\phi(r)$ with $M_{B\leftrightarrow A}\phi(r)$ and integrate in the appendage domain $r \in V_A$. We illustrate the transformation only for the more important mass properties:

$$
\begin{aligned}
m &= \textstyle\int_{V_A} \rho_A \, dV_A = m_A \\
p &= \textstyle\int_{V_A} (M_{B\leftrightarrow A} r + r_A)\, \rho_A \, dV_A = M_{B\leftrightarrow A} p_A + r_A m_A \\
L^k &= \textstyle\int_{V_A} M_{B\leftrightarrow A} \phi^k \rho_A \, dV_A = M_{B\leftrightarrow A} {L_A}^k \\
S^k &= \textstyle\int_{V_A} [M_{B\leftrightarrow A} r + r_A, M_{B\leftrightarrow A} \phi^k] \rho_A \, dV_A = M_{B\leftrightarrow A} {S_A}^q + [r_A, M_{B\leftrightarrow A} {L_A}^q] \\
I &= \textstyle\int_{V_A} ((M_{B\leftrightarrow A} r + r_A)^T (M_{B\leftrightarrow A} r + r_A) - (M_{B\leftrightarrow A} r + r_A)(M_{B\leftrightarrow A} r + r_A)^T) \rho_A \, dV_A \\
&= M_{B\leftrightarrow A} I_A M_{A\leftrightarrow B} - m_A ({r_A}^T r_A - r_A {r_A}^T) + p^T r_A - p {r_A}^T + {r_A}^T p - r_A p^T
\end{aligned}
\tag{3.122}
$$

Clearly mass, static momentum and inertia after the transportation must be summed over all the appendages with that of the body, while the modal participation factors vectors add a new column in the modal participation matrix of the satellite.

Using the transformation $\varepsilon = W\eta$ and setting $\omega_0 = 0$, Eqs. 3.118, 3.119 become:

$$I_G\delta\ddot{\phi} + S_G\ddot{\eta} + \delta\dot{h}_w + [\delta\dot{\phi}, \overline{h_w}] = \delta c - [p + L\eta, f/m] \tag{3.123}$$

$$S_G^T\delta\ddot{\phi} + \ddot{\eta} + 2\Sigma\Omega\dot{\eta} + \Omega^2\eta = Q^\eta - L^T f/m \tag{3.124}$$

Note that: $\delta c = c_d + \delta c_c$ includes both the exogenous disturbing torque and the control torque.

Equations 3.123, 3.124 are typically used in the linear design of the attitude controller (see Chaps. 5, 6), to define the AOCS response δc_c, δh_w when the attitude and flexible dynamics are subjected to the disturbing inputs c_d, f_d and f_c (because the last produce typically an effect on the attitude when orbital maneuvers are performed).

3.12.1 Properties of the Linear Model

We investigate some properties of the linear system which will be used later in the frame of the control analysis. Equations 3.123, 3.124 can be set in homogeneous form in order to evaluate the free oscillating modes of the satellites.

$$\begin{aligned} &I_G\delta\ddot{\phi} + S_G\ddot{\eta} = 0 \\ &S_G^T\delta\ddot{\phi} + \ddot{\eta} + 2\Sigma\Omega\dot{\eta} + \Omega^2\eta = 0 \end{aligned} \tag{3.125}$$

The diagonal matrix Ω, lists in its diagonal the so called *cantilevered angular frequencies* of all appendages, which are the angular frequencies of the oscillations when forcing to zero the satellite rigid motion ($\delta\phi = 0$), these are the eigen-values of the following system:

$$\ddot{\eta} + 2\Sigma\Omega\dot{\eta} + \Omega^2\eta = 0 \tag{3.126}$$

The system of Eq. 3.125 can also be written:

$$\begin{aligned} &\delta\ddot{\phi} + I_G{}^{-1}S_G\ddot{\eta} = 0 \\ &(E_{N_f} - S_G^T I_G{}^{-1}S_G)\ddot{\eta} + 2\Sigma\Omega\dot{\eta} + \Omega^2\eta = 0 \end{aligned} \tag{3.127}$$

The second of Eq. 3.127 has a mass matrix different from unity, its eigenvalues provide the angular frequencies of the so called *free oscillations*. The mass matrix of this equation is a second order form with the following property:

$$\eta^T(E_{N_f} - S_G^T I_G{}^{-1}S_G)\eta \leq \eta^T\eta, \forall\eta.$$

This means that the mass matrix of the free oscillations weights less than that of the cantilevered oscillations (which is here the identity matrix) while the two systems share the same *stiffness matrix* Ω^2.

In this case, if we consider the damping negligible, from the theory of small elastic oscillations (see the theorems in [19]) it follows that $\Omega \leq \Omega_f$ where Ω_f lists in its diagonal the *free angular frequencies*, $\Omega_f = diag(\omega_{f,1}, \omega_{f,2}, \omega_{f,3}, ...\omega_{f,N_f})$ of the system of Eq. 3.125.

We call now $\eta_k,\ k = 1, 2...N_f$ the eigen-modes of

$$(E_{N_f} - S_G^T {I_G}^{-1} S_G)\ddot{\eta} + 2\Sigma\Omega\dot{\eta} + \Omega^2\eta = 0 \tag{3.128}$$

disregarding damping[10] from the properties of the mass and stiffness matrices they form a complete real orthonormal system.

We choose the eigen-modes in order to satisfy the mass normalization:

$$\eta_k{}^T (E_{N_f} - S_G^T {I_G}^{-1} S_G)\eta_j = \delta_{ij} \tag{3.129}$$

The eigen-modes of the total system of Eq. 3.125 must include, in addition to η_k, also three rigid rotations with null angular frequency.

The modal matrix W of Eq. 3.125 can be easily written considering that rigid modes are normalized with the inertia matrix and that elastic ones must satisfy the first of Eq. 3.125 that is $\delta\phi_k = -{I_G}^{-1} S_G \eta_k$:

$$W = \begin{pmatrix} {I_G}^{-\frac{1}{2}}, & -{I_G}^{-1} S_G \eta_1, & -{I_G}^{-1} S_G \eta_2, & ....., & -{I_G}^{-1} S_G \eta_{N_f} \\ 0_{N_f x3}, & \eta_1, & \eta_2, & ....., & \eta_{N_f} \end{pmatrix} \tag{3.130}$$

with the $N_f + 3$ angular frequencies $(0, 0, 0, \omega_{f,1}, \omega_{f,2}, \omega_{f,3}, ...\omega_{f,N_f})$.

We easily verify that the matrix W transforms the system matrices of Eq. 3.125 in the following way:

$$W^T \begin{pmatrix} I_G, & S_G \\ {S_G}^T, & E_{N_f} \end{pmatrix} W = E_{N_f+3} \tag{3.131}$$

The stiffness matrix is transformed in:

$$W^T \begin{pmatrix} 0_{3x3}, & 0_{3xN_f} \\ 0_{N_f x3}, & \Omega^2 \end{pmatrix} W = \begin{pmatrix} 0_{3x3}, & 0_{3xN_f} \\ 0_{N_f x3}, & {\Omega_f}^2 \end{pmatrix} \tag{3.132}$$

---

[10] From a mathematical point of view a generic damping matrix precludes the existence of a real modal matrix. A sufficient condition to have a real modal matrix requires a damping matrix proportional to the stiffness matrix. We will assume this hypothesis from now in order not to use complex modes. It is usual engineering practice to specify directly the damping factor for each free mode and to create the modal matrix from the cantilevered mode description without damping in order to avoid complex modes.

If we assume the cantilevered damping matrix proportional to the cantilevered stiffness one, also the damping matrix is diagonalized in:

$$W^T \begin{pmatrix} 0_{3x3}, & 0_{3xN_f} \\ 0_{N_f x3}, & 2\Sigma\Omega \end{pmatrix} W = \begin{pmatrix} 0_{3x3}, & 0_{3xN_f} \\ 0_{N_f x3}, & 2\Sigma_f \Omega_f \end{pmatrix} \tag{3.133}$$

Then, we write Eqs. 3.123, 3.124 in terms of modal amplitudes z:

$$Wz = \begin{pmatrix} \delta\phi \\ \eta \end{pmatrix} \tag{3.134}$$

Defining $u = \delta c_c - \delta \dot{h}_w - [\delta\dot{\phi}, \overline{h_w}]$ the control input torque that includes also the effect of the *momentum management system* and assuming $f = 0$, $Q^\eta = 0$, $c_d = 0$ Eqs. 3.123, 3.124 become:

$$\ddot{z} + \begin{pmatrix} 0_{3x3}, & 0_{3xN_f} \\ 0_{N_f x3}, & 2\Sigma_f \Omega_f \end{pmatrix} \dot{z} + \begin{pmatrix} 0_{3x3}, & 0_{3xN_f} \\ 0_{N_f x3}, & {\Omega_f}^2 \end{pmatrix} z = \begin{pmatrix} {I_G}^{\frac{1}{2}} \\ -{\eta_1}^T {S_G}^T \\ \ldots\ldots \\ -{\eta_{N_f}}^T {S_G}^T \end{pmatrix} {I_G}^{-1} u \tag{3.135}$$

which are easily integrable.

Another important consideration comes from the analysis of the transfer function of the forced system, we analyze first the case without damping.

$$\begin{aligned} I_G \delta\ddot{\phi} + S_G \ddot{\eta} &= u \\ S_G^T \delta\ddot{\phi} + \ddot{\eta} + \Omega^2 \eta &= 0 \end{aligned} \tag{3.136}$$

If we Laplace transform Eq. 3.136, multiplying it by e^{st} and integrating in time from $0 \rightarrow \infty$, the transfer function from $u \rightarrow \delta\ddot{\phi}$ results to be $\delta\ddot{\phi} = M(s)^{-1}u$, with $M(s)$ given by

$$M(s) = I_G - s^2 \sum_{k=1 \ldots N_f} \frac{{S_G}^k {S_G}^{kT}}{s^2 + {\Omega_k}^2} \tag{3.137}$$

If we include the damping in the *dynamic mass matrix* $M(s)$ we get:

$$M(s) = I_G - s^2 \sum_{k=1 \ldots N_f} \frac{{S_G}^k {S_G}^{kT}}{s^2 + 2\zeta_k \Omega_k s + {\Omega_k}^2} \tag{3.138}$$

In this equation we have decomposed the modal participation factors in the modal component vectors: $S_G = [{S_G}^1, {S_G}^2 \ldots {S_G}^{N_f}]$.

We observe from Eq. 3.138, that when $s \rightarrow j\Omega_k$, it follows that $u = M(s)\delta\ddot{\phi} \rightarrow -j\frac{{S_G}^k {S_G}^{kT} \delta\ddot{\phi}}{2\zeta_k}$, if as usual $\zeta_k \approx 0$ then, ${S_G}^{kT} \delta\ddot{\phi} \approx 0$ in order to keep their ratio finite and close to u.

Thus, when the input torque oscillates at the cantilevered frequencies we have a null response on the rigid body motion in the direction of the vector ${S_G}^k$.

This property explains why in the bode plots of the torque to attitude transfer function we observe for each flexible mode a couple zero-pole, with the zero anticipating the pole; the zero is placed in the position of the cantilevered angular frequency and the pole in the position of the free angular frequency (see the Bode plots in Sect. 6.3).

3.12.2 Typical Satellite Flexible Modes

We close the chapter reviewing the more important typical flexible elements on a satellite. We analyze their main characteristics in terms of frequency and damping, reporting shortly the way they interact with the control design.

- *sloshing*. The fuel for chemical or plasmic propulsion is stored on-board satellites typically in the form of pressurized liquid inside titanium or aluminum containers. Under micro-gravitational conditions the liquid tends to form bubbles attached to the container walls and kept together by the tensional surface forces. When the acceleration field f/m (refer to Eq. 3.124) is greater than the liquid tensional forces, then we have sloshing, and the liquid dynamics can be considered as a pendulum (see the fundamental work of [20]) with a sloshing mass m_s and length L_s; in a linearized dynamic, this pendulum can be modeled as a flexible mode with a concentrated mass and a spring.
  The pendulum mass is charged with potential energy at the beginning of a firing phase (by the term $L^T f/m$), oscillates at a relatively low frequency and damps down very slowly. We can speak of sloshing when there is enough acceleration during a firing phase so that the Bond number is higher than unity: $Bo = gR^2\rho/\sigma > 1$, in such a case the effect of the liquid surface tensional forces is negligible and the dynamic of the liquid is more predictable in the linear range.
  In the Bond number formula $g = f/m$, R $[m]$ is the dimension of the container, ρ $[Kg/m^3]$ is the density of the liquid, σ $[N/m]$ is the tensional force of the liquid. The frequency of the oscillation is derived by the pendulum formula $\sqrt{g/L_s}$, the normalized rotational participation factor $S = \sqrt{m_s}L_p$ (with L_p the distance between the pendulum attachment point and the CoM projected along the acceleration direction), the sloshing mass m_s participating to the mode is a certain fraction of the liquid mass.
  These modes are characterized by very small damping factors—in the range of 0, 001...0, 005-, very low frequencies—for example with $f = 400\,\text{N}$, $m = 5000\,\text{Kg}$, $L_s = 0,5\,\text{m}$ we get $(\sqrt{g/L_s}/(2\pi) = 0,06\,\text{Hz}$—and large participation factors (because when using chemical propulsion large tanks m_s can be 10 % of the satellite mass) that typically interact with the controller design, giving problems in terms of control saturation and stability.
  Unfortunately to solve the pendulum characteristics L_s, L_p, m_s it is necessary to solve an elliptic PDE (partial differential equation) for all tank filling ratios.

During long maneuvers the flexible characteristics of these modes may change significantly, requiring a very robust design. An excellent reference text for modeling sloshing in containers is [20], a more recent reference for the theory of the sloshing is [21], an appropriate mathematical setting for the sloshing PDE problem is provided in [22];

- *solar arrays*. Solar arrays integrated on high power satellites are deployed as very long wings formed by stiff panels attached to each other with springs. The length of these wings can be up to 10 m or more. The solar array frequencies are typically in the range of some tenths of a Hertz with damping coefficients in the range of 0, 01...0, 02. A lot of effort is normally developed to characterize the solar array flexible dynamics during ground testing and in flight. When excited by the control loop dynamic or by the maneuver acceleration, the solar array oscillations produce a disturbance on satellite's dynamics which may lead to instabilities. Even if there is little concern for the solar array pointing, the effect of the solar array flexible modes has to be carefully checked in the design of the rotational AOCS controller;
- *antennas*. Antenna frequencies are very often higher than the solar array ones, because their dimension is normally less than 3...4 m. This fact in principle makes these modes less important for the AOCS design. However, in modern AOCS design the antennas' flexibilities are becoming more and more important for various reasons:
  - recent projects of antenna reflectors deal with antennas of more than 5 m which have frequencies similar to solar arrays. However, while a stable solar array excitation is not of concern, a large antenna dynamic deformation can produce a loss of performance service;
  - antennas for large band services (i.e. Ka-Band) require a better pointing accuracy than satellite thermo-elastic deformations (few times 0, 01 deg). Thus even with a very fine AOCS pointing system, it is necessary to implement a decentralized antenna pointing controller. The coupling of the dynamics between AOC controller, antenna controller and antenna flexible modes must be carefully studied, typically the AOCS controller and the antenna one are designed uncoupled in the control bandwidth;
  - platform agility. When the satellite implements an agile attitude maneuver, the antennas' flexible modes are excited by the resulting angular acceleration (by the term $S_G^T\dot{\omega}$), the resulting oscillation must be tranquillized in order to use the antenna with good performances. It is very effective to shape the slope of $S_G^T\dot{\omega}$ such that its Fourier Transform is as decoupled as possible from the antenna frequencies. The higher the frequency of the antenna, the faster the allowed agility.

References

1. A.E. Bryson, *Control of Spacecraft and Aircraft* (Princeton University Press, Princeton, 1994)
2. J.R. Wertz, *Spacecraft Attitude Determination and Control* (D. Reidel Publishing Company, Dordrecht, 2002)
3. M.J. Sidi, *Spacecraft Dynamics and Control* (Cambridge University Press, Cambridge, 1997)
4. P. Santini, Stability of flexible spacecraft. ACTA Astronaut. **3** (1976)
5. T.R. Kane, P.W. Likins, D.A. Levinson, *Spacecraft Dynamics* (McGraw-Hill, New York, 1983)
6. P.W. Likins, R.E. Roberson, J. Wittenburg, *Dynamics of Flexible Spacecraft*, Department of General Mechanics, Course held in Dubrovnik (Springer, 1972)
7. L. Mazzini, P. Santini, Non linear equations of flexible systems for control applications. ASI Technical Report financed, Contract N. ARS-96-19 (1996)
8. L. Meirovitch, *Methods of Analytical Dynamics* (McGraw-Hill, New York, 1970)
9. J.L. Lagrange, *Analytical Mechanics*, Translated from the 'Mecanique analytique, nouvelle edition' of 1811 (Springer, 1997)
10. H. Goldstein, *Classical Mechanics* (Zanichelli, 1980)
11. A.A. Shabana, *Dynamics of Multibody Systems* (Cambridge University Press, Cambridge, 2005)
12. O.A. Bauchau, *Flexible Multibody Dynamics* (Springer, Dordrecht, 2010)
13. B. Wie, Singularity analysis and visualization for single-gimbal control moment gyro systems. J. Guidance, Control Dyn. **27**(2) (2004)
14. V. Tamilia, L. Mazzini, Artemis satellite analysis and performance verification of the antenna pointing control systems using flexible multibody dynamics Lagrangian equation modelization IAF-98-A.5.07, in *49th International Astronautical Congress*, Sept. 28–Oct. 2, Melbourne, Australia (1998)
15. O. Montenbruck, E. Gill, *Satellite Orbits, Models, Method and Applications* (Springer, Berlin, 2005)
16. B.N. Agrawal, R.S. Mac Lelland, G. Song, Attitude control of flexible spacecraft using pulse width pulse frequency modulated thrusters. Space Technol. **17**(1), 15–34, Elsevier (1997)
17. H. Bittner, H.D. Fischer, M. Surauer, Reaction jet attitude control systems for flexible spacecraft, in *IFAC Proceedings, Automatic Control in Space*, Noordwijkerhout (1982)
18. K.T. Alfriend, S.R. Vadali, P. Gurfil, J.P. How, L.S. Breger, *Spacecraft Formation Flying, Dynamics Control and Navigation*. Elsevier Astodynamics Series (2010)
19. V.I. Arnold, *Matematiceskie metody klassiceskoj mechaniki* (MIR, Moscow, 1979)
20. H. Abramson, The dynamic behavior of liquids in moving containers NASA SP-I06 (N67-15884), Scientific and Technical Information Division, National Aeronautics and Space Administration, Washington, DC. (1961)
21. R.A. Ibrahim, *Liquid Sloshing Dynamics* (Cambridge University Press, Cambridge, 2005)
22. V. Komornik, P. Loreti, L. Mazzini, Null controllability of the free surface of a liquid in a container. SIAM J. Control Optim. **43**(2), 685–696 (2004)

Chapter 4
The AOCS Functions

Abstract This chapter presents the operating modes of a typical modern satellite attitude control system (AOCS). The various modes are presented in terms of functional requirements, H/W architecture, functional block diagram and controller characteristics. The control mode dynamics are described through various simulations.

4.1 Introduction

This chapter presents the operating modes of a typical modern satellite attitude and orbit control system (AOCS). After a description of the mission phases and the AOCS functional blocks, the various modes are presented in terms of functional requirements, H/W architecture, functional block diagram and controller characteristics. The control mode dynamics are described through various simulations.

Over the last 40 years many different AOCS architectures have been proposed and developed for different applications (i.e. telecommunications, scientific, Earth observation...).

A full review of all possible design being impossible, we have decided to present a *multi-mission* architecture, conceived to be flexible to implement most of the requirements in existing applications. This architecture is not a real AOCS, specific adaptations in terms of equipment and design must be introduced to cope with the different orbits or some specific requirements. However this concept allows to introduce the discussion on a wide class of design issues and any specificity will be highlighted as necessary. The design solutions presented are not unique, we can also have different design implementations with respect to those presented to respond to specific needs.

In the last years, due to the high level of flexibility, autonomy and processing capability embedded in the avionic sensors and on-board software (OBS), the avionic solutions converge to a small number of possibilities which can be reused in different contexts. It is then possible to talk about a *multi-mission avionic bus*, able to fulfill all most important AOCS requirements.

The three axis stabilized attitude and orbit control system architecture shown in Sect. 4.3 is able to re-orient the spacecraft in a predefined target attitude from a

© Springer International Publishing Switzerland 2016

L. Mazzini, *Flexible Spacecraft Dynamics, Control and Guidance*,
Springer Aerospace Technology, DOI 10.1007/978-3-319-25540-8_4

nominal Nadir pointing attitude (that is to maintain **BRF** rotated with respect to **ORF** of some target Euler angles).

Depending on the P/L different spacecraft attitudes can be desired:

- *Earth* pointing: *geocentric* or *geodetic*;
- *Sun* pointing;
- *inertial* pointing;
- *roll/pitch/yaw offset* and bias capability.

Moreover this AOCS architecture can implement different steering pointing maneuvers. Specific steering maneuvers are required to support the payload or the spacecraft according to different mission needs such as:

- maintain the solar array wings orthogonal to the Sun direction (in order to maximize the generation of power);
- steer the yaw angle during the orbit (to maximize the sun input on the solar arrays in inclined orbits or to compensate the doppler shift due to the Earth motion for a SAR satellite);
- bias attitude in roll (to re-point the bore-sight of an optical/radar/transmitter payload);
- bias attitude in pitch (to re-point the bore-sight of an optical/radar/transmitter payload);
- steer attitude to point different ground targets;
- maintain an inertial attitude (e.g. long firing maneuver, star imaging);

The torque produced by the reaction wheels (or CMG when high agility is required) mounted in pyramidal configuration (see Chap. 9 for the different RW configurations), allows the spacecraft to perform re-pointing maneuvers. The attitude is measured with star trackers and the angular rate with three axes gyroscopes. Low noise sensors and actuators, a stiff structure and a properly designed control logic guarantee a very stable platform with precise pointing (mandatory characteristics for missions adopting high performance optical sensors or synthetic aperture radar).

In telecommunication missions that usually require less demanding performances the same architecture may be used by changing the class and the cost of sensors and actuators.

The AOC system is designed to perform the orbit correction maneuvers required by the mission. The first orbital maneuvers are those to correct orbital errors due to the launcher's mis-performance in releasing the satellite in the nominal separation orbit, after those to acquire/achieve/maintain the specified operational orbit are performed, to finish with de-orbiting maneuvers at the end of the satellite's life.

These maneuvers are implemented using the on-board propulsion subsystem. Different architectures and propulsion technologies can be used, see Chap. 8 for a survey on these technologies.

4.2 Typical Mission Profiles for LEO and GEO Mission

In most LEO missions (typically Earth observation missions) the launch vehicle (LV), after lift-off and after its cruise mode, releases the satellite in an injection orbit close to the operational one with a predefined attitude and a residual angular rate agreed upon by the launch authority and the owner of the satellite.

In the GEO mission case (typically a telecommunication mission), the launch vehicle after lift-off and after its cruise mode releases the satellite in a parking orbit (i.e. GTO—geostationary transfer orbit) that is not the operational orbit and therefore the satellite has to perform some long thruster firings with its own propulsion system (chemical or electrical), in order to reach the operational target orbit.

The mission profiles for LEO and GEO have a different *launch and early orbit phase* (LEOP) approach and AOCS has to be designed to support the specific LEOP profile. Moreover, AOCS has to be designed to support all mission's needs from LEOP to satellite disposal. The satellite must implement at least the following operational phases: *launch phase*, *transfer phase*, *deployment phase*, *mission phase*, *deorbit phase*. Each phase is supported by one or more AOCS modes. The transition between different modes is carried out by the on-board software (OBS) upon receiving a ground telecommand or when triggered by a timer.

Launch Phase

The launch phase is supported by the AOCS implementing an idle operational mode that is the only AOC mode allowed during the satellite launch mode. In idle mode no attitude control is realized. Only satellite survival activities are carried out during this phase (e.g. active thermal control). The avionic must implement multiple H/W barriers in order to prevent dangerous unwanted events in the launcher fairing, like thrusters activations or deployments.

A robust logic has to be implemented to detect and confirm the satellite's separation from the launcher.

Transfer Phase

Transfer phase is a pre-operational phase lasting from the launcher separation to the acquisition of the operational orbit. The transition from launch phase to transfer phase is typically autonomously realized by software as soon as separation from the launcher is detected. In this phase AOCS functions are required to damp the initial LV separation rates,[1] acquire an attitude with positive power balance (Sun acquisition) and realize if necessary a series of orbital maneuvers to reach the final operational orbit.

The rate damping and *safe* attitude pointing can be achieved according to a specific AOC design. Different schemes and different sensors/actuators can be used by AOC control loop to reach this objective. In this chapter two possible designs are presented

[1] The separation rates can be quite high such as 1–2 deg/s.

such as the *safe hold mode* (SHM) and the *Sun acquisition mode-emergency mode* (SAM-EM).

The main goal of this mode is to be simple and reliable. This implies using a minimal hardware and a simple logic that guarantees a convergence of the loop starting from *any reasonable* initial attitudes and rates.

In case of GEO missions the transfer phase is maintained until reaching the target orbit. Then deployment phase is entered upon receiving a telecommand.

The classic approach of the orbital transfer to GEO (for a description of the control modes and phases of a classic telecom satellite see [1]) implements a sequence of firings of a liquid apogee engine (LAE). The attitude in this mode is controlled such to point the LAE thrust vector along the optimal direction. The number of firings depends on the transfer orbit strategy selected by mission operation engineers and usually requires more than one transition into *orbit control mode* (usually two or three apogee boosts).

The theory of optimal orbital transfers presented in Sect. 7.8 can be used to determine the optimal firing direction.

In more recent design approaches these transfers are realized using low thrust plasmic thrusters continuously firing for very long durations. The theory of this kind of optimal transfer is studied in Sect. 7.9.

During firings that last many orbits the AOCS will continuously align the firing thruster along the optimal direction (which rotates continuously in **ORF**) and will rotate the satellite and the solar arrays in order to maximize the solar power input. In these missions the deployment phase is anticipated before the transfer phase.

Deployment Phase

The deployment phase is normally entered when the satellite has finally reached the target orbit (i.e. after the LV separation in LEO/MEO missions or after the transfer orbit in GEO missions). In this phase all satellite's appendages (i.e. solar arrays, antennas, instruments) will be deployed. The AOC mode used to deploy all appendages is selected considering the dynamic conditions and the thermal and power constraints.

Mission Phase

The mission phase is entered after the deployment and it is maintained up to the *end of mission* before satellite disposal. In this phase AOC system has to support all nominal operations required by the satellite. Basically the *nominal pointing mode* (i.e. mission mode) is designed to achieve a good attitude pointing accuracy and knowledge necessary for the correct payload operations. Specific functions linked to the payload requirements are implemented in this phase like agile pointing or steering capability. Moreover, AOCS is also required to support station keeping maneuvers necessary to maintain the satellite in the correct orbital box (i.e. GEO mission) or close to the correct reference orbit (i.e. LEO mission). This objective is achieved through an orbit control mode capable to implement the required DV maneuvers.

Deorbit Phase

This phase is entered when the satellite reaches its *end of mission* because consumable resources are going to finish and the owner decides to de-orbit the satellite. The AOC has to implement the DV maneuver to transfer the satellite into its graveyard orbit (GEO missions) or decaying orbit (LEO missions), following the rules of the *international space law*.

We have briefly seen the AOCS modes and functions that must be designed. Each AOCS mode can be present in different mission phases. The following section discusses the control modes mainly in terms of their software architecture and data flow.

4.3 Control Modes Architecture

Generally a control mode is designed to support different mission phases and each mode can be divided into several sub-modes. Each mode is characterized by its own set of H/W on control (H/W that is actively used by the on-board software to implement the control loop).

Transitions from one control mode to another are commanded either by dedicated ground telecommands (TC) or by automatic recovery actions activated by the detection of an on-board equipment failure.

The functions fulfilled by the AOCS control system are coded inside periodic tasks executed by the on-board S/W in a time cycle that is generally in the range of 50–500 ms. The time cycle must be fast enough to represent the dynamics and the bandwidth of the controller, taking into account the constraints of the available processing resources and the sensor and actuators measurement and execution delays. The discretization of the control algorithms is therefore an important subject for any on-board real-time control system like AOCS.

The control mode periodic tasks (see Fig. 4.1) can be grouped into:

- *pre-processing function*. It is in charge of processing the raw data acquired from sensors in order to make them available to the control system every time cycle.
- *orbital pre-processing function*. It is in charge of providing orbital and time information to all on-board subsystems.
- *attitude determination function*. It is in charge of providing both S/C attitude and attitude error information. The attitude error will be used by the operational mode controller to implement the control law required by the mode.
- *mode management function*. Depending on the current operational phase mode management function is in charge of implementing the actual control law defining the control command.
- *post-processing function*. This function is in charge of representing the control command in raw data suitable for the AOCS actuators (RW motor voltages, RCT valves on-time…).

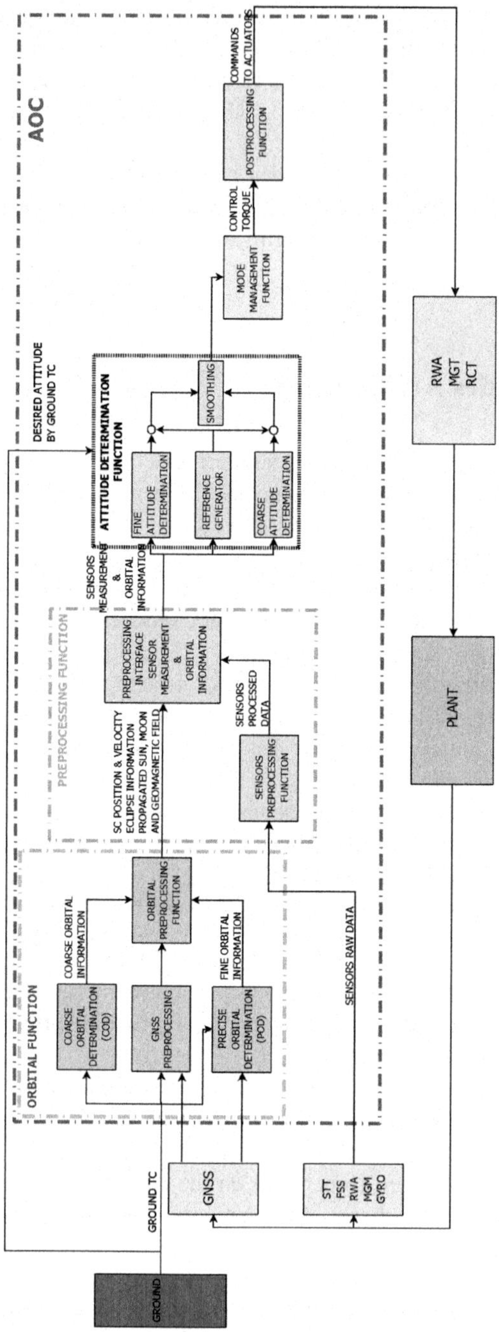

Fig. 4.1 AOCS functional data flow

4.3.1 Mode Management Function

The mode management function is in charge to implement AOC operating modes and submodes and the appropriate AOC transition logic between the current AOC mode/submode and the new AOC mode/submode.

All AOC transitions can be triggered either during the nominal mission operation using ground telecommands, or following a recovery action due to a software or a hardware alarm. Moreover, the mode management function is in charge of implementing the actual control laws that define the control actions. In Sect. 4.4 a description of AOC modes' algorithms is given.

A possible operating mode chart is shown in Fig. 4.2. As already said, the transitions between the various AOCS modes can start due to a ground telecommand (TC) or the detection of a on-board failure (FDIR). The operating mode used during the launch phase is the *idle mode*. This mode is autonomously entered by the software as soon as the separation from the launcher is detected by a hardware switch.

This mode is a stand-by mode in which the on-board Computer (OBC) is active but actuators are inhibited and no attitude control is applied. The internal states of all subsystems including the avionic are packed and formatted in a digital telemetry stream by the on-board Computer (OBC) software and down-linked to the ground station for analysis. This telemetric process is common to all operating modes which must provide an overall observability such to guarantee suitable reaction times to

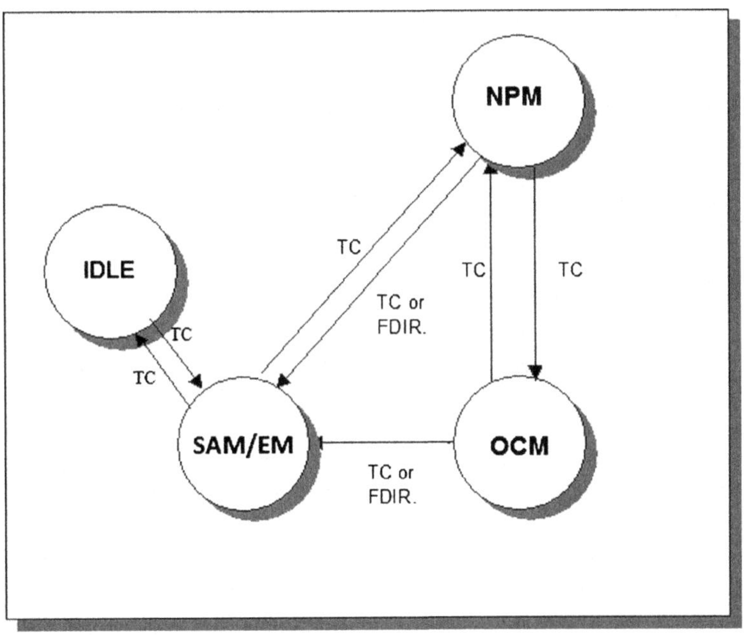

Fig. 4.2 Typical AOCS modes transition logic

the ground station in case anomalies or failures occur. The telemetric data rate is an important difference between LEO and GEO satellites. LEO satellites, in general, have short and periodic contact periods (a few minutes per passage) with the ground stations and therefore need higher telemetry data rates (in the range of many Mbits/s) while the GEO satellites are characterized by continuous contact periods and need much slower data rates (in the range of 10–100 Kbits/s). In addition, LEO satellites normally store the on-board states in dedicated memories in order to make available to ground a historic file of the satellite internal states.

In the *multimission avionic* two operating modes, the *Sun acquisition mode-emergency mode* (SAM-EM) and the *safe hold mode* (SHM)—this second mode only for LEO satellites—are foreseen for the pre-operational phase, when the objective is to charge the satellite's batteries by pointing the solar arrays towards the Sun and to maintain an attitude which allows a good telecommand and telemetric flow with the ground station in order to check the health of the on-board instruments. The same modes are typically used also in contingency cases (i.e. after a failure).

While the SAM-EM mode is used in all operational orbits, the SHM is a mode used only in LEO Sun Synchronous orbit (SSO) because it needs to operate in quite a strong environmental magnetic field.

The SAM-EM is designed to orient the spacecraft in a safe Sun-pointing three-axes controlled attitude. This mode is nominally entered by a ground telecommand (TC) during the pre-operational phase. Moreover, it is the first recovery mode entered autonomously by the software when a hardware failure occurs. The SAM-EM mode design is carried out to minimize the transient acquisition phase and provide a fast, steady state attitude acquisition. This mode uses the Sun sensor, magnetometer (only in LEO orbits), gyro and a coarse orbit determination (COD) algorithm. The actuation is carried out with the reaction wheels (RW). The magneto-torquers (MGT) are used in LEO orbits to unload wheel momentum when reaction wheels reach a certain maximal speed. In GEO orbits the same momentum unload is realized using reaction control thrusters (RCT).

The SHM is a robust mode specific of LEO SSO orbit conditions that guarantees extremely long survival periods using a minimum set of actuators and sensors: only magnetometer (MGM), MGT and RW.

The *normal pointing mode* (NPM), is designed to achieve a fine pointing attitude with a specified accuracy during the operational phase. This mode implements a fine three axes attitude control, suitable for payload imaging operations using Star trackers and gyro sensors.

Some new NPM mode designs may implement also gyro-less controllers using only star trackers for angular and rate feedback. With the present star tracker technology the gyro less approach can be widely applied unless there are very demanding stability and agility requirements.

The precise orbit determination (POD) is used in order to transform the inertial attitude information of the star tracker into **ORF** attitude error. As in the SAM-EM, reaction wheels are used for torque actuation and MGT (in LEO) or RCT (in GEO) for wheel momentum unloading. The NPM is the nominal mode during the payload operations and all features and requirements needed for the achievement of the operational mission conditions dictated by the specific payload characteristics must be considered in the design phase.

In order to perform the orbital maneuvers a dedicated operational mode is designed: the *orbit control mode* (OCM) which is able to handle arbitrary attitude slews in order to align the propulsion system thrust vector in the desired direction according to the requested maneuver acceleration profiles (see Chap. 7).

All these modes are nominally entered by ground TCs. At mode entry point, AOCS autonomously re-configures the equipment requested for the new mode. A possible attitude control mode transition logic is shown in Fig. 4.2.

4.3.2 Pre-processing Function

The pre-processing function is in charge to process all sensors and actuators sampled raw data for any AOC mode. The pre-processing algorithms transform all data coming from sensors (i.e. gyro, GPS receiver, Star tracker, MGM, FSS) and actuators (i.e. RWA) into data usable by the controller.

The sensor raw data will be subjected to at least format conversion, reference frame transformation (from sensor frame to **BRF**), calibration, filtering and any other processing necessary to provide a proper input to the algorithms implementing attitude and orbit determination and operational mode management.

It distributes the necessary sensor information to the other on-board functions (i.e. telemetry or FDIR functions).

4.3.3 Post-processing Function

The post-processing function has in charge to process all command data produced by the controller and to transform them into data accepted by the relevant sensors and actuators used in each AOC mode.

The post-processing function elaborates the high level actuation information produced by the mode management function in each AOC mode and converts them into the actuators' input format. The actuators high level commands produced by the controller will be subject to at least reference frame transformation (from **BRF** to actuator frame), calibration and format conversion plus any other processing necessary to provide a proper input.

4.3.4 Orbital Pre-processing Function

The orbital pre-processing function is designed to provide the orbital information (i.e. at least satellite position, velocity and time) to all other on-board AOC functions.

The orbit determination task can be implemented autonomously on-board using a *global navigation system* (GNSS) receiver (see Chap. 9), or by the ground station (G/S) using the two way ranging measurements produced by the telemetry telecommand link. The first methodology is preferred in the LEO missions where the visibility of the GNSS constellations is good while the visibility of the G/S is short and infrequent, the second methodology is used mainly in GEO missions where the conditions are opposite. It is only recently that GEO missions using GNSS receivers receiving the signals of the GNSS satellites visible behind the Earth have been proposed in order to augment the autonomy of the satellite in the case of very long LEOP with low thrust transfer.

When G/S calculates the orbit an on-board propagator is periodically updated by TC in order to align it with the real orbit.

When the orbit determination is realized through the on-board processing of a global navigation system (GNSS) receiver data, AOCS implements a real time precise orbit determination (POD) based on a navigation *extended Kalman filter* (EKF), using a prediction correction approach (see [2] which deals also with navigation filters). The POD has access to the output of the GNSS receiver and uses an EKF in order to improve the accuracy of the GNSS measurement by precisely modeling the reference orbit, including perturbation models for atmospheric drag and geo-potential. This precise orbital propagator produces the best orbit knowledge available on-board.

The POD calculates the orbital data in terms of spacecraft orbital position and velocity. As a reference, typical position and velocity errors about all **ECI** axes are shown in Figs. 4.3 and 4.4 for a real time on-board estimator in a LEO SSO orbit.

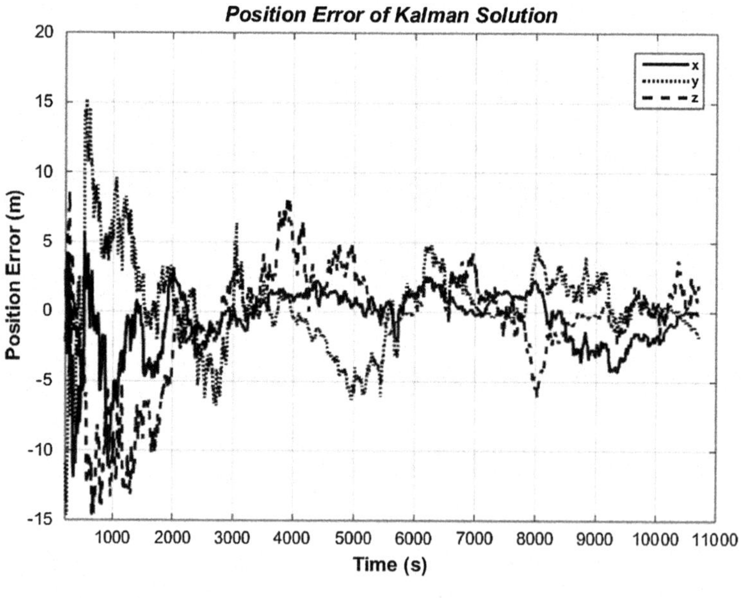

Fig. 4.3 POD EKF 3 axes position error

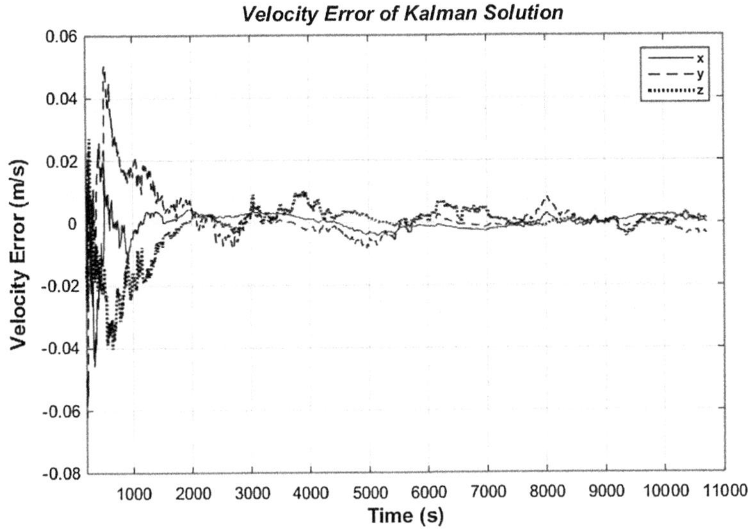

Fig. 4.4 POD EKF 3 axes velocity error

In case of double failure of the GNSS receiver, the POD can be used in a degraded mode in which it produces its output, propagating a set of orbit states periodically up-linked by ground. This back up use of POD in LEO is similar to that of GEO missions and the orbit state is produced with the available G/S ranging measurements.

The AOCS may implement also a coarse orbit propagator (COD) which propagates a coarse spacecraft's position and velocity starting from the initial information uploaded periodically by ground. It is used when there is no much need of accuracy or a limited availability of OBC processing resources (CPU time, memory).

4.3.5 Attitude Determination Function

The attitude determination function implements the algorithms (see [3]) that are designed to provide the attitude estimation necessary to feed attitude controllers. These algorithms, when optimal estimates are necessary can be improved using the method that is briefly presented in Sect. 5.3, where kalman filters introduce the satellite dynamic information to implement an optimal estimation (see [4, 5]). We can have coarse as well as fine attitude estimators, depending on the mission phase and the required attitude pointing performance. The fine and the coarse attitude determination algorithms are implemented typically in a single module together with a reference attitude generator.

By comparing the attitude estimate with the reference attitude this module provides the attitude error estimate to the operational mode controller (see Fig. 4.5).

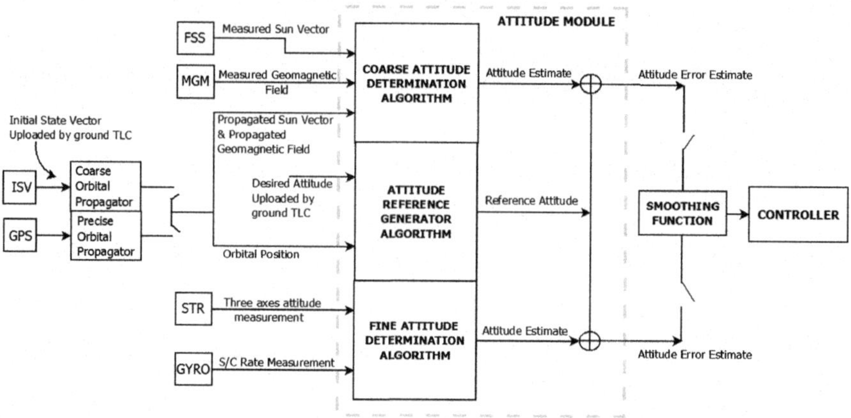

Fig. 4.5 Attitude determination functional flow

The coarse and fine attitude determination algorithms are fully interchangeable. The first one is used during the SHM or SAM-EM when the main goal is to re-orient the spacecraft in a safe attitude, while the second one is used during the mission phases when it is necessary to reach a high pointing accuracy and a high level of maneuverability relying on a very agile and flexible attitude control system.

In the *multimission avionic* the coarse attitude algorithm is a *Triad-like* algorithm (see for example [3, 6, 7]) that estimates the spacecraft attitude using the low accuracy sensor (Sun sensor and magnetometer) as well as the COD.

The fine attitude determination algorithm is an attitude estimation algorithm based on the gyro and star tracker measurement. Combining the gyro and the star tracker measurements, using a prediction correction approach, this algorithm smooths the noisy measurements and provides the best attitude estimate available on-board.

The fine attitude determination algorithm is designed to work with one, two or even three star tracker sensors both in hot and cold redundancy, the sensor configuration depending on the level of the desired accuracy and the study of the star tracker blinding conditions from the Sun, the Earth and the Moon (see Chap. 9). In case of star tracker unavailability (temporary blinding conditions), the algorithm autonomously propagates the last attitude estimate with the gyro rate until the star tracker measurement is available again.

The reference attitude is calculated by the attitude reference generator algorithm which uses the orbital information, coming from the COD/POD. This algorithm, properly configured, is able to generate a variety of attitude maneuver profiles. Depending on the mission needs, minimum jerk maneuver, minimum time maneuver, roll, pitch, yaw offsets and periodic maneuvers can be implemented. Moreover a fully customized maneuver made up of polynomial and harmonic profiles can be configured by ground telecommand.

The interchangeability of coarse and fine algorithms, the presence of a coarse and fine orbital propagator and the possibility of a temporary star tracker unavailability lead to the design of a smoothing module. Every time a change in the orbital reference or in the measurement occurs, the software autonomously performs a smooth transition, using a dedicated weighting function, to minimize the attitude transient.

4.3.5.1 Coarse Attitude Determination Algorithm

The coarse attitude determination function provides a coarse attitude estimate with the construction of two triads of orthonormal unit vectors using the Sun sensor and magnetometer measurement as well as the coarse orbit propagator information. This method is the so-called *Triad* algorithm. This kind of implementation is typical of LEO orbits; in GEO orbits, where the magnetic field is too low for a reliable attitude measurement, it is typical to implement a different algorithm that uses the Sun sensor and the gyro data to compute the attitude in the two directions orthogonal to the Sun's one, while the attitude along the Sun's direction, that cannot be resolved in absolute terms, is determined with respect to some initial constant. The partial attitude observability that results must be taken into account in the design of GEO acquisition modes.

The attitude determination problem is equivalent to determine in real time the attitude rotation matrix $M_{B \hookleftarrow O}$ that describes the orientation of the spacecraft **BRF** with respect to the **ORF**.

Let's indicate

- s_{BRF} Sun unit vector measured by the Sun sensor;
- m_{BRF} magnetic field unit vector measured by the magnetometer sensor;
- s_{ORF} Sun unit vector provided by orbital propagator;
- m_{ORF} magnetic field unit vector provided by orbital propagator.

where s_{BRF} and m_{BRF} refer to **BRF** components, while s_{ORF} and m_{ORF} refer to **ORF** components.

The following relations follow from the definition of the attitude matrix, if we assume for the moment a null measurement error:

$$\begin{aligned} s_{BRF} &= M_{B \hookleftarrow O} s_{ORF} \\ m_{BRF} &= M_{B \hookleftarrow O} m_{ORF} \end{aligned} \tag{4.1}$$

We define now the following orthogonal unit vectors of a new orthogonal triad **TRD**, in **BRF** and **ORF** components:

$$
\begin{aligned}
t1_{BRF} &= s_{BRF} \\
t2_{BRF} &= \frac{s_{BRF} \times m_{BRF}}{|s_{BRF} \times m_{BRF}|} \\
t3_{BRF} &= t1_{BRF} \times t2_{BRF} \\
t1_{ORF} &= s_{ORF} \\
t2_{ORF} &= \frac{s_{ORF} \times m_{ORF}}{|s_{ORF} \times m_{ORF}|} \\
t3_{ORF} &= t1_{ORF} \times t2_{ORF}
\end{aligned}
$$

Two rotation matrices can be constructed bringing from the triad **TRD** to **BRF** and **ORF**:

$$
\begin{aligned}
M_{B\leftarrow T} &= [t1_{BRF}, t2_{BRF}, t3_{BRF}] \\
M_{O\leftarrow T} &= [t1_{ORF}, t2_{ORF}, t3_{ORF}]
\end{aligned} \tag{4.2}
$$

Then the attitude matrix $M_{B\leftarrow O}$ can be easily calculated:

$$
M_{B\leftarrow O} = M_{B\leftarrow T}[M_{O\leftarrow T}]^T \tag{4.3}
$$

Having defined the sequence of Euler angles rotations to move from the **ORF** to the **BRF**, the three attitude angles can be easily derived (for example inverting the trigonometric expressions in Eq. 2.12 or using the simple quaternions' inversion formulas described in Sect. 2.3.1).

If we assume now, that measurement is affected by random Gaussian error, the expression of the attitude error covariance matrix can be computed and shows that it is inversely proportional to the angle between the Sun and magnetic field vectors also called co-linearity angle. In general, orbit transient phases where the co-linearity is low or null may exist. In such periods of time this algorithm cannot be used and it is customary to propagate the attitude with a gyroscope.

The expression of the attitude error covariance is found to be (see [8, 9]):

$$
\begin{aligned}
P = \sigma_S^2 I + \frac{1}{||s_{BRF} \times m_{BRF}||^2}[(\sigma_B^2 - \sigma_S^2) s_{BRF} s_{BRF}^T \\
+ \sigma_S^2 (s_{BRF}{}^T m_{BRF})(s_{BRF} m_{BRF}^T + m_{BRF} s_{BRF}^T)]
\end{aligned} \tag{4.4}
$$

$\sigma_{S_{ref}}$ = Sun reference half cone error covariance
$\sigma_{S_{meas}}$ = Sun measured half cone error covariance
$\sigma_{B_{ref}}$ = B field reference half cone error covariance
$\sigma_{B_{meas}}$ = B field measured half cone error covariance

$$
\begin{aligned}
\sigma_S^2 &= (\sigma_{S_{ref}})^2 + (\sigma_{S_{meas}})^2 \\
\sigma_B^2 &= (\sigma_{B_{ref}})^2 + (\sigma_{B_{meas}})^2
\end{aligned} \tag{4.5}
$$

In case of more reference and observation vectors, a better attitude estimate can be achieved using the *Quest* algorithm that finds an optimal weighted overlap of measurement and references (see [10]).

For a coarse attitude estimate using Sun sensor and magnetometer measurement the *Triad* algorithm is a good compromise between the computational effort and the achieved accuracy.

4.3.5.2 Fine Attitude Determination Algorithm

The fine attitude determination algorithm is based on gyro and star tracker measurement. The gyro rate is used to propagate the attitude kinematic equation (2.11); the residual between star tracker measurement and propagated attitude estimate, properly weighted by filter gains, is added to the gyro measured rate to correct the attitude prediction and to estimate the gyro rate drift.

The attitude dynamics in terms of quaternions is described by the kinematic equation (2.11):

$$\dot{q} = \frac{1}{2}(0, \omega) \otimes q \tag{4.6}$$

where q is the attitude quaternion that describes the spacecraft attitude orientation providing the transformation matrix $M_{B \leftarrow E}$ from **ECI** to **BRF** using Eq. 2.9, while $\omega = \omega_{BE,B}$ is the spacecraft angular rate in **BRF** components.

The Star tracker sensor will provide a noisy attitude quaternion measurement, the noise being higher in the direction of the sensor bore-sight. This noise will produce noisy dynamics of the controlled attitude and poor pointing performances. The sensor error is not actually made up only of random noise: important bias and low frequency error components also affect the measurement quality.

However, it is customary to calculate the attitude determination algorithms using a random noise hypothesis because this method produces good results and simplifies the calculations of the filter gains.

It is also possible to treat the low frequency errors by introducing a suitable dynamic model of these errors, though this approach will not be presented here.

The purpose of the attitude determination filter is to fuse the measurements from other star trackers and a gyro in order to minimize the noise of the final estimated quaternion.

Let us indicate the measured quaternion of the ith star tracker q_i and $q_{n,i}$ the error quaternion of i-th star tracker, subscript n indicates that this quaternion depends only on the measurement noise. The true spacecraft attitude can be obtained by multiplying the small error quaternion with the star tracker measured attitude quaternion (see 2.10):

$$q = q_{n,i} \otimes q_i \tag{4.7}$$

Neglecting the bias and low frequency error components, the error quaternion $q_{n,i}$ is supposed to be due only to random zero mean Gaussian white noise. It can be assumed that $q_{n,i}$ represents a near identity transformation, so that it is possible to consider the scalar part q_0 equal to one and the vector part $||\mathfrak{q}_{n,i}|| \ll 1$ so that, to the first order, the norm of the quaternion is still unitary: $q_{n,i} \approx (1, \mathfrak{q}_{n,i})$. This is how small rotations are represented in the quaternion theory for linearization purposes.

Given $\mathfrak{n}_i$ the ith sensor noise in the star tracker reference frame, $\Theta_i = E[\mathfrak{n}_i \mathfrak{n}_i^T]$ the star tracker covariance matrix (the matrix containing the PSD of the noise in the star tracker reference frame, for the definition of PSD see Sect. 5.3) and $M_{B \leftarrow S_i}$ the transfer matrix describing the orientation of the ith star tracker reference frame with respect to the spacecraft **BRF**, the error quaternion $\mathfrak{q}_{n,i} = M_{B \leftarrow S_i} \mathfrak{n}_i$ PSD is obtained as follows:

$$E[\mathfrak{q}_{n,i}\mathfrak{q}_{n,i}^T] = E[M_{B \leftarrow S_i} \mathfrak{n}_i \mathfrak{n}_i^T M_{B \leftarrow S_i}^T] = M_{B \leftarrow S_i} \Theta_i M_{B \leftarrow S_i}^T \tag{4.8}$$

It can be assumed that the spacecraft's angular rate is related to the gyro ω_g measurement according to the following equation:

$$\begin{cases} \omega = \omega_g + \mathfrak{b} + \mathfrak{n}_g \\ \dot{\mathfrak{b}} = 0 \end{cases}$$

where:

ω is the true spacecraft rate referred to the body fixed reference frame
ω_g is the gyro measured spacecraft rate already converted to body reference frame
$\mathfrak{b}$ is the gyro drift supposed to be almost constant in time ($\dot{\mathfrak{b}} = 0$)
$\mathfrak{n}_g$ is the gyro random noise modeled as zero mean Gaussian noise

Let us define the state of our process given by $\begin{pmatrix} q \\ b \end{pmatrix}$ and its estimate by $\begin{pmatrix} \hat{q} \\ \hat{b} \end{pmatrix}$ where $\hat{q}$ is the attitude quaternion estimate and $\hat{b}$ is the gyro drift estimate.

As shown in Sect. 5.3, the state estimation filter copies the state dynamics equations augmented by a forcing term which is linear in the error between measurements and estimate. Under certain assumptions the estimated state converges to the true state. We will follow this approach. Let us now indicate q_e the small error quaternion estimate so that:

$$q = q_e \otimes \hat{q} \tag{4.9}$$

The error quaternion estimate q_e is supposed to be very small so that it is possible to consider the scalar part almost equal to one $q_e \approx (1, \mathfrak{q}_e)$.

Combining Eqs. 4.7 and 4.9 it is possible to calculate the residual quaternion between the star tracker measurement quaternion and the attitude quaternion estimate. Let us define $q_{r,i}$ the residual quaternion:

$$q_{r,i} = q_i \otimes \hat{q}^{-1} \tag{4.10}$$

Using the quaternion multiplication rule defined in Sect. 2.10 we have:

$$q_{r,i} = q_{n,i}^{-1} \otimes q \otimes \hat{q}^{-1} = q_{n,i}^{-1} \otimes q \otimes q^{-1} \otimes q_e = q_{n,i}^{-1} \otimes q_e.$$

Neglecting the second order terms $O(\mathfrak{q}_{n,i}^T \mathfrak{q}_e)$ the ith star tracker residual can be written as:

$$q_{r,i} = (1, \mathfrak{q}_e - \mathfrak{q}_{n,i}) \tag{4.11}$$

Considering that the residual quaternion is very small, defining $\mathfrak{e} = (e_1, e_2, e_3)$ the angular residual error between the star tracker measurement and the quaternion estimate, it is possible to assume the angular residual error equal to twice the quaternion's vector components:

$$e_i = 2\mathfrak{q}_{r,i} = 2(\mathfrak{q}_e - \mathfrak{q}_{n,i}) \tag{4.12}$$

The following equations can now be used for the state prediction:

$$\begin{aligned} \dot{\hat{q}} &= \tfrac{1}{2}(0, \hat{\omega}) \otimes \hat{q} \\ \dot{\hat{b}} &= \textstyle\sum_{i=1}^{n} K_{b,i}\mathfrak{e}_i \\ \hat{\omega} &= \omega_g + \hat{\mathfrak{b}} + \textstyle\sum_{i=1}^{n} K_{m,i}\mathfrak{e}_i \\ \mathfrak{e}_i &= 2\mathfrak{q}_{r,i} = 2(q_i \otimes \hat{q}^{-1})|_{1,2,3} \end{aligned} \tag{4.13}$$

The $K_{b,i}$ and $K_{m,i}$ are the filter gains matrices. They represent the weights given to the measurement error. In order to optimize the gains, the dynamics of the error estimate is derived.

From Eq. 4.9 the estimate error quaternion can be derived:

$$q_e = q \otimes \hat{q}^{-1} \tag{4.14}$$

Differentiating Eq. 4.14 we get:

$$\dot{q}_e = \dot{q} \otimes \hat{q}^{-1} + q \otimes \frac{d}{dt}\hat{q}^{-1} \tag{4.15}$$

The first addendum can be computed using Eq. 2.11, the second using:

$$\frac{d}{dt}(\hat{q} \otimes \hat{q}^{-1}) = \frac{d}{dt}\hat{q} \otimes \hat{q}^{-1} + \hat{q} \otimes \frac{d}{dt}\hat{q}^{-1} = 0 \tag{4.16}$$

From which it is easy to derive using the fist of Eq. 4.13:

$$\frac{d}{dt}\hat{q}^{-1} = -\frac{1}{2}\hat{q}^{-1} \otimes (0, \hat{\omega}) \tag{4.17}$$

Therefore the error dynamics equation becomes:

$$\dot{q}_e = \frac{1}{2}(0, \omega) \otimes q_e + \frac{1}{2} q_e \otimes (0, -\hat{\omega}) \tag{4.18}$$

Assuming that:

$$\omega - \hat{\omega} = \delta\omega \tag{4.19}$$

$$q_e = (1, \mathfrak{q}_e) \tag{4.20}$$

We get:

$$\begin{aligned} &\tfrac{1}{2}((0, \omega) \otimes q_e + q_e \otimes (0, -\omega)) = -(0, \omega \times q_e) \\ &\tfrac{1}{2} q_e \otimes (0, \delta\omega) \approx (0, \delta\omega) \end{aligned} \tag{4.21}$$

Equation 4.18 written in quaternion vector component becomes:

$$\dot{\mathfrak{q}}_e = -\omega \times \mathfrak{q}_e + \frac{1}{2}\delta\omega \tag{4.22}$$

Considering also:

$$\delta\mathfrak{b} = \mathfrak{b} - \hat{\mathfrak{b}} \Longrightarrow \delta\dot{\mathfrak{b}} = -\dot{\hat{\mathfrak{b}}} \tag{4.23}$$

$$\omega \times \mathfrak{q}_e = skew(\omega) q_e \tag{4.24}$$

$$skew(\omega) = \begin{bmatrix} 0 & -\omega_3 & \omega_2 \\ \omega_3 & 0 & -\omega_1 \\ -\omega_2 & \omega_1 & 0 \end{bmatrix} \tag{4.25}$$

we are in condition to write the following error dynamics equation:

$$\begin{cases} \dot{\mathfrak{q}}_e = -skew(\omega)\mathfrak{q}_e - (K_{m,1} + K_{m,2})\mathfrak{q}_e + \frac{1}{2}\delta\mathfrak{b} + \frac{1}{2}n_g + K_{m,1}\mathfrak{q}_{n,1} + K_{m,2}\mathfrak{q}_{n,2} \\ \delta\dot{\mathfrak{b}} = -(K_{b,1} + K_{b,2})\mathfrak{q}_e + K_{n,1}\mathfrak{q}_{n,1} + K_{b,2}\mathfrak{q}_{n,2} \end{cases} \tag{4.26}$$

Assuming that $skew(\omega) \ll (K_{m,1} + K_{m,2})$ Eq. 4.26 can be simplified and in matrix form becomes:

$$\begin{pmatrix} \dot{\mathfrak{q}}_e \\ \dot{\delta\hat{\mathfrak{b}}} \end{pmatrix} = \begin{bmatrix} -(K_{m,1} + K_{m,2}) & \frac{1}{2} \\ -(K_{b,1} + K_{b,2}) & 0 \end{bmatrix} \begin{pmatrix} \mathfrak{q}_e \\ \delta\hat{\mathfrak{b}} \end{pmatrix} + \begin{bmatrix} K_{m,1} & K_{m,2} \\ K_{b,1} & K_{b,2} \end{bmatrix} \begin{pmatrix} \mathfrak{q}_{n,1} \\ \mathfrak{q}_{n,2} \end{pmatrix} + \frac{1}{2}\begin{bmatrix} I \\ 0 \end{bmatrix} \mathfrak{n}_g \tag{4.27}$$

In the case of only one star tracker available, the system of equations becomes:

$$\begin{pmatrix} \dot{\mathfrak{q}}_e \\ \dot{\delta\hat{\mathfrak{b}}} \end{pmatrix} = \begin{bmatrix} -K_{m,1} & \frac{1}{2} \\ -K_{b,1} & 0 \end{bmatrix} \begin{pmatrix} \mathfrak{q}_e \\ \delta\hat{\mathfrak{b}} \end{pmatrix} + \begin{bmatrix} K_{m,1} \\ K_{b,1} \end{bmatrix} \mathfrak{q}_{n,1} + \frac{1}{2}\begin{bmatrix} I \\ 0 \end{bmatrix} \mathfrak{n}_g \tag{4.28}$$

We choose to select diagonal gain matrices in order to simplify the design and uncouple the body measurement axes. The convergence of the error is determined by the roots of the polynomials which identify the eigenvalues of Eq. 4.28, if k_m, k_b, are the generic diagonal element of the $K_{m,1}$, $K_{b,1}$ matrices, the polynomial for the corresponding axis is:

$$s^2 + k_m s + \frac{1}{2} k_b \tag{4.29}$$

so that, being ω_f and ζ_f the filter angular frequency and damping, it follows that

$$\begin{cases} k_m = 2\zeta_f \omega_f \\ k_b = 2{\omega_f}^2 \end{cases} \tag{4.30}$$

The attitude filter is a second order low pass filter for each measurement axis, where k_m and k_b determine to the damping coefficient and the natural frequency. In this case the gain synthesis can be directly carried out by setting the filter poles in the frequency domain.

In case of two star trackers from Eq. 4.27 we derive the same Eq. 4.29 for the eigenvalues, here k_m, k_b, are the generic diagonal element of the $K_{m,1} + K_{m,2}$, $K_{b,1} + K_{b,2}$ matrices. In addition, the steady state error effect due to the star tracker noise can be easily derived from the same equation:

$$\delta \dot{\hat{\mathfrak{b}}} \rightarrow 0 \Longrightarrow \mathfrak{q}_e \rightarrow (K_{b,1} + K_{b,2})^{-1}(K_{b,1}\mathfrak{q}_{n1} + K_{b,2}\mathfrak{q}_{n,2}) \tag{4.31}$$

Assuming that the star tracker error source is a stationary random process, after the filter convergence, the error covariance matrix depends only on the $K_{b,1}$, $K_{b,2}$ filter gains:

$$E[\mathfrak{q}_e\mathfrak{q}_e^T] = (K_{b,1} + K_{b,2})^{-1}(K_{b,1}E[q_{n,1}q_{n,1}^T]K_{b,1} + K_{b,2}E[q_{n,2}q_{n,2}^T]K_{,b2})(K_{b,1}^T + K_{b,2}^T)^{-1} \tag{4.32}$$

For spacecraft rates verifying $skew(\omega) \ll (K_{m,1} + K_{m,2})$, the system gains can be synthesized with an optimization method that minimizes the trace of the error covariance matrix:

$$K_{b,i} = argmin(tr(E[q_e q_e^T])), \ i = 1, 2 \tag{4.33}$$

under constraints on the $K_{m,i}$, $K_{b,i}$ gains, derived taking into account that a reduction of the filter frequency produces a better error smoothness at steady state, but also an increase of the gyro noise effect.

	AOCS MODES REQUIRED HW							
	MGT	RW	RCT	FSS	GYRO	MGM	GPS	STT
SAM-EM	+	x	-	x	x	+		
NPM	+	x			x		+	x
OCM	+	x	-		x		+	x
SHM	+	x		x		+		

(+) HW used only in LEO ; (-) HW used only in GEO ; (x) common HW

Fig. 4.6 AOCS operational modes HW usage

4.4 AOCS Mode Algorithms

As previously mentioned, the design of the *multi-mission* AOCS includes the following operational modes:

- *Sun acquisition mode-emergency mode (SAM-EM)*;
- *normal pointing mode (NPM)*;
- *orbit control mode (OCM).*
- *safe hold mode (SHM) specific for LEO-SSO missions*;

Figure 4.6 specifies the hardware used in each AOCS operational mode.

4.4.1 Sun Acquisition Mode-Emergency Mode

The Sun acquisition mode-emergency mode (SAM-EM) is used to perform a robust and fast acquisition of the Sun pointing attitude.

The SAM-EM may be used in different S/C platform configurations and is applied in some nominal phases (i.e. to point solar arrays towards the Sun in order to reload the batteries) as well as in contingency scenarios. The emergency mode is made up of the following submodes:

- rate damping;
- Sun search;
- coarse Sun pointing;
- fine Sun pointing;
- eclipse.

The SAM-EM may be entered by ground TC or at any time during the mission according to FDIR logic. The SAM-EM mode is able to guarantee the stability and performance requirements in any of the nominal S/C configurations (from stowed to fully deployed).

SAM-EM Submodes	MGT	RW	RCT	FSS	GYRO	MGM	GPS	STT
Rate Damping	+	x	-	x	x	+		
Sun Search	+	x	-	x	x	+		
Coarse Sun Pointing	+	x	-	x	x	+		
Fine Sun pointing	+	x	-	x	x	+		
Eclipse	+	x	-		x	+		

(+) HW used only in LEO ; (-) HW used only in GEO ; (x) common HW

Fig. 4.7 SAM-EM submodes

The SAM-EM provides an attitude, controlled either around all three attitude angles (typical in LEO), or around two attitude angles orthogonal to the Sun direction and in rate around the Sun direction (typical in GEO or GTO). The reference attitude is configurable—in a given set defined by the Sun sensors field of view (see Chap. 9)—through the ground TCs.

In LEO, with three axes control, a coarse attitude estimation algorithm like the *Triad* is required (see Sect. 4.3.5.1). To cope with this need the SAM-EM design uses not only the magnetometer (MGM) and the FSS (fine Sun sensor) as AOCS sensors but also a coarse orbital propagator (COD) and a magnetic field model that allow to estimate the magnetic field in **ORF** necessary for the *Triad* algorithm (see Sect. 4.3.5.1).

Figure 4.7 lists for each submode AOCS hardware used in the control loop.

4.4.1.1 Rate Damping

The rate damping submode (Fig. 4.8) is a derivative loop with a zero angular rate reference. Its goal is to slow down the S/C rates and to take the system to an inertial attitude from which the Sun search phase can be safely and successfully started. The rate damping phase reduces the spacecraft angular rate below a pre-defined threshold on each axis. The SAM-EM is able to quickly despin the spacecraft if the stored angular momentum does not exceed a predefined value compatible with the RW Assembly overall momentum capability.

The primary sensor used during this phase is the gyro package and the actuators used are the RWs.

The exit condition is the achievement of adequately low rates on all three axes.

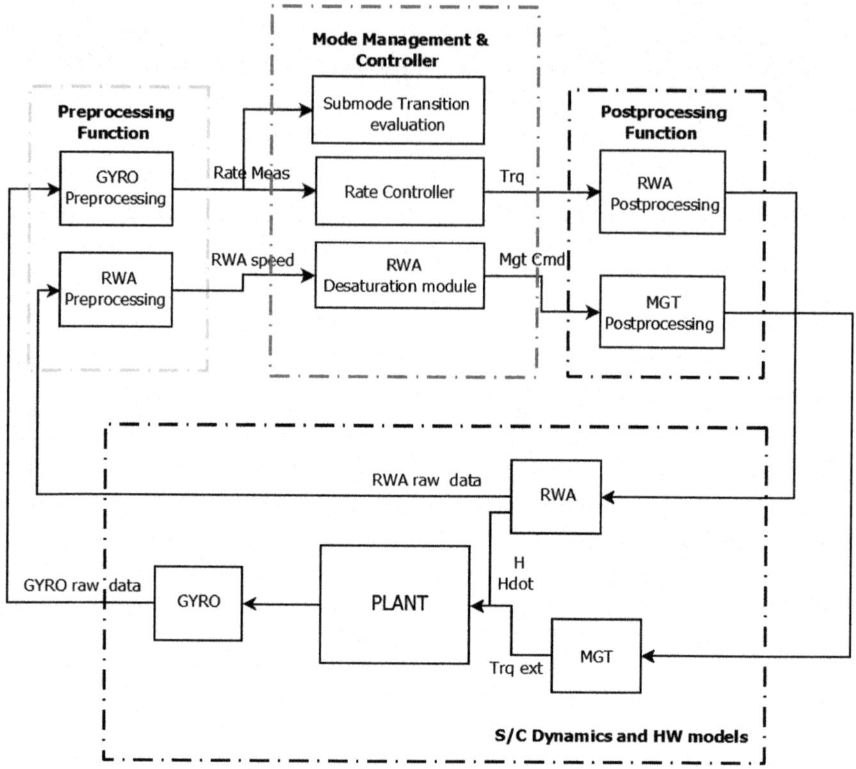

Fig. 4.8 EM rate dumping functional flow

4.4.1.2 Sun Search

The Sun search submode is entered after successful completion of the rate damping phase. Its goal is to maneuver the spacecraft so to perform a rotation about an axis fixed in **BRF** to bring the Sun in the FOV of a Sun sensor. The on-board collocation and orientation of the FSSs (fine Sun sensors) must provide a full coverage of the space after a 360 deg. rotation around this fixed axis, such that the Sun occurrence in one of the FSS FOV is guaranteed. When said condition is achieved (i.e. Sun in FOV), the coarse Sun pointing submode is automatically entered.

The primary sensor used during this phase is the gyro package and the actuators are the RWs while the Sun sensors are only used to trigger the Sun's presence.

At submode entrance, if the Sun is already present in the FSS field of view, the sense of rotation is chosen in order to move towards the Sun. In this case the Sun search will last only the time needed to confirm the Sun's presence.

The Sun search functional flow diagram is shown in Fig. 4.9, where it is possible to see that the core structure of the controller is a rate controller which commands the rotation rate desired to have the Sun entering in the FSS field of view.

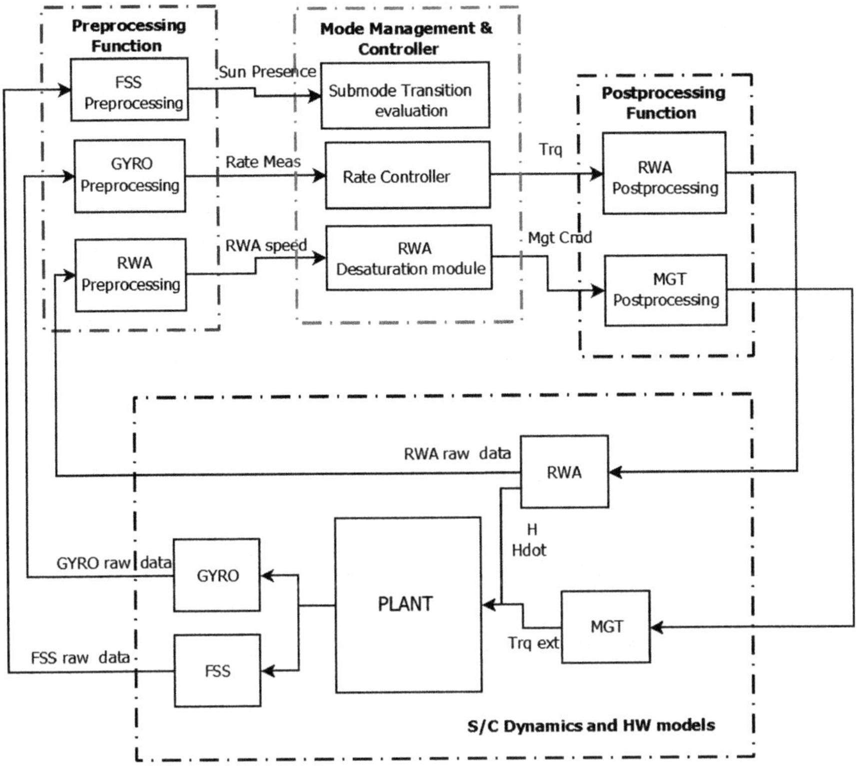

Fig. 4.9 EM sun search functional flow

4.4.1.3 Coarse Sun Pointing

The coarse Sun pointing (CSP) submode is entered after the accomplishment of the Sun search submode. This submode has two specific implementations: one applicable to LEO mission and one applicable to GEO mission.

4.4.1.4 Coarse Sun Pointing in LEO

It consists of a sequence of three large orthogonal rotations, realized for example according to the Euler angles rotation sequence 1-3-2 (see [6]). At the start of the sequence an estimate of the satellite attitude errors with respect to a desired attitude is obtained using a coarse attitude estimation algorithm that combines, in a *Triad-like* algorithm, the low accuracy Sun sensor and magnetometer measurement with the related reference provided by the available orbital propagator. After a sequence of roll, yaw and pitch rotations is defined to reach the desired Sun pointing attitude.

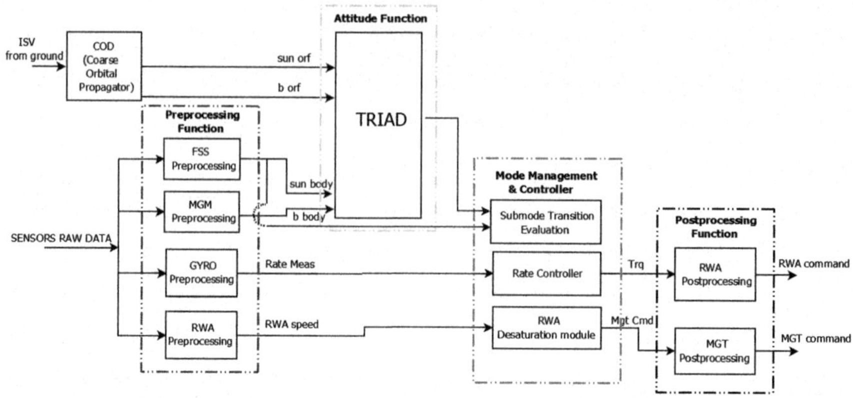

Fig. 4.10 Coarse sun pointing functional flow in LEO

During each rotation of the sequence, the control error is provided by the gyro measured rate so that two axes are kept at null rate and in the third the desired angle of rotation is integrated with the gyro rate until the desired rotation is obtained.

The combination of the first two maneuvers will bring the Sun in the X_{BRF}/Z_{BRF} plane. At this point it is necessary only to implement the pitch maneuver.

In case of no eclipse, as soon as the three attitude errors have been reduced, the transition to fine Sun pointing submode will be performed. In Fig. 4.10 the coarse Sun pointing functional flow is shown.

4.4.1.5 Coarse Sun Pointing in GEO

In GEO AOCS we typically do not have a coarse three axis measurement (because the magnetic field is too low), therefore the acquisition cannot be based on a full three axis maneuver computation.[2]

The main sensors used in CSP submode in GEO are gyro and FSS, while in LEO also the MGM was used to have a three axis attitude determination.

In GEO a different algorithm based on a two axis control instead of a three axis control can be used. The error signal is computed as the cross vector product between the Sun unit vector measured **BRF** and the target Sun unit vector in the **BRF** attitude. The target Sun unit vector in body axis is typically selected in order to maximize the solar array power generation. This error signal is then projected in the body reference frame and it is used as position error while the gyro measurement is used as rate error (i.e. the controller is a PD control loop).

The Sun acquisition in GEO is split in two parts: coarse acquisition and fine acquisition. These two phases use the same control loop structure, but with differ-

[2]The use of star trackers to perform Sun Pointing modes with three axes attitude measurement has been proposed and is becoming a possibility considering the good reliability of current STTs. However the question is much debated considering that most designers prefer to use different sensors for contingency (the SAM-EM is also a contingency mode) and nominal modes.

ent gains. This choice provides a smooth and gradual acquisition process without undesired transients. In fine Sun pointing, ground may command angular bias values and rate bias values around the Sun line if they are required (e.g. to have gyroscopic stiffness around that axis).

4.4.1.6 Fine Sun Pointing

The fine Sun pointing submode is entered after the successful completion of the coarse Sun pointing and guarantees that the attitude errors are lower than a predefined threshold.

This submode consists of a proportional derivative loop which is fed-in by attitude errors signal (proportional channel) and by the gyroscope measurements (derivative channel).

The sensors used during this phase are the gyro, the FSS and the MGM (only in LEO mission).

In LEO missions using the MGM, we can implement a three axis position control, without the MGM, like in most GEO missions, the control around the Sun direction is merely derivative and may allow for a slow drift of the satellite attitude.

The main actuators in the loop are the RWs. The MGMs and MGTs are used in LEO to desaturate the reaction wheels (RWs) stored angular momentum while the RCTs are used in GEO to desaturate the reaction wheels (RWs) stored angular momentum.

The emergency fine Sun pointing functional flow (for LEO mission) is shown in Fig. 4.11.

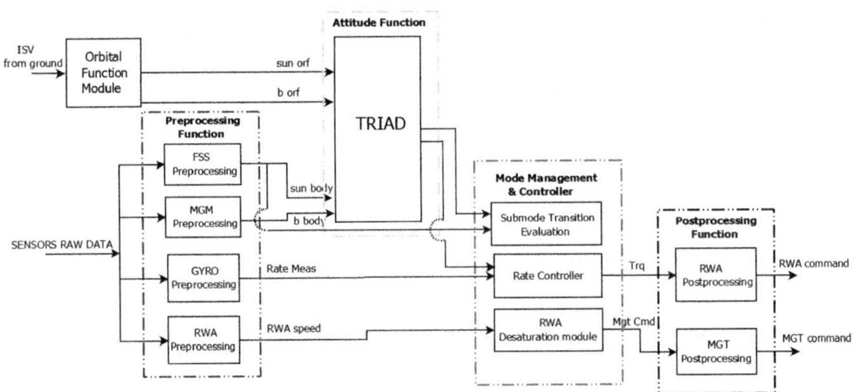

Fig. 4.11 EM fine sun pointing functional flow in LEO

4.4.1.7 Eclipse

The eclipse submode is automatically entered during the eclipse periods from Sun search, coarse Sun pointing and fine Sun pointing submodes. It consists in a derivative loop (see Fig. 4.12). The eclipse can be tackled in different ways depending on the orbital conditions (LEO or GEO) and the entry submode.

Specifically:

1. When the previous submode was the Sun search, the loop is fed-in with a null angular rate reference. This is done to resume the Sun search exactly from the point in which it was interrupted.
2. When the eclipse is entered from a fine Sun pointing or coarse Sun pointing submode in the LEO, the loop is fed-in with the orbital rate along the direction normal to the orbital plane.

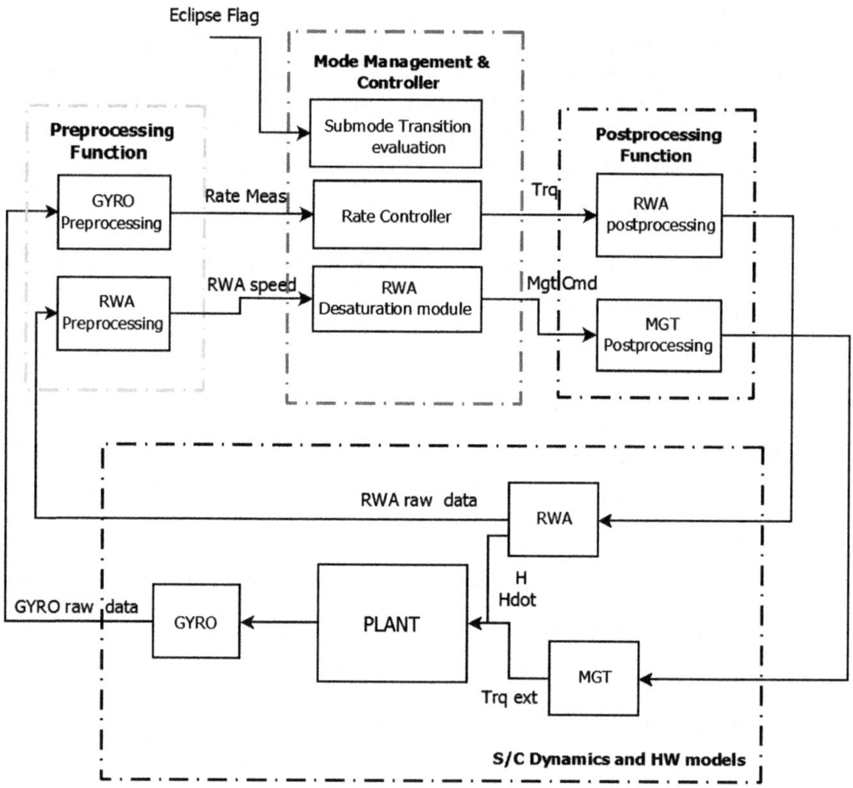

Fig. 4.12 EM eclipse functional flow

3. When the eclipse is entered from a fine Sun pointing or coarse Sun pointing submode in the GEO, a bias or null angular rate around the Sun line is commanded. This small rate bias provides some gyroscopic stiffness, so that at the end of the eclipse the Sun is expected to reappear close to its reference. At eclipse exit the rate damping submode, as well as the following Sun search and coarse Sun pointing submodes, should find all forward conditions verified, and should last only few seconds, before the new transition into fine Sun pointing.

For attitude control the sensor used during this phase is the gyro package, and the actuators are the RWs. The MGM and MGT are also used to desaturate the RWs' stored angular momentum in LEO mission, RCT are used in GEO with the same scope. The exit from the eclipse is triggered by the eclipse state provided by the on-board propagator or by a timer set by ground.

4.4.1.8 SAM-EM Control Loop

4.4.1.9 Rate Control Loop Structure

The SAM-EM rate control loop is a pure proportional rate controller. This control loop is used in rate damping, Sun search and coarse Sun pointing submode, to implement open loop S/C rotations.

The proposed control structure is shown in Fig. 4.13 where $H_p(s)$ is the plant, $H_g(s)$ is the gyro pre-filtering and $K_d H_c(s)$ is the controller.

Where:

$$H_c(s) = K_f \frac{1 + \tau_f s}{s^2 + 2\zeta_f \omega_f s + \omega_f^2} \tag{4.34}$$

$$H_g(s) = \frac{1}{1 + \tau_g s} \tag{4.35}$$

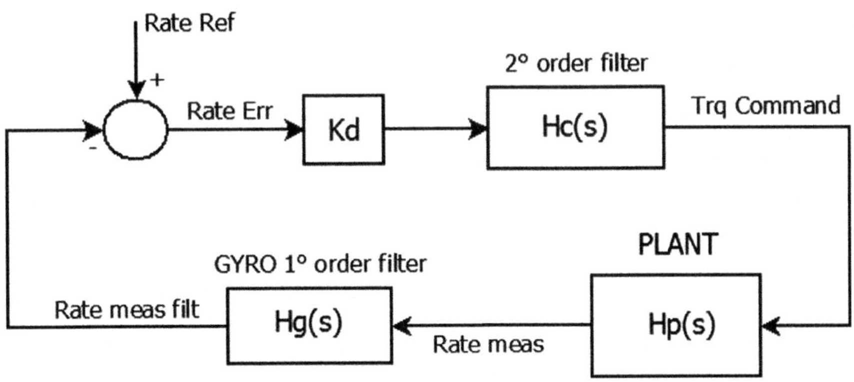

Fig. 4.13 EM rate control loop

The Plant is made up of a satellite with flexible appendages (e.g. solar array and antennas). The complete set of linear differential equations that can be used to describe the dynamics of a body with flexible appendages can be found in Chap. 3 Eqs. 3.136 and 3.137.

In a more general form, the plant is a multi Input multi Output (MIMO) system that implies that the control design has to be approached with control techniques (e.g. LQG/LQR, H_∞,...) suitable for this class of dynamic systems (see Sect. 6.3).

However, in many cases like the one we consider here, the plant has a highly diagonally dominant matrix of inertia and modal participation factors aligned to the principal axes. The plant can be assumed as three separate Single Input Single Output (SISO) systems obtained through the transfer function built around the diagonal elements of the matrix of inertia. Thus, it is possible to use a more simple control technique in frequency domain or time domain (e.g. poles place, root locus, PID controller synthesis, phase-lag and phase lead compensator...) to synthesize the controller (many text are available to study these basic techniques see among the others [11–13]).

Following this consideration, we recall that Eq. 3.136, bring to the following plant transfer function:

$$\delta\phi = (I_G - s^2 \sum_{k=1 \ldots N_f} \frac{{S_G}^k {S_G}^{kT}}{s^2 + {\Omega_k}^2})^{-1} \frac{u}{s^2} \tag{4.36}$$

where $u = \delta c_c - \delta \dot{h}_w - [\delta\dot{\phi}, \overline{h_w}]$ is the control torque. In this case we assume $\delta c_c = 0$, and $\overline{h_w} = 0$, so that $u = -\delta \dot{h}_w$.

If we specialize Eq. 4.36 to a single axis transfer function $H_p(s)$ from the control torque to the attitude angle output, considering only a two iso-frequency dominant flexible modes composed of two symmetric appendages (see also [14]), we obtain:

$$H_p(s) = \frac{s^2 + {\Omega_1}^2}{Js^2((1 - 2\delta^2/J)s^2 + {\Omega_1}^2)}$$

where J is the diagonal component of the matrix I_G in the selected axis, $(\Omega_1 = \Omega_2)$ is the cantilevered angular frequency of the first dominant flexible mode while the pole $(\frac{\Omega_1}{\sqrt{1-2\delta^2/J}})$ is the free-free frequency of this flexible mode. The factor δ is the participation factor, corresponding to the square root of the diagonal component of the matrix ${S_G}^1{S_G}^{1T} = {S_G}^2{S_G}^{2T}$ on the selected axis. It is counted twice because we are assuming to have two identical symmetric appendages.

In the proposed AOCS architecture, we have direct access to the rate measurement using a gyro as a rate sensor. The filter $H_g(s)$ is introduced as a pre-filter to reduce the output noise of the rate sensor. To simplify the control design and gain selection, the cut-off frequency of the gyro pre-processing filter has to be selected at least one decade higher than the control bandwidth in order not to limit the bandwidth of the controller.

Figure 4.13 shows a second order filter $H_C(s)$ introduced to filter the flexible appendages (i.e. solar arrays). To design this filter we can use different techniques mainly related to the level of amplitude reduction and phase shaping needed at a predefined frequency. A possible structure of this filter is shown in Eq. 4.34.

Due to the fact that in this mode a very large bandwidth is not necessary, the controller is simply a gain Kd that has been tuned for all S/C configurations (i.e. in stowed and deployed configuration). Its value is selected in order to obtain the desired static error in the rate loop.

This simple solution is enough to achieve the desired bandwidth with sufficient margins both for the phase and amplitude without using a more complicated structure like the phase-led/phase-lag compensator [11, 12].

A typical bode diagram for this type of control loop chain is shown in Fig. 4.14.

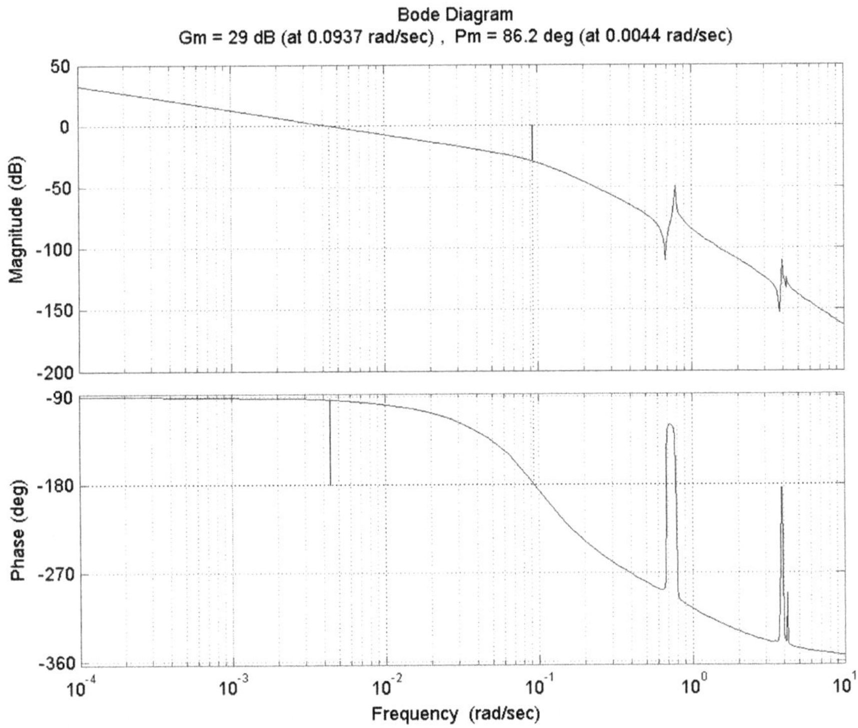

Fig. 4.14 Bode open loop chain—rate control loop

In this picture, it is possible to recognize that the control loop has a derivative action up to the selected bandwidth (0.004 rad/s), high gain margin and phase margin and it has a strong gain reduction (up to 50 db) on the peak of the first frequency of the solar array.

4.4.1.10 Rate and Position Control Loop Design

The rate and position control loop is a proportional-derivative (PD) controller used in the Sun pointing submode to perform three axis attitude control. The plant structure, pre-filtering and flexible mode filter presented in the previous chapter can be reused for this new control chain.

Figure 4.15 shows a rate and position control loop structure.

where:

$$H_c(s) = K_f \frac{1 + \tau_f s}{s^2 + 2\zeta_f \omega_f s + \omega_f^2} \tag{4.37}$$

$$H_g(s) = \frac{1}{1 + \tau_g s} \tag{4.38}$$

The main difference between this control structure and the rate control loop is in the selection of a PD controller instead of only derivative controller. The K_p and K_d gain selection can be carried out using different techniques like the Ziegler-Nichols method (see [12]) or an equivalent approach for the design of a phase-lead compensator (see [11]).

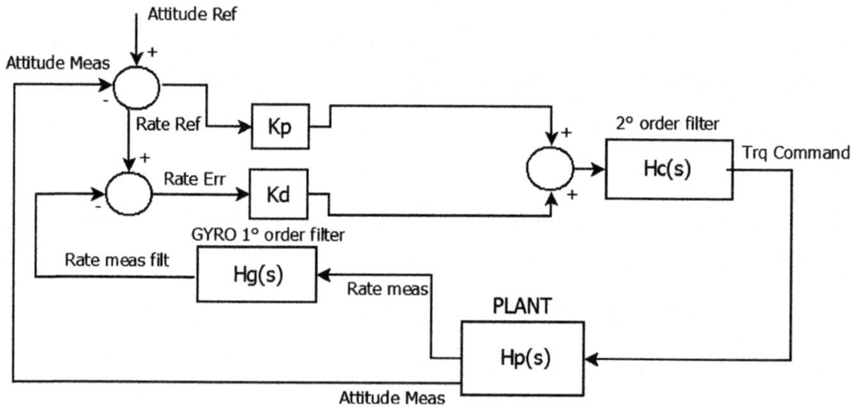

Fig. 4.15 EM rate and position control loop

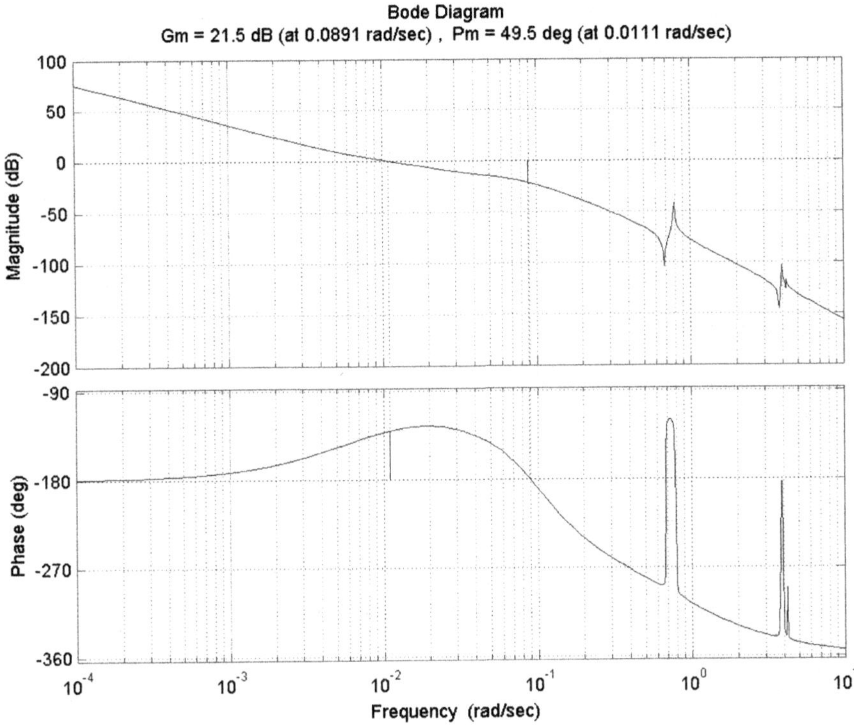

Fig. 4.16 Bode open loop chain—position rate control loop

All the control parameters have to be selected according to a design process able to guarantee the desired performance and appropriate stability margin and gain margin that ensure a robustness with respect to gain loop reduction or time delay in the loop.

The open loop bode diagram of the rate and position loop is shown in Fig. 4.16.

4.4.1.11 SAM-EM Simulation

This section shows some simulation results of SAM-EM obtained using the design solution presented in the previous sections. In particular, the simulation results are related to a S/C in fully deployed condition (i.e. all appendages deployed) with a not-null initial attitude angle and angular rates during a Summer Solstice in a SSO orbit.

The spacecraft data used for the simulations are shown in the following tables.

The inertia tensor of the fully integrated satellite including the appendages is shown in Fig. 4.17.

The appendages considered are two identical solar array wings (SAW1 and SAW2), connected to the main body in two hinges positioned in:

$r_{SAW1} = (0.6829,\ 0.7908,\ 1.2711)^T\ [m]$, and $r_{SAW2} = (-0.6829,\ -0.7908,\ 1.2711)^T\ [m]$.

Inertia tensor is a 3x3 matrix [kgm^2] w.r.t a body fixed frame centred in the S/C COG parallel to the mechanical build reference.**[kgm^2]**						
DESCRIPTION	**I_{xx}**	**I_{yy}**	**I_{zz}**	**I_{xy}**	**I_{xz}**	**I_{yz}**
DEPLOYED Conf.	3467.0	17193.0	14566.0	-512.0	56.3	35.3

Fig. 4.17 Spacecraft inertia tensor

Diagonal component [Hz]	**Minimal**	**Tolerance**
1° Out of plane	0.1039	10%
1° In plane	0.5356	10%
2° Out of plane	0.6635	10%
1° Torsional	0.8259	10%

Fig. 4.18 Modal natural frequencies of the SAW in deployed configuration

SAW modal damping coefficients The Solar Panel properties are reported in the fully deployed configuration.	
1° Out of plane	0.003
1° In plane	0.003
2° Out of plane	0.003
1° Torsional	0.003

Fig. 4.19 Modal damping coefficients of the SAW in deployed configuration

SAW Modal Rotation Coupling Factors The Solar Panel modal rotation coupling factors are reported in the fully deployed configuration.			
[m$\sqrt{kg}$]	X	Y	Z
1° Out of plane	-8.946340E-03	1.250615E-03	-4.596133E+01
1° In plane	4.854559E-01	-4.651433E+01	-4.144528E-03
2° Out of plane	7.994077E-02	-1.549985E-02	5.530611E+00
1° Torsional	-4.096249E+00	2.708088E-01	1.149119E-02

Fig. 4.20 Modal rotation participation factors of the SAW

The transformation matrix from the SAW2 reference frame to the **BRF** is the identity matrix, the transformation matrix from the SAW1 reference frame to the **BRF** is a 180 deg rotation around the X **BRF** axis which means that $Y_{BRF} = -Y_{SAW1}$ and $Z_{BRF} = -Z_{SAW1}$.

SAW Modal Translation Coupling Factors The Solar Panel modal translation coupling factors are reported in the fully deployed configuration.			
$[\sqrt{kg}]$	X	Y	Z
1° Out of plane	-4.635139E-03	6.583623E+00	1.610784E-04
1° In plane	-1.332722E-02	1.718402E-03	-6.991334E+00
2° Out of plane	1.645563E-02	-3.466420E+00	-2.312155E-03
1° Torsional	5.056233E-04	2.638664E-02	3.697665E-02

Fig. 4.21 Modal translation participation factors of the SAW

Controller Parameters	Roll	Pitch	Yaw
K_d [N*m*sec/rad]	20	115	70
τ_g [sec]	1	1	1
K_f [rad2/sec3]	3.6e-03	3.6e-01	0.01
τ_f [sec]	0	0	0
ω_f [rad/sec]	0.06	0.6	0.1
ς_f	0.7	0.7	0.7

Fig. 4.22 SAM-EM rate controller parameters

Controller Parameters	Roll	Pitch	Yaw
K_d [N*m*sec/rad]	40	175	150
K_p [N*m/rad]	0.3	1.2	1
τ_g [sec]	1	1	1
K_f [rad2/sec3]	3.6e-03	3.6e-01	0.01
τ_f [sec]	0	0	0
ω_f [rad/sec]	0.06	0.6	0.1
ς_f	0.7	0.7	0.7

Fig. 4.23 SAM-EM rate-position controller parameters

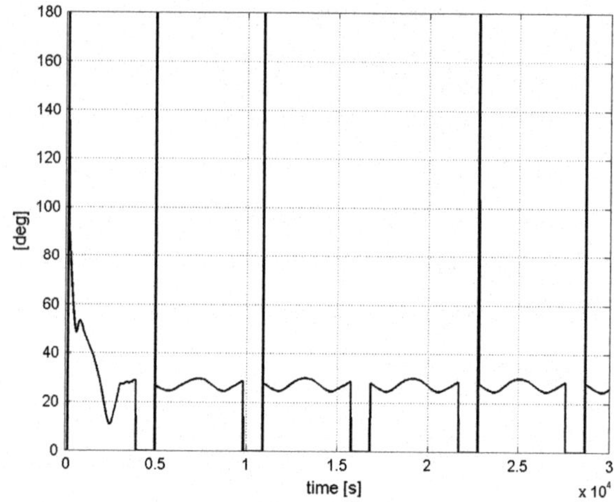

Fig. 4.24 SAM-EM angle between $-Y$ **BRF** axis and sun unit vector

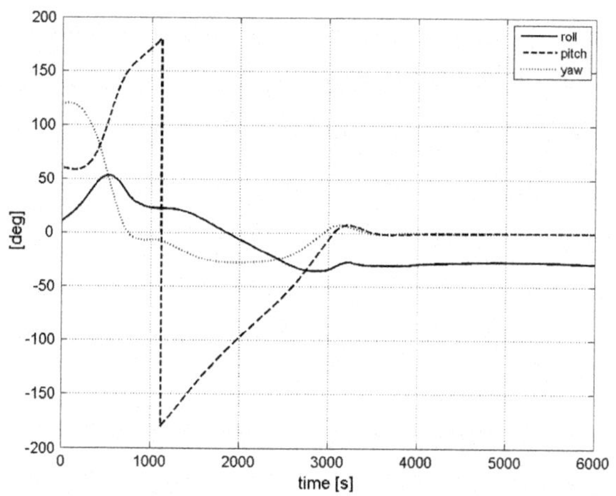

Fig. 4.25 SAM-EM attitude angles according Euler angle rotation sequence 2-1-3

The cantilevered frequencies, damping, rotational and translational participation factors of the two SAW in their SAW reference frames are shown in Figs. 4.18, 4.19, 4.20 and 4.21.

The transformation rules to transfer the modal participation factors to **BRF** has been discussed in Sect. 3.12.

The control gains used for the rate loop and for rate-position loop are shown in Figs. 4.22 and 4.23.

Figure 4.24 shows the behavior of the angle evaluated between $-Y$ **BRF** axis and the Sun unit vector in body axis during 5 orbits. This is an important indicator because the Sun has to be pointed towards the solar array wings (in this case the solar

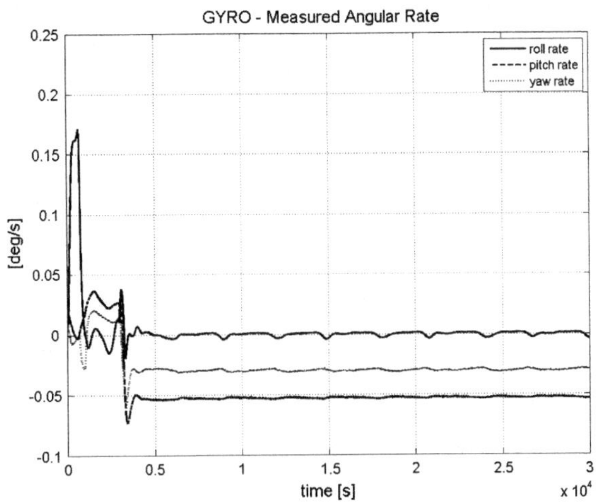

Fig. 4.26 SAM-EM gyro measured angular rates in body axes

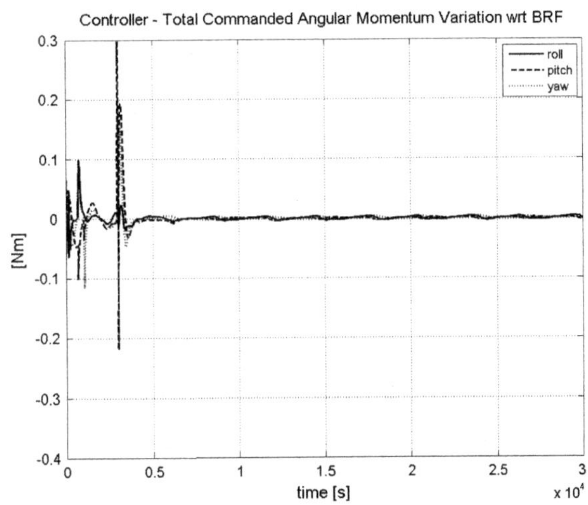

Fig. 4.27 SAM-EM RWA commanded torque in body axes

array normal is parallel to the $-Y$ **BRF** axis) in order to have positive power budget. The plot shows a few holes due to the presence of the eclipse. The SAM-EM mode is a PD 3-axes controlled mode, therefore the steady state pointing is maintained almost constant also during the eclipse holes due to the rate feedback control.

Figures 4.25, 4.26 and 4.27 show the evolution of satellite attitude angles, attitude rates and commanded torques.

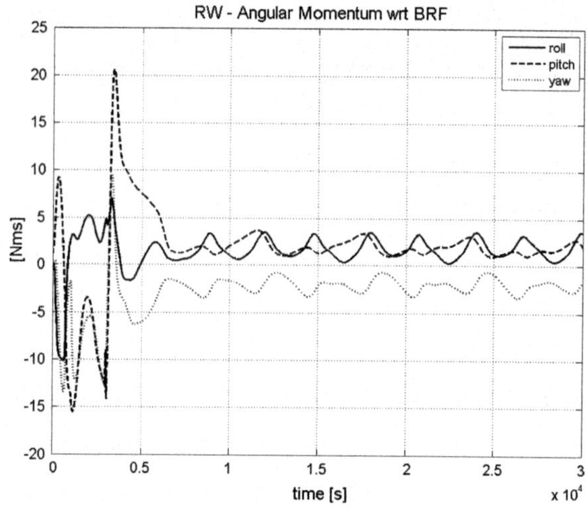

Fig. 4.28 SAM-EM RWA angular momentum in body axes

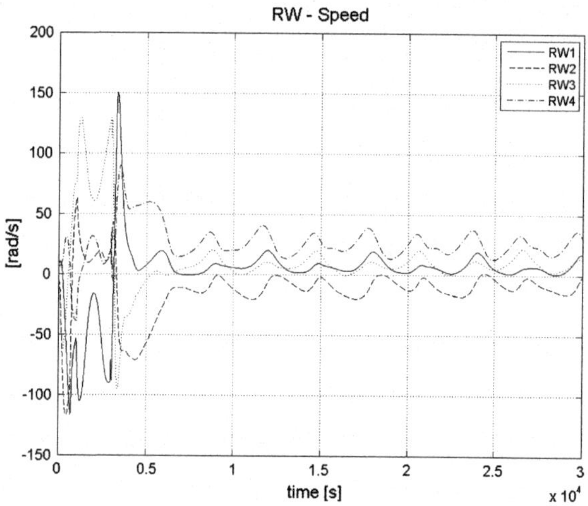

Fig. 4.29 SAM-EM RW velocity in RW axes

Finally the RWA torque and angular momentum in body axes, the RWA velocity in wheel axes and the MGT commanded momentum dipole are shown in Figs. 4.28, 4.29 and 4.30.

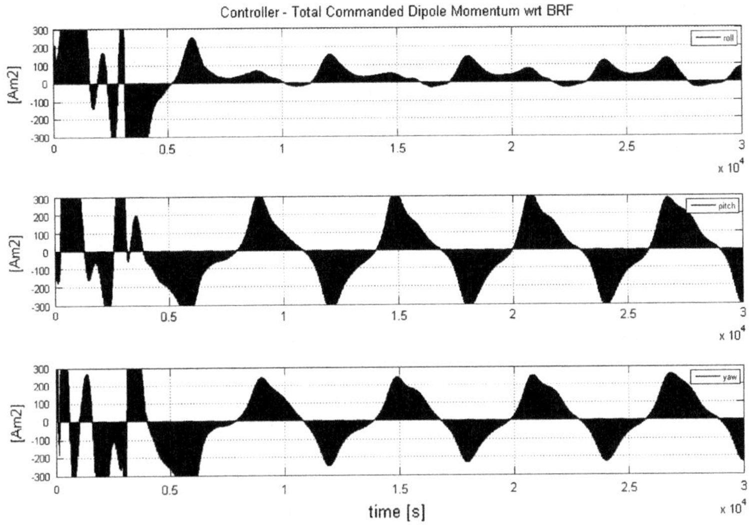

Fig. 4.30 SAM-EM MGT commanded dipole momentum in body axes

4.4.2 Safe Hold Mode

The mode is particularly effective with small satellites, microsatellites and undeployed satellites (with a low gravity gradient). The safe hold mode (SHM) implements an easy to design and test and robust approach to rate damping. This AOC mode is widely used in LEO SSO (i.e. Sun Synchronous orbit) missions because it requires a strong Earth magnetic field.

When the satellite orbit is a LEO Sun Synchronous Dawn-Dusk orbit and when the satellite has a well defined axis of maximum inertia (e.g. $-Y$ **BRF**), the SHM can be designed to provide also a robust Sun pointing attitude acquisition, positive power balance and thermal stability.

In such conditions the steady state attitude of this mode realizes the convergence of the maximum inertia axis to the direction orthogonal to the orbital plane, and in SSO Dawn-Dusk the Sun direction is almost orthogonal to the orbital plane. We will describe in this section the SHM for this specific orbital conditions.

The SHM is designed to support all phases of the mission (e.g. LEOP, nominal mission mode and Disposal) and it can be entered by the FDIR function when an alarm is triggered or by a ground TC when in nominal condition.

This mode uses a *B-dot* control law (see [15]), that is fully autonomous (no position knowledge is required), unlimited in time (there is no propellant consumption), using a limited set of highly reliable units (magnetometers, magneto-torques and reaction wheels) can be easily designed and end to end ground tested (the magnetic field is available also on the Earth surface while other sensors cannot be stimulated naturally in ground AOCS tests).

SHM Submodes	MTQ	RW	RCT	FSS	GYRO	MGM	GPS	STT
Rate Damping	x	x		x		x		

Fig. 4.31 SHM mode equipments

The SHM is able to acquire the Sun from a large set of initial conditions, including any initial satellite orientation and a predefined maximum angular rates about any axis in all satellite configurations. If the selection of the magneto-torques (MGT) size is correctly carried out and the energy stored in the battery is sufficient , the Sun acquisition in any satellite configurations will be completed within a predefined worst case time typically expressed in number of orbits (typically 2 or 3 orbits) from the SHM time entry.

The following Fig. 4.31 lists AOCS hardware used to perform the attitude control.

4.4.2.1 SHM Control Loop Design

The SHM is based on a simple control law (the *B-dot*) which consists in measuring the Earth magnetic field B in **BRF** and implementing the following control law

$$m = -k\dot{B} \tag{4.39}$$

where m is the satellite magnetic dipole produced by the torquerods. The *B-dot* is part of the large family of the dissipative controllers used in *non-linear* control, which are very much used in space control.

For these controllers one aims at demonstrating stability—starting from any attitude condition- usually with the methods of the Lyapunov functions whose theory will be illustrated in Sect. 6.1.2.

The physical coupling between the dipole momentum and the external magnetic field generates the control torque that acts on the spacecraft:

$$c_c = m \times B = -k\dot{B} \times B \tag{4.40}$$

The derivative of the magnetic field in **BRF** is given by:

$$\dot{B} = M_{B \leftarrow E}\dot{B}_I + B \times \omega_{BE,B} \tag{4.41}$$

where $\dot{B}_I$ is the Earth magnetic field derivative in **ECI** due to the orbital motion, while the second term is the contribution of the spacecraft spinning motion.

$\dot{B}_I$ depends both on the variation in magnitude $|B|$ and direction e_b of the geomagnetic field along the orbit. Considering

$$B_I = |B|e_b \rightarrow \dot{B}_I = \frac{d|B|}{dt}e_b + |B|\dot{e}_b = \frac{d|B|}{dt}e_b + \omega_{b,E} \times e_b|B| \tag{4.42}$$

where $\omega_{b,E}$ is the magnetic field rotation rate—in **ECI**- during the orbital motion:

$$\dot{e}_b = \omega_{b,E} \times e_b \tag{4.43}$$

Since the first term, related to the change of magnitude, is always parallel to the magnetic field itself, it never contributes to the *B-dot* torque so that (we indicate $M_{B \leftarrow E} \omega_{b,E} = \omega_b$):

$$c_c = -k(M_{B \leftarrow E} \dot{B}_I + B \times \omega_{BE,B}) \times B = -k(B \times (\omega_{BE,B} - \omega_b)) \times B \tag{4.44}$$

From Chap. 3 we know that the energy is an invariant in absence of gravity or magnetic torques when the momentum bias is constant (see Eq. 3.84 when flex mode have negligible effects). We then study the effect of the $\dot{B}$ on the kinetic Energy to see if we can derive general conclusion on the dynamics. The time rate of change of kinetic energy (see 3.4) can be written as

$$\frac{d\mathscr{T}}{dt} = {c_c}^T \omega_{BE,B} = -k[(B \times (\omega_{BE,B} - \omega_b)) \times B]^T \omega_{BE,B} \tag{4.45}$$

For a LEO SSO dawn-dusk orbit, the Earth magnetic field can be assumed always lying in the orbit plane. Its direction variation ω_b is normal to the orbit plane with a magnitude varying in a range around twice the orbital rate, since the dipole field describes two lobes per orbit. When the spacecraft is spinning with a rate much higher than twice the orbital rate, the Earth magnetic field time derivative due to the orbital motion can be neglected.

$$|\omega_{BE,B}| \gg 2\omega_0 \rightarrow |\omega_{BE,B}| \gg |\omega_b| \tag{4.46}$$

Under this assumption, the time rate of change of kinetic energy is always less than zero, so *B-dot* acts like an energy sink, damping out the kinetic energy with a dissipative torque feedback:

$$\frac{d\mathscr{T}}{dt} \cong -k \left\| B \times \omega_{BE,B} \right\|^2 \tag{4.47}$$

Moreover when the spacecraft rate becomes of the order of magnitude of twice the orbital rate, considering that ω_b is normal to the orbit plane, while B lies in it, the time rate of change of kinetic energy $\mathscr{T}$ is

$$\frac{d\mathscr{T}}{dt} = -k \left\| B \times \omega_{BE,B} \right\|^2 + k|B|^2 {\omega_b}^T \omega_{BE,B} \tag{4.48}$$

When the spacecraft angular rate is normal to the magnetic field rate, that is $\omega_{BE,B}$ lies in the orbital plane (${\omega_b}^T \omega_{BE,B} = 0$), the *B-dot* law provides dissipative effects damping out the energy with the torque feedback. This energy dissipation property provides a damping effect on the transverse dynamics (motion outside the orbital plane).

For a simple rigid body or in case of additional gyroscopic stiffness, this dissipative effect tends to damp the nutation dynamics, bringing the spacecraft to rotate around the direction of ω_b with the maximum inertia axis.

When the spacecraft rate is finally normal to the orbit plane, aligned with ω_b, the energy variation rate is given by:

$$\frac{d\mathcal{T}}{dt} = -k[(B \times (\omega_{BE,B} - \omega_b)) \times B]^T \omega_{BE,B} = -k|B|^2 \omega_{BE,B}{}^T (\omega_{BE,B} - \omega_b) \tag{4.49}$$

For the longitudinal dynamics the *B-dot* control law provides a torque feedback driven by the $(\omega_{BE,B} - \omega_b)$ rate error. It tends to bring the error to zero and then to bring the spacecraft rate $\omega_{BE,B}$ to be equal to the magnetic field rate ω_b (if $\omega_{BE,B} < \omega_b$ the kinetic energy begins to increase).

However, when the spacecraft is affected by a gravity gradient torque larger than the torque rods (MGT, see Chap. 9) capability, the transverse dynamics, induced by the gravity gradient, are damped by *B-dot* control law and the residual motion tends to be in the orbit plane. In this configuration the residual longitudinal motion keeps it close to the gravity gradient equilibrium attitude, with an average rate equal to the orbital one. Moreover the *B-dot* control torque variation along the orbit, because of the change of the geomagnetic field, induces a swinging motion around that equilibrium condition.

In order to achieve the sun pointing attitude with the $-Y$ **BRF** body axis sun pointing, the *B-dot* control law is used together with a fixed angular momentum polarized along the $-Y$ **BRF**. In this case the *B-dot* control low aligns the internal angular momentum to the magnetic field orbital rate. At steady state the *B-dot* control law drives the spacecraft rate to be almost equal to the magnetic field rotation rate (twice the orbital rate, two lobes per orbit).

4.4.2.2 Rate Damping Submode

The rate damping submode is used to slow down the angular rates around the three axes leaving the satellite spinning at a small rate, with the $-Y$ **BRF** body axis (maximum inertia axis) pointing toward the Sun.

As shown in the previous section a *B-dot* control law is used. The spacecraft rotational rates are damped with a commanded torque that leads the spacecraft body to synchronize with the Earth Magnetic field rate. The Earth Magnetic field has a bi-orbital component in the SSO dawn-dusk orbit, so the *B-dot* control law tries to damp the spacecraft rate to $2\omega_0$, where ω_0 is the orbital rate.

The rate damping *B-dot* control law will be applied in two different situations:

- *satellite with low gravity gradient torque configuration*: The *B-dot* control law is used together with a momentum polarized on wheel assembly. The *B-dot* law induces the alignment of internal angular momentum with the magnetic field rotation rate. Loading a fixed angular momentum on the $-Y$ **BRF** axis (maximum inertia axis) of the **BRF** triad, at the end of the transient phase, the *B-dot* control law guarantees that this axis is aligned with the orbit normal (nominal Sun pointing attitude).

- *satellite with high gravity gradient torque configuration* : The satellite rate de-spin is obtained with a *B-dot* control law, but the large gravity gradient disturbance torque is much higher than the control torque provided by magnetic torque rods. This means the spacecraft is not able to follow the Earth magnetic field, and the gravity gradient forces the spaceraft to point its minimum inertia axis toward the Earth (i.e. $+X$ **BRF** or $-X$ **BRF** axis toward the Earth) with a pitch rotation rate that is almost equal to the ω_0 orbital rate. In this case, instead of trying to control the large gravity gradient disturbance torque, the *B-dot* law is modified, to ensure a reference rotation rate at ω_0.

The primary sensor used during this phase is the magnetometer and the primary actuator is the magnetic torque rod.

The RWA is used only for momentum polarization.

The FSS are used to trigger/signal the Sun's presence but is not actively used in the loop.

Figure 4.32 shows the SHM rate damping functional flow data.

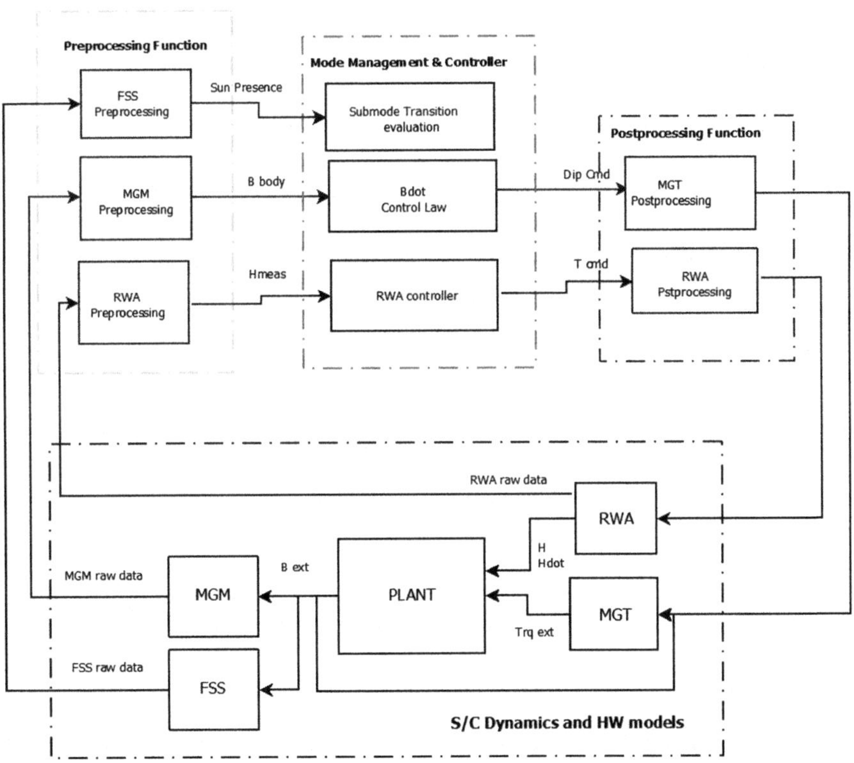

Fig. 4.32 SHM rate damping functional flow

The magnetometer raw data is processed in order to compute the three components of the B field in body axes. The *B-dot* control law realizes a numerical derivative of magnetic field components in order to calculate the magnetic dipole command to the Megneto-torquer (MGT) as follows:

$$m = -k\dot{B} \tag{4.50}$$

$$m_{ctrl} = m - (m^T \hat{B})\frac{\hat{B}}{|\hat{B}|^2} = (I - \frac{\hat{B}\hat{B}^T}{|\hat{B}|^2})m \tag{4.51}$$

where $\dot{B}$ is the derivative vector of the measured magnetic field in body reference frame, $\hat{B}$ is the measured magnetic field unit vector in body reference frame, k is a constant gain and m_{ctrl} is the magnetic dipole momentum commanded to the MGT. In the m_{ctrl} expression the calculated dipole command m is purged from the component parallel to the magnetic field B (the term: $-(m^T \hat{B})\frac{\hat{B}}{|\hat{B}|^2}$ to allow a saving of electric power to be made available to the components of m that are orthogonal to the magnetic field B and thus producing torque.

The physical coupling between the dipole momentum commanded to the MGT and the external magnetic field generates the control torque that acts on the spacecraft.

Due to the sensitivity of the MGM to the MGT generated magnetic field an appropriate duty cycle for their utilization has to be implemented in the real design.

In order to reduce the measurement noise a pre-filtering can be implemented to the B measurement before using this value in the *B-dot* calculation.

Moreover in order to avoid limit cycle between MGM measurement and MGT command a dead-band on command can be added in AOC control loop.

In rate damping submode the reaction wheels are commanded in open loop to load a predefined angular momentum in body axis.

The RWA pre-processing provides the on-board computer with accumulated angular momentum in body axes, the RWA controller calculates the RWA body axes control torque, while the RWA post-processing evaluates the torque command for each wheel starting from the torque request in body axes.

4.4.2.3 SHM Simulation

This section shows a simple simulation result for SHM as per the design solution presented in the previous sections. In particular, the simulation results are related to a S/C in stowed condition with not-null initial angular rates (i.e. after separation from launcher vehicle) during a Summer Solstice in a SSO (Sun Synchronous orbit) at an altitude of about 710 km.

The spacecraft is in stowed configuration and its inertia tensor used for the simulation is shown in Fig. 4.33.

Inertia tensor is a 3x3 matrix [kgm^2] w.r.t a body fixed frame centred in the S/C COG parallel to the mechanical build reference.**[kgm^2]**						
DESCRIPTION	**I_{xx}**	**I_{yy}**	**I_{zz}**	**I_{xy}**	**I_{xz}**	**I_{yz}**
STOWED Conf.	2558.6	3135.4	1492.9	-18.50	86.39	28.17

Fig. 4.33 Spacecraft inertia tensor in stowed configuration

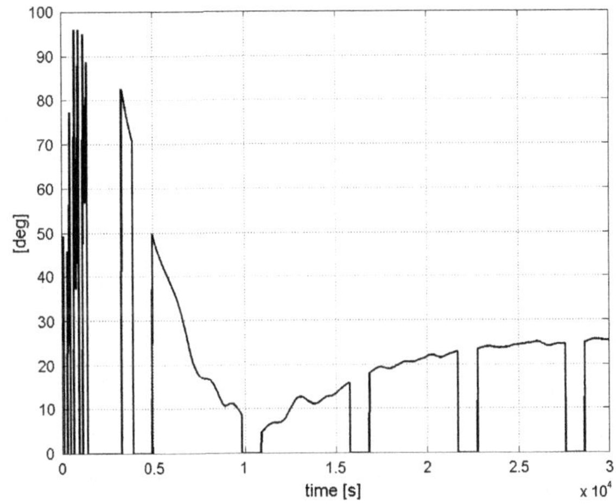

Fig. 4.34 SHM angle between $-Y$ **BRF** axis and sun unit vector

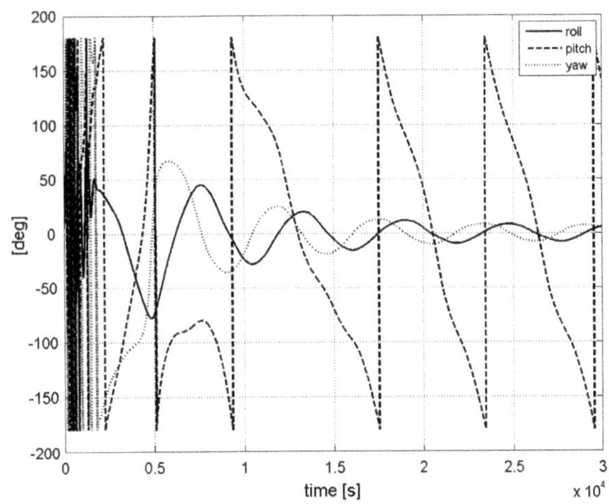

Fig. 4.35 SHM attitude angles according to Euler angle rotation sequence 2-1-3

The relevant data for the *B-dot* control law are: $k = 4nT/(\text{s Am}^2)$ and the MGT saturation value for the dipole momentum $m_{sat} = 300\,\text{Am}^2$.

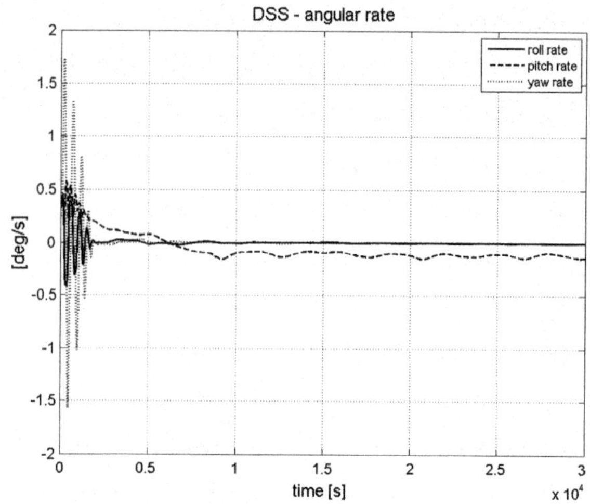

Fig. 4.36 SHM S/C angular rate

Figure 4.34 shows the behavior of the angle between $-Y$ **BRF** axis and the Sun unit vector. This is an important indicator because the Sun has to be pointed towards the solar array wings (in this case solar array normal is parallel to the $-Y$ **BRF**) in order to have positive power budget. The plot shows a few holes due to the presence of an eclipse repeated in each orbit (i.e. summer solstice simulation).

The next two Figs. 4.35 and 4.36 show the time history of attitude angles and angular rates.

Figure 4.37 shows the reaction wheel angular momentum biased at a predefined value according to the design in order to polarize a *preferred stability attitude* along $-Y$ **BRF** axis according to the design.

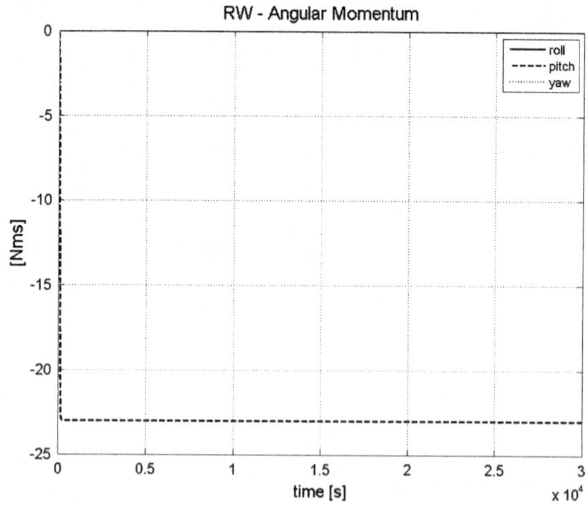

Fig. 4.37 SHM RW angular momentum in body axes

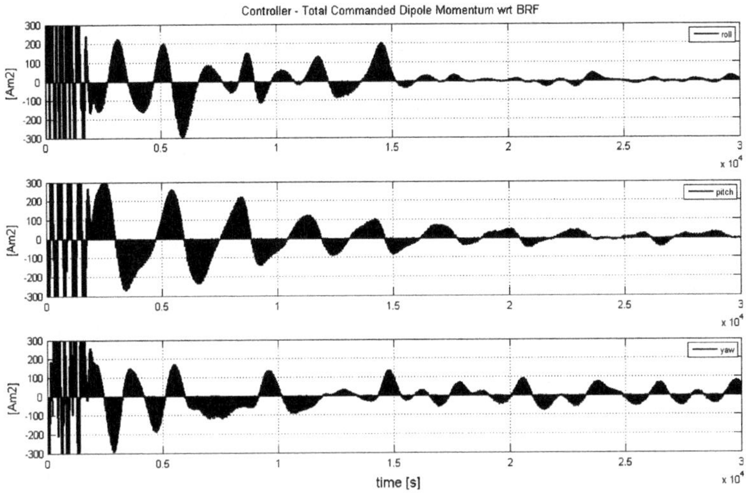

Fig. 4.38 SHM MTG commanded dipole momentum in body axes

The MGT dipole momentum behavior is shown in Fig. 4.38. It is possible to recognize the high command activity at the beginning due to the rate damping phase while when the steady state condition is reached the behavior is smoother.

Finally Fig. 4.39 shows the result of a one day simulation of SHM with a typical evolution of the Sun unit vector projected in the elevation azimuth plane.

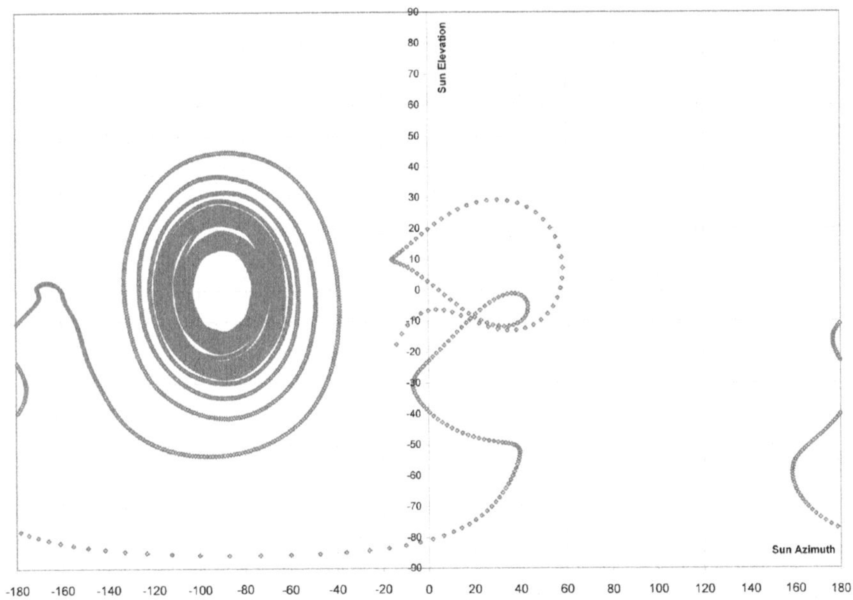

Fig. 4.39 SHM sun elevation versus azimuth evolution (1 day)

4.4.3 Normal Pointing Mode (NPM)

The NPM is designed in order to provide an attitude performance suitable for the P/L operations. The mode is entered in nominal scenarios by a ground TC from SAM-EM.

The normal mode is composed of the following submodes:

- tranquillization;
- fine pointing.

The tranquillization phase allows AOC subsystem to acquire the stable target nominal attitude as required by the mission. The controller is exactly the same in both submodes, but the bandwidth of the tranquillization submode is typically much shorter in order to avoid the controller saturation when the initial conditions at NPM are quite higher then the ones required in fine pointing.

The fine pointing submode is entered only when the attitude and rate errors are lower than the forward transition thresholds for a proper confirmation time, which avoid actuator saturation. Moreover, if in fine pointing the controller detects large attitude or rate errors, then it switches back to the tranquilization submode.

During the fine pointing submode AOC subsystem provides fine pointing accuracy and stable performance through fine attitude sensors, low disturbance actuators, stiff structure and properly designed control logic.

Figure 4.40 lists for each submode AOCS hardware used in the NPM control loop.

The NPM uses, as attitude sensors, the STT and gyro, whose data is combined in a fine attitude estimation filter. This filter smooths the measurement noise and provides the best attitude estimate available on-board. The required pointing accuracy is also ensured by the use of a precise orbit determination (POD) algorithm which provides a high precision orbital position and velocity through the on-board processing of the GNSS receiver solutions.

The NPM uses the RW for attitude control. An inertial constant external disturbance torque produces an accumulation of *angular momentum* in the RWs. This momentum must be unloaded to prevent the RW saturation. This task is realized with the use of MGT in LEO orbits and with the RCT in GEO orbits.

Moreover, the normal pointing mode provides a steering attitude ability. The steering attitude is obtained through a sequence of three elementary rotations (2-1-3 sequence) from **ORF** to **BRF**.

NPM SubModes	MGT	RW	RCT	FSS	GYRO	MGM	GPS	STT
Tranquillization	+	x			x		+	x
Fine Pointing	+	x			x		+	x

(+) HW used only in LEO ; (-) HW used only in GEO ; (x) common HW

Fig. 4.40 NPM sub mode

The Earth pointing attitude is nominally defined with roll, pitch and yaw angle equal to zero degrees.

Ground is allowed to enable/disable the Steering attitude. When the steering attitude is disabled the S/C's attitude is the Earth pointing attitude.

It is also possible to define different nominal reference attitudes depending on the specific mission (i.e. inertial reference frame, geodedic or geocentric reference frame, …).

The NPM is designed to be robust to star sensor blinding condition (Sun, Moon, Earth) providing fine attitude control even during such events.

4.4.3.1 Fine Attitude Function

The fine attitude function is part of the attitude determination function shown in Sect. 4.3.5. This function is designed to continuously provide the NPM and OCM controllers with both attitude/rate estimate and attitude/rate reference.

The fine attitude function includes an attitude estimation filter, a smoothing function and a reference attitude generator function.

A possible implementation of an *attitude estimation filter* is described in Sect. 4.3.5.2.

4.4.3.2 Attitude Smoothing Function

The *attitude smoothing function* is in charge of ensuring a smooth transaction after a temporary unavailability of STT measurement using a dedicated weighting function to minimize the transient condition.

Given a quaternion $q_1(t)$ and a quaternion $q_2(t)$, where $q_2(t) = q_1(t) \otimes q_2(t)$, and given a predefined time period T, the smoothing function is in charge of calculating a third quaternion that verifies the following conditions:

$$\begin{cases} q_3 = q_1 & t = 0 \\ q_3 = q_2 & t = T \end{cases} \tag{4.52}$$

Providing at the same time a smooth transition from $q_1(t)$ to $q_2(t)$.

4.4.3.3 Attitude Reference Generator Function

The *attitude reference generator function* provides the NPM mode controller with reference attitude, rate and acceleration for fine pointing attitude control and S/C maneuvering (see Fig. 4.41).

S/C maneuvering is always performed using a feed forward torque compensation algorithm. The torque control law consists of two parts: the commanded feed forward torque and the angular and rate feedback.

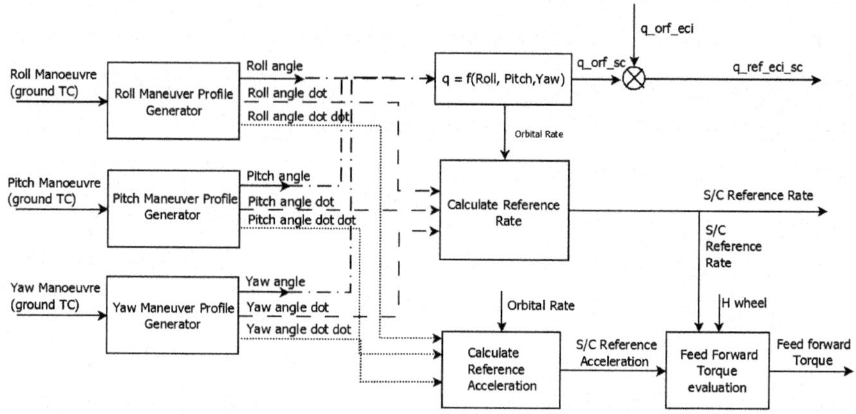

Fig. 4.41 NPM reference attitude generator

The maneuver function calculates the desired angular acceleration, rate and angular position as functions of time. The last two functions are used to feed the linear feedback system. The feed forward torque is calculated starting from the desired angular acceleration and desired angular rate.

The ground station is normally in charge of uploading, for any desired maneuver, a suitable data set which is used by the on-board software to cyclically generate, using a predefined interpolation algorithm, the following desired angular position, rate and acceleration as functions of time:

$$\begin{cases} \varphi(t), \dot{\varphi}(t), \ddot{\varphi}(t) \\ \theta(t), \dot{\theta}(t), \ddot{\theta}(t) \\ \psi(t), \dot{\psi}(t), \ddot{\psi}(t) \end{cases} \tag{4.53}$$

using, for example, the following interpolation formulas:

$$f(t) = \sum_{i=0}^{5} b_i t^i + \sum_{j=1}^{5} s_j \sin(\omega_j^s t) + \sum_{k=1}^{5} c_k \cos(\omega_k^c t) \tag{4.54}$$

$$\dot{f}(t) = \sum_{i=0}^{5} i b_i t^{i-1} + \sum_{j=1}^{5} s_j \omega_j^s \cos(\omega_j^s t) - \sum_{k=1}^{5} c_k \omega_k^c \sin(\omega_k^c t) \tag{4.55}$$

$$\ddot{f}(t) = \sum_{i=0}^{5} i(i-1) b_i t^{i-2} - \sum_{j=1}^{5} s_j {\omega_j^s}^2 \sin(\omega_j^s t) - \sum_{k=1}^{5} c_k {\omega_k^c}^2 \cos(\omega_k^c t) \tag{4.56}$$

The data set which is uploaded for each maneuver and for each angle will be:

$$\begin{array}{ll} b_i & \text{Polinomial coefficients from the 0-th to the 5-th order} \\ s_j & \text{Sine amplitudes} \\ \omega_j^s & \text{Sine angular frequencies} \\ c_k & \text{Cosine amplitudes} \\ \omega_k^c & \text{Cosine angular frequencies} \end{array} \tag{4.57}$$

Then the on-board software can calculate the reference angular rate and its derivatives using the kinematic equation presented in Sect. 2.3.2.1:

$$\omega = \mu(\varphi, \theta, \psi) \begin{pmatrix} \dot{\varphi} \\ \dot{\theta} \\ \dot{\psi} \end{pmatrix} - M_{213}(\varphi, \theta, \psi)\omega_0 Y_{ORF} \tag{4.58}$$

where $-\omega_0 Y_{ORF}$ is the orbital rate in **ORF** which is calculated together with its derivative as a function of time using a numerical routine included in the on-board orbital propagator. Following Sect. 2.3.2.1 the angular acceleration is also obtained:

$$\begin{pmatrix} \dot{\omega}_1 \\ \dot{\omega}_2 \\ \dot{\omega}_3 \end{pmatrix} = \dot{M}_{213}(\varphi, \vartheta, \psi) \begin{pmatrix} 0 \\ -\omega_0 \\ 0 \end{pmatrix} + M_{213}(\varphi, \vartheta, \psi) \begin{pmatrix} 0 \\ -\dot{\omega}_0 \\ 0 \end{pmatrix}$$
$$+ \begin{pmatrix} \ddot{\varphi}\cos(\vartheta) - \dot{\varphi}\dot{\vartheta}\sin(\vartheta) - \ddot{\psi}\cos(\varphi)\sin(\vartheta) + \dot{\psi}\dot{\varphi}\sin(\varphi)\sin(\vartheta) - \dot{\psi}\dot{\vartheta}\cos(\varphi)\cos(\vartheta) \\ \ddot{\vartheta} + \ddot{\psi}\sin(\varphi) + \dot{\psi}\dot{\varphi}\cos(\varphi) \\ \ddot{\varphi}\sin(\vartheta) + \dot{\varphi}\dot{\vartheta}\cos(\vartheta) + \ddot{\psi}\cos(\varphi)\cos(\vartheta) - \dot{\psi}\dot{\varphi}\cos(\varphi)\sin(\vartheta) - \dot{\psi}\dot{\vartheta}\sin(\varphi)\cos(\vartheta) \end{pmatrix}$$

While applying the closed loop controller output torque, calculated from the attitude error with respect to the reference attitude, it is customary to apply, at the same time (open loop feed-forward), the torque requested for the maneuver to obtain a rapid response.

Setting $\omega = \omega_{REF}$ we can calculate the torque which can be applied to the satellite in order to produce the desired motion, the so-called *feed forward torque*. Generally this torque is applied with a momentum management system based on reaction wheels (RW) or CMGs. In the RW case we command the RW momentum derivative so that (see Sect. 3.7 and Eq. 3.74) the satellite follows the reference kinematics:

$$-\dot{h}_{wheel} = J\dot{\omega}_{REF} + \omega_{REF} \times J\omega_{REF} + \omega_{REF} \times h_{wheel} \tag{4.59}$$

This open loop control law will be more effective if $\omega(t = 0) \approx \omega_{REF}(t = 0)$, if the inertias used in the software are close to the actual ones and if the wheel speed measurement error is small. The attitude errors that remain after the open loop compensation will be recovered in real time through the closed loop control.

The reference attitude generator is a general purpose function that can implement a different steering law according to the mission's needs, simply by customizing the set of parameters of generic function. This structure is suitable, for example, to implement common steering laws like:

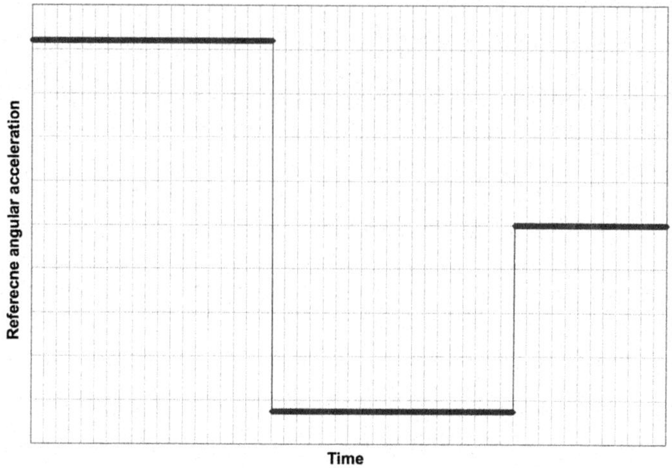

Fig. 4.42 NPM reference angular acceleration minimum time

- minimum jerk slew;
- minimum time slew.

The minimum jerk slew is used to minimize the load on actuators and avoid the actuation system to stimulate the S/C flexible modes while the minimum time slew is used to minimize the maneuver time.

As an example, hereafter we will show how the reference generator steering law can be configured to implement a minimum time slew angular acceleration profile on a single axis as shown in Fig. 4.42.

Given the acceleration function, the angular position and angular rate function can be derived, see Figs. 4.43 and 4.44:

$$\begin{cases} \alpha = a_0 + a_1 t + a_2 t^2 \\ \dot{\alpha} = a_1 + a_2 t \qquad 0 \le t \le \frac{T}{2} \\ \ddot{\alpha} = a_2 \end{cases} \tag{4.60}$$

$$\begin{cases} \beta = b_0 + b_1 t + b_2 t^2 \\ \dot{\beta} = b_1 + b_2 t \qquad \frac{T}{2} \le t \le T \\ \ddot{\beta} = b_2 \end{cases} \tag{4.61}$$

Assuming that the following conditions are verified:

$$\begin{cases} \alpha(0) = 0, \qquad \dot{\alpha}(0) = 0 \\ \beta(T) = \alpha_{max}, \quad \dot{\beta}(T) = 0 \\ \dot{\alpha}(\frac{T}{2}) = \dot{\beta}(\frac{T}{2}) \\ \ddot{\alpha} = -\ddot{\beta} = \frac{c_{max}}{J} \end{cases} \tag{4.62}$$

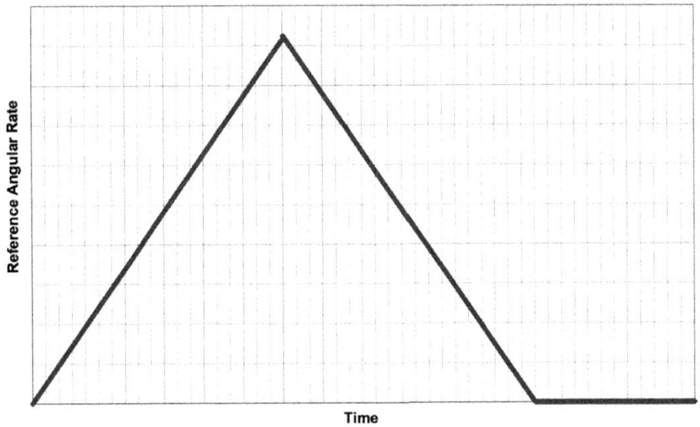

Fig. 4.43 NPM reference angular rate minimum time

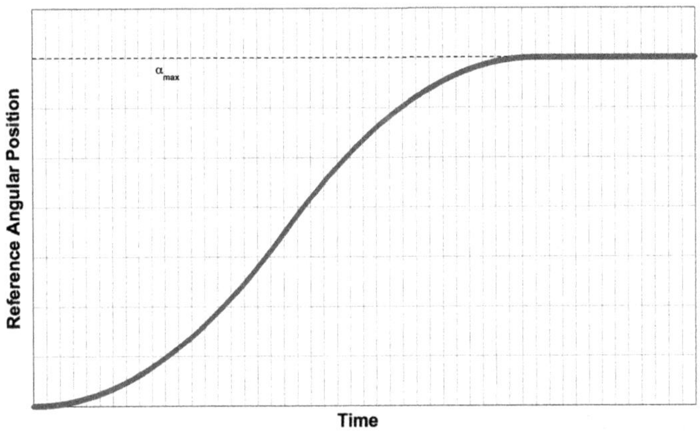

Fig. 4.44 NPM reference angular position minimum time

where J is the satellite inertia, c_{max} is the maximum torque available, α_{max} is the maximum angular position that have to be reached in the maximum time T.

The angular position coefficient can be easily calculated

$$\begin{cases} a_0 = 0, & \alpha_1 = 0, & \alpha_2 = \frac{c_{max}}{2J} \\ b_0 = \frac{c_{max}}{2J}\left(\frac{T}{2}\right)^2, & b_1 = 2(\alpha_{max} - b_0)\left(\frac{T}{2}\right)^{-1}, & b_2 = -b_1\left(\frac{T}{2}\right)^{-2} \end{cases} \tag{4.63}$$

4.4.3.4 Tranquillization/Fine Pointing

The tranquillization submode and fine pointing submode use the same hardware and functional logic (Figs. 4.40 and 4.45).

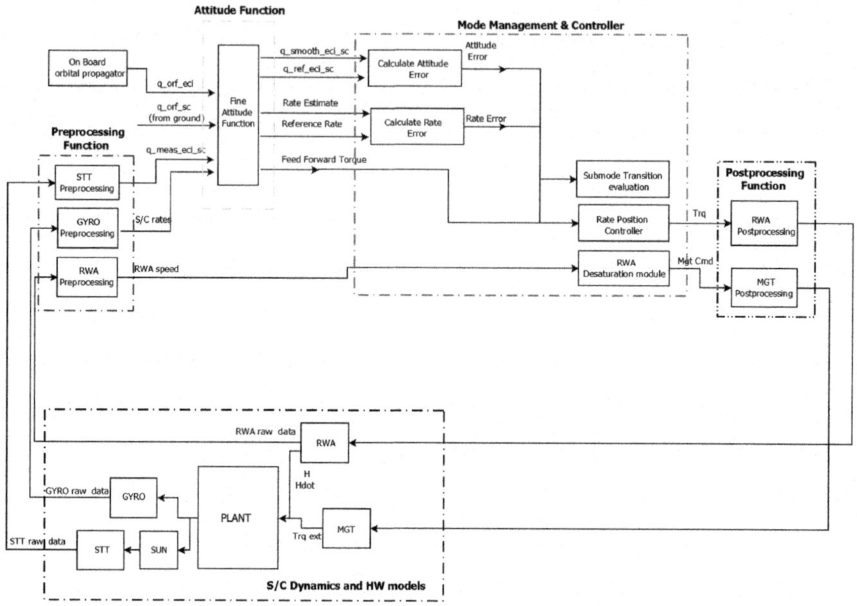

Fig. 4.45 NPM tranquillization fine pointing functional flow

The control loop gains of the two submodes are different because the NPM has to manage large initial errors and at the same time has to guarantee the fine accuracy required by the payload. To cope with these conflicting requirements, two sets of control gains have been selected: the tranquillization gains and fine pointing gains.

The first one has to guarantee the NPM convergence to the target attitude also in the presence of large initial errors maintaining the RWs in the linear torque range (no torque saturation), the steady state error of this loop is typically relaxed with respect to the required NPM performances. The second set of gains are selected to guarantee the fine attitude pointing performance starting from relatively small initial error conditions maintaining again the RWs in their linear torque range.

The fine pointing submode is entered only when attitude and rate errors are confirmed to be permanently below a given threshold, otherwise it is exited. During the transient phases the controller error normally increases, consequently the tranquillization submode is entered. When the controller operates nominally, at a steady state, the controller error remains in the fine error band and the fine pointing submode is kept.

The primary sensors used during these phases are the gyro package and the star tracker while the actuators used are the RWs. The magnetic torque rods are used only for momentum unloading.

4.4.3.5 NPM Controller Design and Stability

The NPM controller is detailed in Figs. 4.46 and 4.47. The design methods here presented will be SISO, they are applicable as already said when there is much coupling between the control axes.

Frequency methods, as already shown in the previous Sect. 4.4.1.9, are based on a combination of PID controller plus a phase-lead/phase-lag compensator and filtering, shaping the open loop gain and phase bode diagram of the overall transfer function. With reference to Fig. 4.47, calling T_{in} the input of the phase lead compensator and T_{out} the output of the structural filter—described in Fig. 4.46 by the cascade of two second order filters in space state representation $\mathbf{A_f}$, $\mathbf{B_f}$, $\mathbf{C_f}$, $\mathbf{D_f}$—the shaping is performed with the transfer function $H(s)$, such that:

$$T_{out} = H(s)T_{in} \tag{4.64}$$

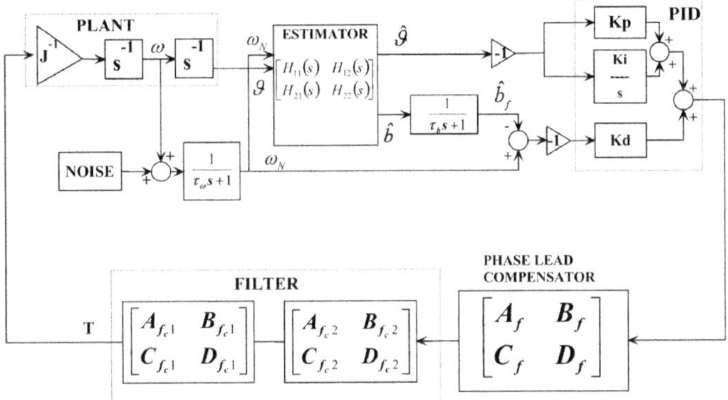

Fig. 4.46 NM single axis model

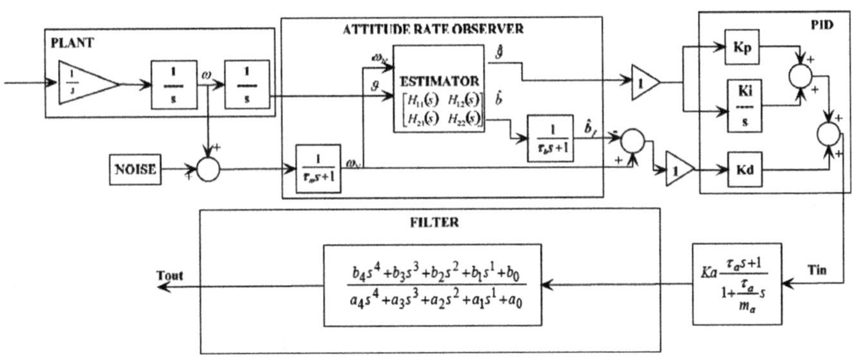

Fig. 4.47 NPM open loop system transfer function

The design goal is to guarantee at least ≈6 db gain margin and 30° phase margin for each body axis. The NPM controller is composed of a proportional-integral-derivative controller (PID controller), a first-order phase-lead compensator and a structural filter used to filter the flexible appendages (e.g. solar array).

All the considerations made on the plant structure in Sect. 4.4.1.9 are applicable also for the plant that has to be stabilized by NPM.

The PID controller attempts to correct the error between the attitude and rate measurement and the related references, by calculating and then exerting a torque corrective action. The PID gain has been tuned in order to provide an adequate system bandwidth to ensure good error responsiveness to fulfill the attitude pointing stability requirement. Very often, the choice to introduce an integral action is made in order to have a nominally zero error in steady state conditions in the presence of a constant disturbance torque.

The structural filter is used to filter the frequencies of the solar arrays. In this particular design where the desired bandwidth of the control loop is close to the solar array frequency (less than one decade), the design of the filter is more complicated than a classic second order filter. In general the order of the proposed filter is higher than 2nd and it does not only have poles but also zeros. In this specific case, the selected filter is a 4th order filter.

Different filter techniques can be used to achieve the desired performance in the filtering. The filter has to be synthesized with a proper selection of poles and zeros that allow it to achieve a strong amplitude reduction (more than with a simple second order filter) at a desired frequency with very low amplitude gain reduction before that frequency. The associated drawback is a strong phase reduction around the frequency bandwidth. For this reason, it is necessary to introduce a phase-lead compensator to properly manage the phase around the control bandwidth in order to achieve the desired phase margin. The typical design filter techniques are based on elliptic filters theory.

Attitude Filter Transfer Function

The attitude estimation filter, combining the gyro and the star tracker measurement (respectively ω and θ), smooths the noisy measurements and provides the best gyro drift $\hat{b}$ as well as attitude estimate $\hat{\theta}$.

The attitude filter pass-band has been chosen in order to ensure the required pointing performance at a steady state.

Phase Lead Compensator

The phase-lead compensator transfer function can be written as follows

$$H_1(s) = K_a \frac{1 + s\tau_a}{1 + s\frac{\tau_a}{m_a}} \tag{4.65}$$

The K_a, τ_a, m_a parameters have to minimize the gain increase while ensuring a satisfactory system phase margin.

Filter Synthesis

All the identified parameters have to be selected according to a design process able to guarantee the desired performance. In particular, the low pass filter is a fourth order one introduced to filter out the flexible appendages (i.e. solar arrays). Different techniques can be used to design it, mainly related to the level of amplitude reduction and phase shaping needed at a predefined frequency. The order of this low pass filter is higher than that of the one used in the SAM-EM controller due to the fact that the desired control bandwidth is close to the flexible mode frequency, and the loop shaping must be more accurate.

In Fig. 4.48, it is possible to see a strong gain reduction (up to 27 db) at the peak of the first frequency of the solar array that guarantees a good robustness with respect to undesired flexibility effects like the spill-over that can induce loop instability.

The selection of a PID controller has been proposed to have a good steady state performance also in presence of constant or quasi-constant disturbance torques. This solution, applied to a plant that at low frequency can be considered a double integrator, requires some design attentions.

In fact, this control structure realizes a control loop chain with a good phase margin (>30 deg) but introduces a zero-dB gain crossing at low frequency (i.e. *system with conditioned stability*). In this case, a possible gain reduction in the control actuation can provide instability.

The loop design presented in Fig. 4.48 has been optimized and shows a gain margin at low frequency better than 14 db. This gain margin implies a robustness to saturation five times over the maximum torque capability of the actuator, which is normally considered sufficient.

The normal pointing mode control loop can also be designed with alternative techniques to achieve performance and stability when there is a strong coupling between the different axes (MIMO synthesis): in this case the linear control theory and linearized equations (see Sect. 3.11) can be used. For this kind of control, the linear methods are extremely powerful (see Sects. 5.4.3 and 5.5).

4.4.3.6 NPM Pointing Budget

The NPM pointing budget is carried out with the intent to evaluate the pointing attitude error during the normal pointing mode.

The NPM pointing accuracy is the deviation of the actual pointing vector from the desired pointing vector at any instant. The desired pointing vector typically refers to a payload antenna or a telescope line of sight.

The driving issues of AOCS pointing accuracy budget in this mode are the STT and gyro measurements errors and the coupling of these errors when combined in the estimation filter; moreover, the controller errors impact the pointing accuracy.

The pointing accuracy budget reports the 3σ value of the Attitude Performance Error (APE) that is defined as a difference between the commanded attitude and the actual attitude in a specified reference frame (see [16]).

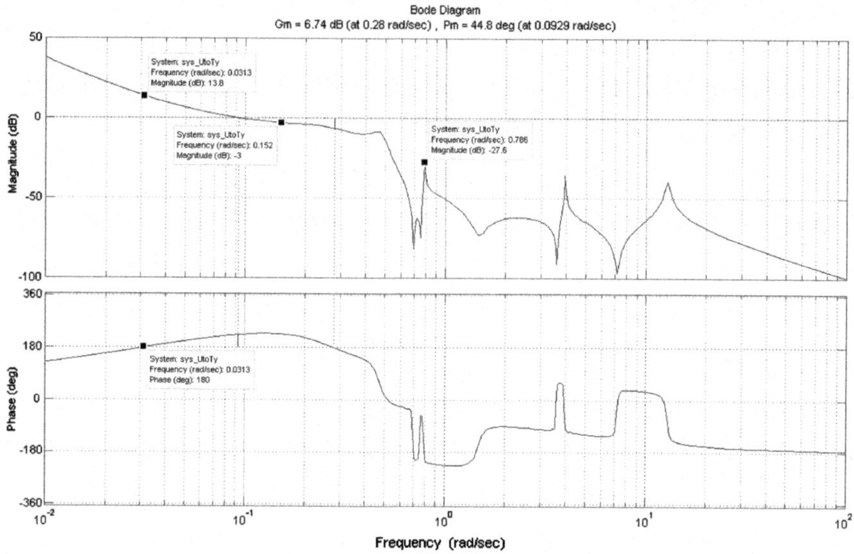

Fig. 4.48 Bode open loop chain—NPM control loop

The issues that will be dealt with in this section can also be applied to all other control loops with specific adaptations.

In order to get to the pointing error, the different error sources must be analyzed carefully.

According to [16, 17], it is possible to follow a step-by-step engineering process to obtain an attitude control pointing errors budget. The process starts with an unambiguous formulation of pointing error requirements and leads step-by-step to the evaluation of the system pointing error.

A complete and exhaustive approach can be found in [16, 17] that are used by system engineers as guidelines in all European Space Agency projects.

Usually three pointing error categories can be identified for each satellite's **BRF** axis represented by $i = 1, 2, 3$, the error sources are identified with the index j:

Bias and Drift

The bias error category includes all error sources that can be considered constant during the spacecraft's life time, and it typically includes errors like misalignment. The drift category includes all errors that vary at low rates during the mission, and errors related to aging effects are typically part of this category. Biases and drifts are summed quadratically according to [16] providing that there are no correlated errors:

$$b_i = \sqrt{3 \sum_j B_{ij}^2} \tag{4.66}$$

It is important to underline that the B_j values have to be intended as the worst case, or maximum absolute, values for each error contribution, while b_i is a statistical 3σ effective contribution of the bias class.

Harmonic

This category contains all error sources that vary with a frequency that is in the order of an orbital period. This category typically contains effects related to thermal variation during the orbit.

Harmonic errors are summed quadratically according to [16] providing that there are no correlated errors:

$$h_i = \sqrt{\frac{9}{2}\sum_j H_{ij}^2} \tag{4.67}$$

It is important to underline that the H_j values have to be intended as the worst case, or maximum absolute, values for each error contribution, while h_i is a statistical 3σ effective contribution of the harmonic class.

Random

This category contains error sources that are characterized by a high frequency behavior, typically sensor noises and actuator torque noise and vibrations are part of this category.

Random errors are summed quadratically according to [16] providing that they are all uncorrelated errors:

$$r_i = \sqrt{9\sum_j \sigma_{ij}^2} \tag{4.68}$$

It is important to underline that the σ_j values have to be intended $1\,\sigma$ of the j error source.

In the total pointing budget all errors of different categories are summed quadratically:

$$\sqrt{b_i^2 + h_i^2 + r_i^2} \tag{4.69}$$

The pointing accuracy error sources of the normal pointing mode budget can be grouped into three error categories: attitude measurement errors, control errors and targeting errors.

Attitude Measurement Errors

The attitude measurement error must be calculated considering the fusion of STTs and gyro data. The AOC design foresees the implementation of the attitude estimation filter which combines the measurement provided by the attitude sensors (STTs and gyro), thereby allowing a reduction of the STT measurement noise and the gyro drift effects.

The error sources that affect attitude measurement are presented hereafter:

- STT measurement noise and gyro measurement noise (ARW and readout noise, random),
- STT relativistic aberration, star catalog error, STT optical distortion, STT focal length error (low frequency, harmonic),
- STT focal length variation temperature (orbital frequency, harmonic),
- STT focal length variation temperature (aging, drift),
- STT internal and external misalignment (bias),
- STT ground to orbit bore-sight stability (bias),
- gyro rate bias (bias) and rate bias thermal stability (aging).

Control Errors

This term is obtained through a dynamic simulation of the S/C attitude during one orbit. The possible main sources of disturbance torques are solar disturbance, gravity gradient, MGT actuation, solar array disturbance induced by movement of the solar array to track the Sun, RWA micro-vibration induced by static and dynamic RWA imbalance, RWA torque ripple effect and dynamic disturbances due to P/L re-orientations.

Targeting Errors

The dominant contribution to this term comes from the error in the knowledge of the exact spacecraft position and time along its orbit.

When all error sources have been evaluated and classified according to the rules defined above, the overall pointing budget for NPM can be calculated.

The same approach can be extended to all other AOC modes.

4.4.3.7 NPM Simulation

This section shows a simple simulation result for NPM as per the design solutions presented in the previous sections. In particular, the simulation results are related to a S/C in fully deployed condition (i.e. all appendages deployed) during a Summer Solstice on a SSO orbit.

The spacecraft relevant data (spacecraft inertia tensor, solar array frequencies, damping, rotational and translational participation factors) used for the simulations are shown in Figs. 4.17, 4.18, 4.19, 4.20 and 4.21.

The control gains structure is shown in Fig. 4.47. The selected values are shown in Fig. 4.49.

Fig. 4.49 NPM controller parameters

Controller Parameters	Roll	Pitch	Yaw
K_P [N*m/rad]	21.6537	119.3402	66.2799
K_D [N*m*sec/rad]	274.2259	1679.2799	900.9856
K_I [N*m/(rad*sec)]	0.3047	1.6793	0.901
τ_m [sec]	1	1	1
τ_b [sec]	600	600	600
b4	0.009846	0.009917	0.00555
b3	0.001774	0.001448	0.001079
b2	0.03182	0.02632	0.01883
b1	0.0008833	0.0006341	0.0005321
b0	0.01442	0.01035	0.00864
a4	1.0	1.0	1.0
a3	0.432	0.2803	0.4076
a2	0.3203	0.3021	0.2512
a1	0.07824	0.05353	0.05593
a0	0.01618	0.01462	0.009473
K_a	1	1	1
τ_a [sec]	7.0273	4.17	7.0273
m_a	10	24.9701	10

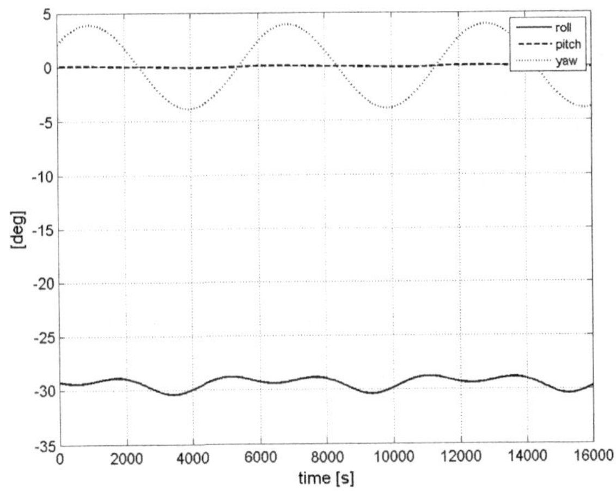

Fig. 4.50 NPM attitude angles in nominal right looking attitude with roll/yaw steering laws

Figure 4.50 shows the behavior of NPM at steady state when the satellite is in nominal attitude (i.e. Right Looking (RL) attitude—about −30 deg in roll axis). It is possible to see that a yaw steering maneuver has been superimposed over the Right

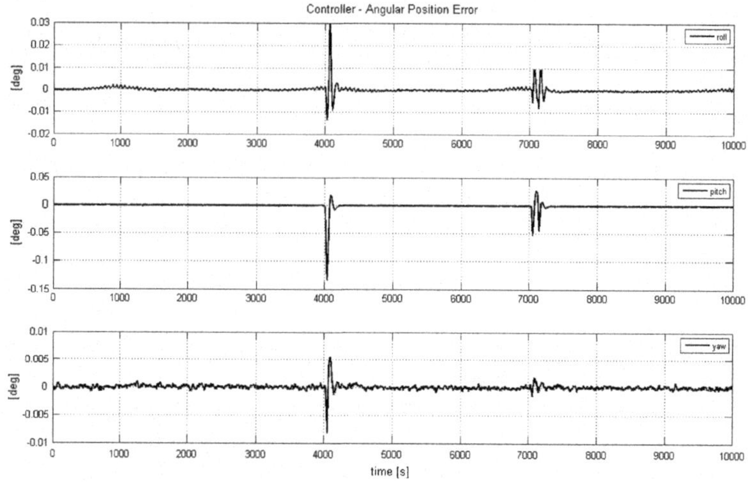

Fig. 4.51 NPM S/C attitude angle errors

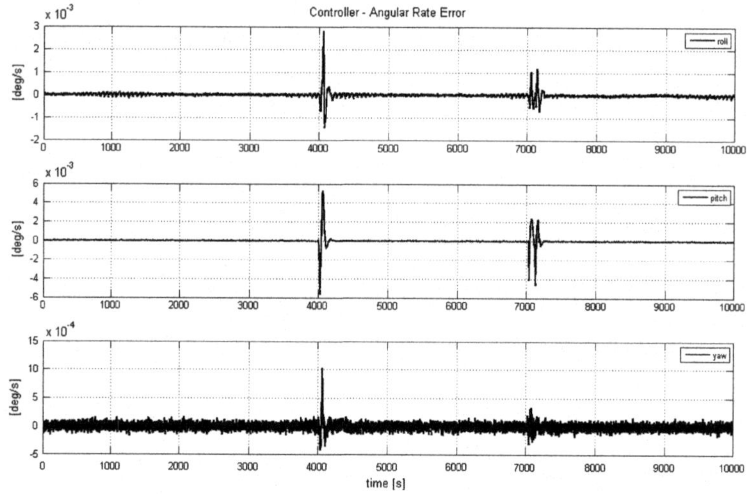

Fig. 4.52 NPM S/C angular rate errors

Looking attitude in order to support a payload need (e.g. doppler compensation). The other two Figs. 4.51 and 4.52 are related to the attitude angle errors and rate angle errors. In these two figures it is possible to see several spikes in attitude angles and angular rates due to disturbance torque induced by fast payload repointings.

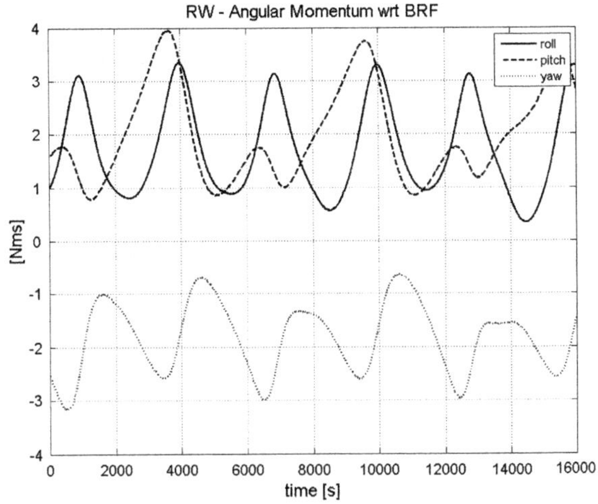

Fig. 4.53 NPM RWA angular momentum in body axes

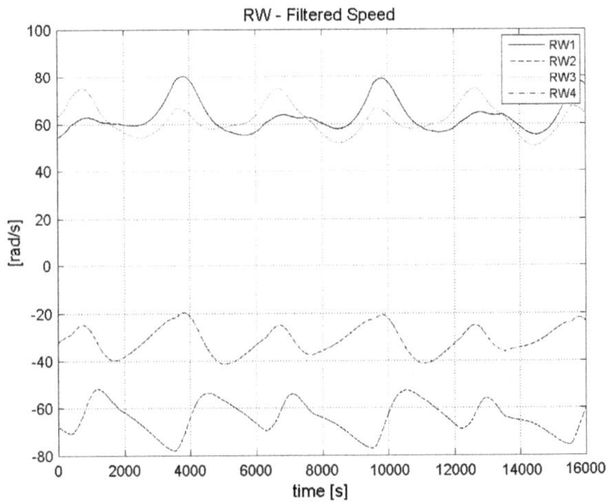

Fig. 4.54 NPM RW velocity in RW axes

Finally the RWA angular momentum in body axes, RWA velocity in wheel axes and MGT commanded momentum dipole are shown in Figs. 4.53, 4.54 and 4.55.

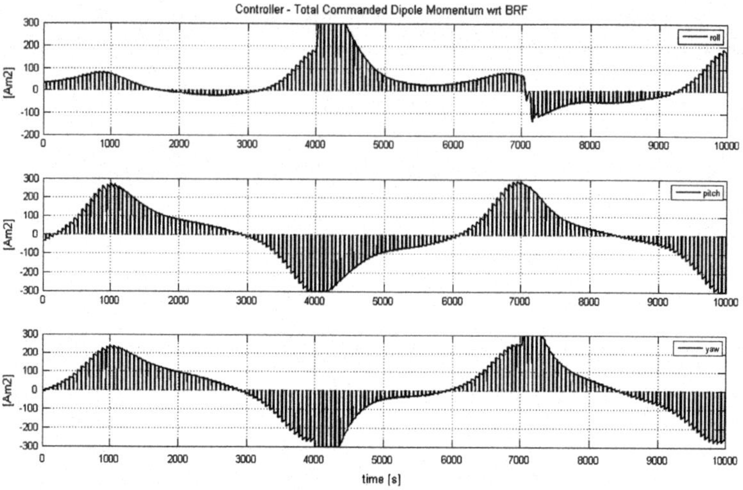

Fig. 4.55 NPM MGT commanded dipole momentum in body axes

4.4.4 Orbit Control Mode (OCM)

The OCM is used to:

- acquire the nominal orbit after launch;
- re-phase the spacecraft position on the orbit;
- correct the effect of the orbital perturbation on the desired orbital parameters;
- implement the de-orbiting maneuver at the end of the mission, in order cause the ultimate spacecraft re-entry in the Earth atmosphere. To this end extra fuel mass is budgeted in order to guarantee the necessary ΔV.

The maximum ΔT for the duration of a maneuver following a single firing has to be limited to a specific value due to the momentum build-up that results from the unavoidable misalignment of the thrust direction with respect to the spacecraft's center of mass, variable during the mission as fuel is consumed. The attitude estimator in OCM is the same used in NPM, and in fact also the attitude controller closely traces the structure of the controller used in NPM, though specific features have been included to address sloshing disturbances and RCT disturbance torques.

The orbit control mode is composed of the following submodes:

- hold mode;
- maneuver;

Figure 4.56 lists AOCS hardware used in the OCM control loop for each submode.

Hardware and functional logic of OCM submodes are the same used in NPM submodes.

The OCM mode supports the in-plane and out-of-plane orbital correction maneuvers starting from the separation from the launch vehicle, through the entire mission

OCM SubModes	MGT	RW	RCT	FSS	GYRO	MGM	GPS	STT
Hold Mode	+	x	-		x		+	x
Manoeuvre	+	x	-		x		+	x

(+) HW used only in LEO ; (-) HW used only in GEO ; (x) common HW

Fig. 4.56 OCM sub mode

life up to the de-orbit phase. An increase in pointing errors is generally allowed during maintenance thrusting when they have a short duration.

The OCM has been designed in order to re-gain the pointing performance within 2 min from the end of the thrust.

4.4.4.1 Hold Mode

This sub-mode is automatically entered upon transition to OCM. The entry conditions are: attitude error below a selected threshold for each axis, initial rates not exceeding a predefined threshold rate.

This sub-mode will set the spacecraft attitude in accordance to the reference quaternion up-linked from ground station, reducing the attitude and angular rate errors below predefined thresholds.

The control loop gains used in this submode can be typically the same as those used in tranquillization submode of NPM because the two submodes fulfill the same function: to damp the initial conditions to an acceptable entry condition for NPM.

4.4.4.2 Maneuver

Transition from hold mode submode to maneuver submode is triggered by a ground telecommand in order to fulfill the out-of-plane or in-plane maneuver to implement the required ΔV.

The control loop gains used in maneuver submode must be tuned to lower the attitude sensitivity to the disturbance torques induced by RCT firing during the orbital maneuver phase. Such tuning must take into account the need of stability and performance robustness (see Sect. 5.5.4).

4.4.4.3 OCM Simulation

This section shows some simulation results for OCM as per the design solution shown in the previous sections. In particular, the simulation results are related to a S/C in fully deployed condition (i.e. all appendages deployed) during the Summer Solstice in a SSO orbit with two burns performed by RCTs to simulate an orbit correction.

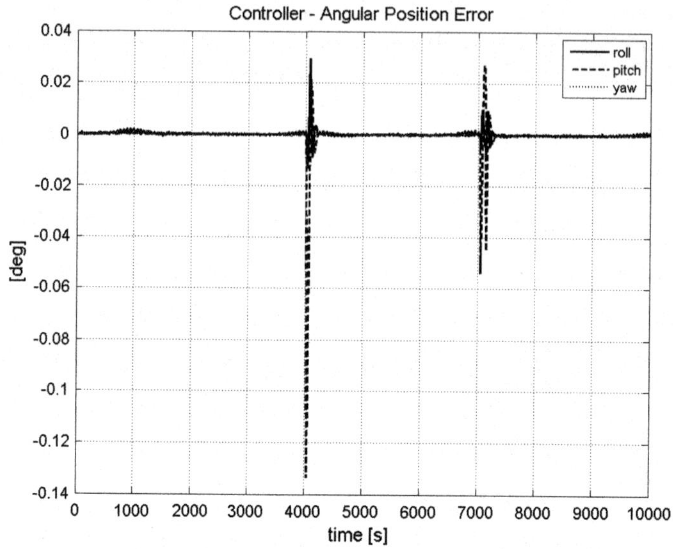

Fig. 4.57 OCM S/C attitude angle errors during RCT firings

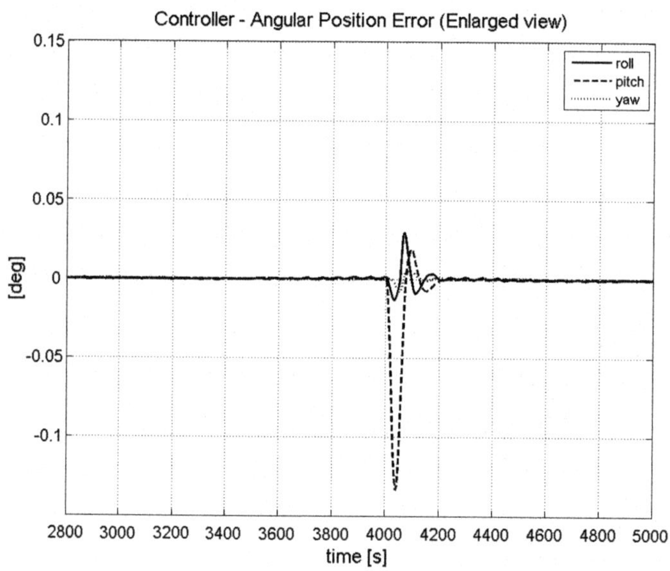

Fig. 4.58 OCM S/C attitude angle errors during RCT firing—zooming

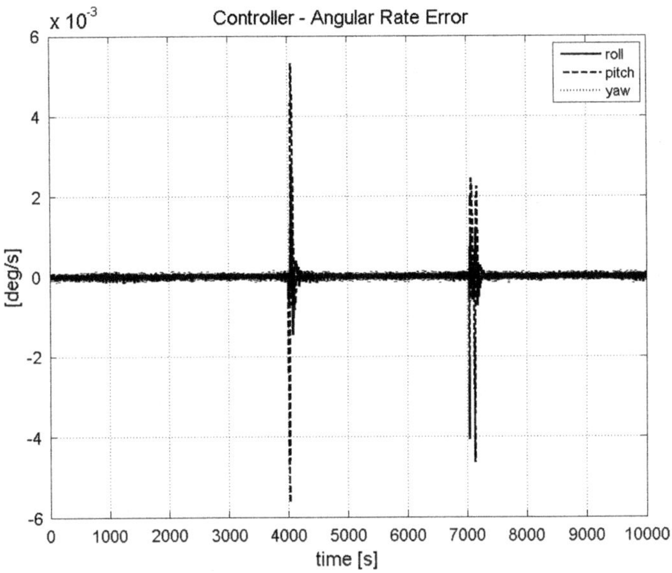

Fig. 4.59 OCM S/C angular rate errors during RCT firings

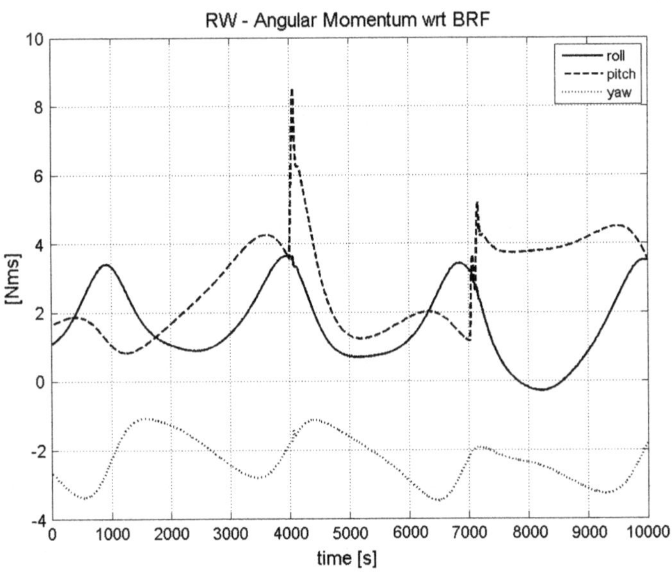

Fig. 4.60 OCM RWA angular momentum in body axes during RCT firings

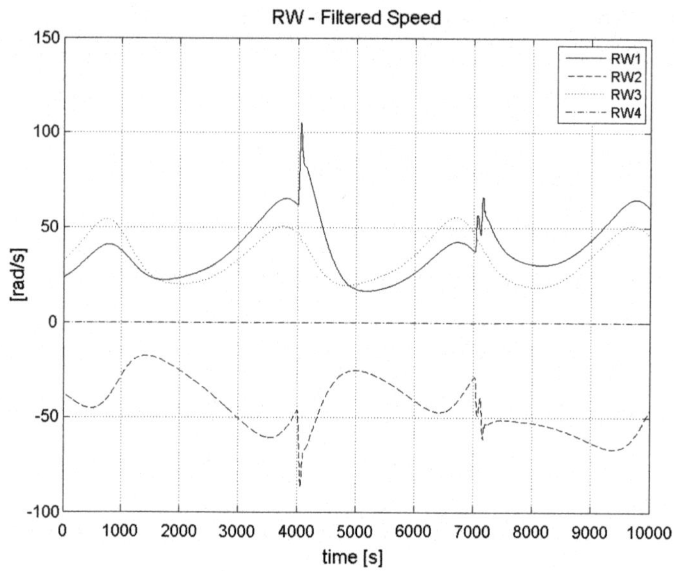

Fig. 4.61 OCM RW velocity in RW axes during RCT firings

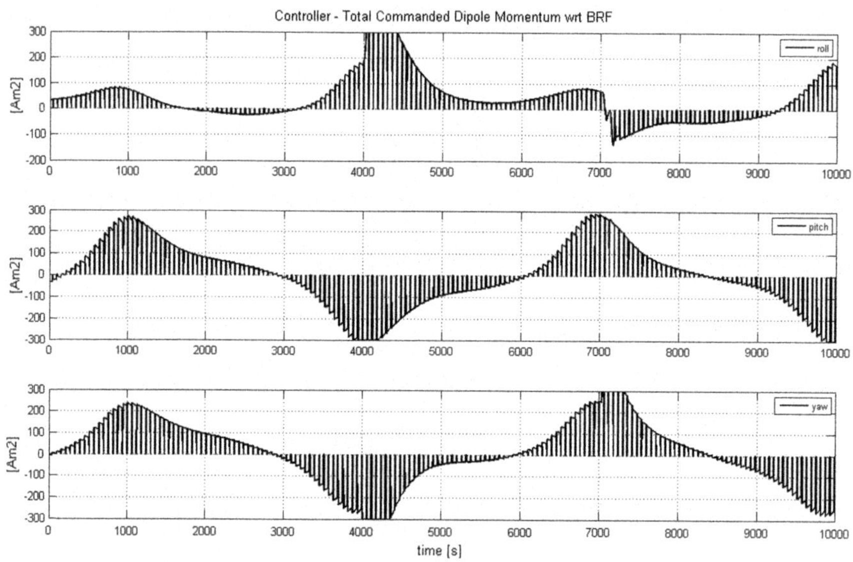

Fig. 4.62 OCM MGT commanded dipole momentum in body axes

The spacecraft relevant data (spacecraft inertia tensor, solar array frequencies, damping, rotational and translational participation factors) used for the simulations are shown in Figs. 4.17, 4.18, 4.19, 4.20 and 4.21.

The control gains used in this simulation are the same used for the NPM simulation shown in Fig. 4.49.

Figures 4.57 and 4.58 show the attitude error dynamics in OCM during a RCT firing. The second figure provide a zoom of the dynamics. Figure 4.59 shows the attitude rate errors in the same simulation.

Finally the RWA angular momentum in body axes, RWA velocity in wheel axes and MGT commanded momentum dipole are shown in Figs. 4.60, 4.61 and 4.62.

References

1. L. Mazzini, M. Surauer, in *The prospective of Integrated on-board Data System (ICDS) for Attitude Control Data Management and Platform Service: the ICDS for the ARTEMIS Satellite*, 15th International Communicatons Satellite Systems Conference, AIAA-94-1194-CP
2. B.P. Gibbs, *Advanced Kalman Filtering. Least Squares and Modeling* (Hoboken, Wiley, 2011)
3. G. Campolo, F. Cesaro, A. Intelisano, M. Pantaleoni, in *Design of the Integrated Control System, a Re-configurable Multi-Mission AOC and Data Handling Subsystem*, 6th International ESA Conference on Guidance, Navigation and Control Systems, Loutraki, Greece, 17–20 Oct 2005
4. M.D. Shuster, Three-axis attitude determination via kalman filtering of magnetometer data. J. Guidance Control Dyn. **13**(3), 506–514 (1990)
5. G. Campolo, S. Gerosa, S. Lagrasta, L. Mazzini, S. Ponzi, in *On the design of observer for attitude estimation of geosynchronous satellites*, 44th IAF, Graz, Austria, 16–22 Oct 1993
6. J.R. Wertz, *Spacecraft Attitude Determination and Control* (D. Reidel Publishing Company, Dordrecht, 2002)
7. F. Landis Markley, J.L. Crassidis, *Fundamentals of Spacecraft Attitude Determination and Control* (Springer, New York, 2015)
8. M.D. Shuster, S.D. Oh, Three-axis attitude determination from vector observations. J. Guidance Control **4**(1), 70–77 (1981)
9. M.D. Shuster, S. Tanygin, *The Many Triad Algorithms*. AAS 07-104
10. M.L. Psiaki, Extended quest attitude determination filtering. Flight mechanics symposium, NASA Goddard Space Flight Center, May 1999
11. A. Isidori, *Sistemi di Controllo*, Siderea Vol. II
12. G. Marro, *Controlli Automatici*, Zanichelli
13. B. Friedland, *Control System Design* (Mc Graw-Hill, Singapore, 1987)
14. B. Wie, C.T. Plescia, Attitude stabilization of flexible spacecraft during stationkeeping maneuvers. J. Guidance **7**(4)
15. A.C. Stickler, K.T. Alfriend, Elementary magnetic attitude control system. J. Spacecraft Rockets **13**, (1976)
16. ESSB-HB-E-003 Working Group, *ESA pointing error engineering handbook*, ESA, July 2011
17. ECSS Secretariat, *ECSS-E-ST-60-10C Space Engineering Control Performance*, ESA-ESTEC Requirements & Standards Division Noordwijk, The Netherlands, Nov 2008

Chapter 5
The Optimal Control Theory

Abstract The optimal control theory is the theoretical basis for the generation of control algorithms to reach and maintain the desired orbital and attitude reference trajectory. These algorithms need to have as input a partial or total information regarding the state of the system. In this chapter we will review the classic optimal control theory derived from the *calculus of variations* and then the so-called *Maximum Principle*. We introduce also the control theory applicable to linear systems and the study of the linear control synthesis methods, with particular focus on second order flexible mechanical systems and H_∞ techniques.

5.1 Introduction

The *optimal control theory* is the theoretical basis for the generation of control algorithms to reach and maintain the desired orbital and attitude *reference trajectory*. These algorithms need to have as input a total or partial information regarding the state—orbital, attitude and flexible when relevant—of the system.

It is a quite common design approach applied both in linear and non-linear control—we have seen examples in Chap. 4, for example see Figs. 4.46 and 4.47 to divide the control systems in one algorithm that estimates the state from the available measurements, the so-called *state observer*, and one algorithm which drives the state to the desired value, the so-called *controller*; this splitting in two different components of the control algorithm, is justified by the so-called *Separation Principle*.

After having presented the *Separation Principle*, we go on to review the estimation methods which are used to obtain an optimal state estimation from the measurements: this will be done mainly focusing on linear time invariant systems.

The algorithms defining the controller are often found by solving constrained optimum problems, for example aiming to minimize propellant, time, performance (i.e. in attitude pointing) or a combination of these. This is the subject of optimal control theory and *Maximum Principle* which is based on the full state knowledge; we give a general formulation of the optimal control problem of a non-linear system, introducing the *calculus of variation* on the extended Lagrangian and then the *Maximum Principle*.

© Springer International Publishing Switzerland 2016

L. Mazzini, *Flexible Spacecraft Dynamics, Control and Guidance*,
Springer Aerospace Technology, DOI 10.1007/978-3-319-25540-8_5

We will see how the elaboration of an optimal control requires the computation of auxiliary information called co-state. When this computation can be done by the on-board computer (OBC), a state feedback controller can be implemented, vice-versa when this computation is too complex for the on-board processing capability we will have to count on the ground station's support. When the system equations are linearized (see Sect. 3.12) the computation of the co-state becomes simpler and standard synthesis techniques can be implemented such as the LQR (linear quadratic regulator) one.

After having reviewed the LQR method we go on to a more general approach to control linear time invariant systems. We pose the optimal control problem in terms of feedback from measurements and synthesize a dynamic controller from measurements. The controllability and observability system features are necessary to design stabilizing controllers and place their dynamic poles at will. However, the performances are still subject to the effect of sensor noise, actuator saturation, sensitivity to changes of satellite parameters: a linear control synthesis technique must be able to minimize these effects on the performances. Among the various techniques that are used today, we will introduce the H_∞ robust optimal control design in a form suitable for second order plants (like the plant described in by Eqs. 3.123, 3.124).

The literature on the subject is wide, there are excellent books where the interested reader can find a full theoretical treatment with many examples, see for example the classic texts [1, 2] or other publications like [3–5] for the optimal control theory, for the linear control synthesis see [6–10], for the H_∞ robust optimal control design see for example [11–14].

In the following chapters (Chaps. 6 and 7) dedicated specifically to AOCS design problems we will apply these basic principles to synthesize adequate control laws in real design examples.

5.2 The Separation Principle

Let us consider a general non linear control problem:

$$\begin{aligned} \dot{x} &= f(x, u, t) \\ y &= h(x, u, t) \end{aligned} \tag{5.1}$$

where:

$$\begin{array}{ll} f : \mathbb{R}^{n+m+1} \rightarrow \mathbb{R}^n & \\ h : \mathbb{R}^{n+m+1} \rightarrow \mathbb{R}^r & \\ u : \mathbb{R} \rightarrow \mathbb{R}^m & \text{represents the input control vector function of time,} \\ y \in \mathbb{R}^r & \text{represents the output measurements vector, and} \\ x \in \mathbb{R}^n & \text{represents the internal state vector of the system.} \end{array} \tag{5.2}$$

$f(x, u, t)$ and $h(x, u, t)$ are typically continuous function of all their arguments and are also continuously differentiable in the variable x, $u(t)$ is a piecewise continuous function of the time.

The output feedback control problem is to find a functional F that defines the input control command at a given time for a given history of past measurements: $u_t = F(y_{T<t})$, such to bring the state to its desired value.

The question we are faced with, is if and when, we can divide the design and implementation of such *feedback controller* in two steps:

(1) a state estimator functional $\hat{x}_t = H(y_{T<t})$
(2) a controller function $u_t = F(\hat{x}_t)$

The *Separation Principle* or the principle of separation of estimation and control, states that the output feedback control problem can be broken into two separate parts, a state feedback control problem and a state estimator design problem.

The state feedback control problem is addressed using the *Maximum Principle*, while the theory of the state estimators for linear time invariant (LTI) systems, is briefly treated in Sect. 5.3. For the general theory of state estimators we refer to the bibliography on Kalman filters (see for example [15] and for non linear systems see in [16] the theory of extended Kalman filters).

The Separation Principle can be used for example in the presence of flexible modes. In this case, usually the on-board S/W may have no direct measurement of the elastic deformation. A possible procedure is to design an optimal robust observer in order to estimate the flexible modes from the rigid body rotations measured by the on board attitude and rate sensors and to feed with this state estimate an optimal state controller.

The Separation Principle suggests designing general purpose estimators which can be used in many different satellite operating modes. These general purpose estimators can be considered, in the state controller design synthesis, in a very simplified way, as providing the full state information in a given low pass bandwidth with low static error and stochastic noise. The overall stability will be guaranteed by the Separation Principle.

It is outside the scope of this book to demonstrate the theorems which justify the Separation Principle. These theorems are important because they help in understanding the structure that a suitable feedback controller should have; broadly speaking there are two classes of theorems which present the *Separation Principle*:

- theorems that state the stability of the global system (plant + estimator + controller) on the basis of the stability of each one of the two parts: the plant + estimator and the controller + plant. These theorems can be extended to many non linear cases (see for example [17, 18] for the quantized measurement case, and [19] which shows that under actuation saturation the Separation Principle is not applicable).
- theorems that consider the output feedback controller problem for a linear system with stochastic Gaussian white noise inputs, such as noise on the measurements or any external disturbance entering in the state equation. These theorems state the conditions under which stability can be provided and how the two parts (estimator +

controller) which constitute the output feedback optimal controller that minimizes a stochastic quadratic function of the control, are calculated.
This problem generates the so called linear quadratic Gaussian (LQG) synthesis technique. In this context, the optimal state feedback controller is called linear quadratic regulator (LQR) and is discussed later in this chapter, and the optimal state estimator also called optimal observer or Kalman filter, is briefly described in Sect. 5.3. This method introduced by Kalman (see [15]) is presented in all control theory text-books like [7, 8] and in a wide paper literature which is well reviewed in [20].

The LQG output feedback controller minimizes the expectation of a quadratic function of the state and the control and is realized by a linear function $u = f(\hat{x})$ where $\hat{x} = E(x|y(\tau < t))$ is the expected state given the history of all the past measurements. The real time calculus of $\hat{x}$ is well established in linear system theory as shown in Sect. 5.3 (Kalman filter). Having solved the state estimation part, we need to solve the optimal control problem, that is to determine the best $u = f(\hat{x})$, the optimal state feedback law.

We will now directly prove an easy theorem of the class (1) for linear time invariant (LTI) systems.

Proposition *Consider the LTI system:*

$$\begin{aligned} \dot{x} &= Ax + Bu \\ y &= Cx \end{aligned} \tag{5.3}$$

where $A \in R^{n \times n}$, $B \in R^{n \times m}$, $C \in R^{r \times n}$. We consider a state observer of the standard form (see Sect. 5.3, Eq. 5.13):

$$\begin{aligned} \dot{\hat{x}} &= A\hat{x} + Bu + L(y - C\hat{x}), \ \textit{or} \\ \dot{\hat{x}} &= (A - LC)\hat{x} + Bu + Ly \end{aligned} \tag{5.4}$$

and a state feedback:

$$u = -K\hat{x} \tag{5.5}$$

The closed loop system eigenvalues, generated putting together the system defined by Eq. 5.3, with the input control defined by Eqs. 5.4 and 5.5, are those of the state observer matrix $A - LC$ and those of the controller matrix $A - BK$ as if the control was applied directly using the full state information; therefore the closed loop system is stable if the observer and the controller matrix are each one individually stable.

In order to demonstrate this proposition it is better to transform the closed loop state from $(x, \hat{x})$ to (x, e) where the error e is defined as:

$$e = x - \hat{x}$$

Then

$$\dot{e} = (A - LC)e$$
$$u = -K(x - e)$$

Now we can write the closed-loop dynamics as:

$$\begin{bmatrix} \dot{x} \\ \dot{e} \end{bmatrix} = \begin{bmatrix} A - BK & BK \\ 0 & A - LC \end{bmatrix}$$

Since this dynamic matrix is triangular, its eigenvalues are those of $A - BK$ together with those of $A - LC$. Thus the closed loop system eigenvalues are those of the observer and the controller as designed independently, and the closed loop system is stable if the two systems are each individually stable.

A typical presentation of class (2) theorems can be found in [20], which provides also a Separation Principle for time delayed systems. We give here a very synthetic format of a typical result for the class (2) theorems following the cited paper.

Let us consider a linear system:

$$\begin{aligned} \dot{x} &= A(t)x + B_1(t)u + B_2(t)w(t) \\ y &= C(t)x + D(t)w(t) \end{aligned} \tag{5.6}$$

Being $w(t)$ a vector of Gaussian white noise signals. The output feedback control problem can be stated as follows:

Problem Find a non-anticipatory map $y_\tau \rightarrow u_t$ with $\tau \in [0, t]$ and $t \in [0, T]$, such that when applied to the system defined by Eq. 5.6, the resulting closed loop system is stabilized and the value of the functional:

$$J(u) = E[\int_T^0 (x^T Qx + u^T Ru)dt + x^T(0)Sx(0)] \tag{5.7}$$

reaches its minimum ($E[\cdot]$ is the expectation operator defined in Sect. 5.3, and Q, R, S are symmetric positive definite matrices of suitable dimensions).

The answer to this problem is that the functional attains its minimum with the following linear controller:

$$\begin{aligned} u &= -K(t)\hat{x} \\ \hat{x} &= E(x|y(s < t)) \\ \dot{\hat{x}} &= A(t)\hat{x} + B_1(t)u + L(t)(y - C(t)\hat{x}) \end{aligned} \tag{5.8}$$

where the two matrix gains $K(t)$ and $L(t)$ are derived by the solution of two Riccati equations (see [20]).

Using the *Separation Principle* we can study the state feedback optimal control problem provided that a state estimator with the following properties is available:

- stable with the plant;
- providing enough large information bandwidth in order not to affect the final performance in terms of disturbance rejection;
- having enough low estimation error not to affect the error budget performance;
- having enough low noise in the estimation error to avoid the controller saturation, which may affect performance as well as stability;
- having enough robustness to unknown parameters (such as the flexible mode characterization) in order to avoid unstable or poorly performing systems.

5.3 Elements of Estimation Theory

The estimation theory is a suite of methods for the state estimation that constitutes a basic tool for the synthesis of attitude and orbit controllers. In fact from the optimal control theory we establish algorithms to control the state that are based on its complete knowledge. In order to support these algorithms it is necessary to have a suitable estimation of the state from the measurements.

The theory of estimation is a very wide field, mostly based on probability theory and signal theory. We will not treat it in detail and we suggest the reader to consult some texts of the huge specialized literature (see for example [15, 16] for a treatment dealing also with many space applications).

The estimation methods are based on two ingredients: a time window of measurements and a model of the dynamics of the system. Given a probabilistic model of the measurement error and of the dynamic model errors, the estimation algorithms provide a state estimate that has the minimal error covariance given the available information.

Deterministic methods exist in which the goal is to characterize *all* the states compatible with the measurements and the dynamic model, these states which form the so called *membership set*.

These last methods are not as used as they should be. The estimation methods can be divided into two branches: the batch estimators like the *least squares* that estimate the state inside the measurements' time window, and the real time estimators where the state is estimated at the end of the measurements' time window. The real time estimators are based principally on the Kalman filter theory and its extension to non linear models.

In this section we will recall the definitions and the results useful to support the understanding of the orbital and attitude state estimation algorithms which have been addressed in this book. We will follow the reference [7].

Let us consider v and w Gaussian zero-mean white-noise process whose mean and covariance matrix are given by

$$\begin{cases} E[w(t)] = 0 \\ E[w(t)w^T(t')] = Q(t)\delta(t-t') \end{cases} \tag{5.9}$$

where $E[w]$ indicates the expectation (i.e. *first-order* statistics), $E[ww^T]$ the covariance matrix (i.e. a *second order* statistic), Q *power spectral density* (PSD) matrix and δ defined as Dirac function (i.e. $\int_{-\infty}^{\infty} \delta(t)dt = 1$).

The symbol $E[\cdot]$ denotes the mathematical expectation that is, the *ensemble* average computed from the probabilistic model of the argument. We define the mean of a scalar random signal $x(t)$

$$E[x(t)] = \int_{-\infty}^{\infty} xpdf[x, t]dx \tag{5.10}$$

where $pdf[x, t]$ represents a probability density function, $pdf[x, t]dx$ gives the probability of finding $x \in [x - dx/2, x + dx/2]$ at some time t.

Moreover, it is also possible to define the correlation function as

$$E[x(t)x(\tau)] = \int_{-\infty}^{\infty}\int_{-\infty}^{\infty} x_1 x_2 pdf[x_1, x_2; t, \tau]dx_1 dx_2 \tag{5.11}$$

where $x_1 = x(t)$, $x_2 = x(\tau)$. Both *first-order* and *second-order* statistics can be extended in a straightforward way to vector random processes.

If we introduce a vector of processes

$$x(t) = \begin{bmatrix} x_1(t) \\ \dots \\ x_n(t) \end{bmatrix}$$

we define the expectation of the state as:

$$E[x(t)] = \begin{bmatrix} E[x_1(t)] \\ \dots \\ E[x_n(t)] \end{bmatrix}$$

and the state correlation matrix as:

$$E[x(t)x^T(\tau)] = \begin{bmatrix} E[x_1(t)x_1(\tau)] & \dots & E[x_1(t)x_n(\tau)] \\ \dots & \dots & \dots \\ E[x_n(t)x_1(\tau)] & \dots & E[x_n(t)x_n(\tau)] \end{bmatrix}$$

Let us consider a dynamic system in a usual linear state representation:

$$\begin{cases} \dot{x} = Ax + Bu + Fv \\ y = Cx + w \end{cases} \tag{5.12}$$

where x is the state, u the control action, y the measurement while v and w are random signals that can be introduced to consider the noise of actuators and the noise on the sensors.

In the case that v and w are white-noise processes, having $E[v(t)v^T(t)] = V(t)\delta(t-\tau)$ and $E[w(t)w^T(t)] = W(t)\delta(t-\tau)$ with $V(t)$ and $W(t)$ defined as power spectral density (PSD) matrices, Kalman and Bucy have demonstrated that the *optimal state estimator* has a structure equal to the structure of an observer, that is (see [7]):

$$\dot{\hat{x}} = A\hat{x} + Bu + \hat{K}(y - C\hat{x}) \tag{5.13}$$

According to this approach, the dynamic equation of the estimation error $e = \hat{x} - x$ is determined by subtracting Eq. 5.12 from 5.13:

$$\dot{e} = (A - \hat{K}C)e - Fv + \hat{K}w \tag{5.14}$$

We assume that the covariance matrix $E[v(t)w^T(t)]$ between v and w is zero (i.e. v and w are uncorrelated signals). The covariance matrix of the error depends on the gain matrix K that is selected and on its initial value.

From the Kalman theory the gain matrix that provides the minimal error covariance matrix is:

$$\hat{K} = \hat{P}C^T W^{-1} \tag{5.15}$$

where P is the error covariance that is propagated with the following matrix Riccati equation:

$$\dot{\hat{P}} = A\hat{P} + \hat{P}A^T - \hat{P}C^T W^{-1} C\hat{P} + FVF^T \tag{5.16}$$

The state estimate which minimizes the covariance matrix of the error can be demonstrated to be the same as the expected state conditioned to the measurements, so that the state estimated by Eq. 5.13 with the gain defined by Eq. 5.15 and P propagated by Eq. 5.16 satisfies also the relationship $\hat{x} = E(x|y(\tau < t))$. The optimal observer is not an LTI system because of the time variation of the covariance matrix P. It is therefore usual to use the steady state solution of Eq. 5.16. It can be demonstrated that a non negative steady state solution of this Riccati equation exists if the couple of matrices (A, C) is detectable and the couple $(A, FV^{1/2})$ is stabilizable. We will give the definitions of these two properties in Sect. 5.4.3.

All the details on this topic are shown in [7] as well as in all basic textbooks in control theory. Extensions to non linear systems and discrete systems can be found in [16].

5.4 Optimal Control

We follow in this section the variational methods as presented in [2, 21].

The following problem, known as the *problem of Mayer*, deals with the research of an optimal control law $u(t)$ for a non-linear dynamical system subject to equality constraints.

Problem (*Optimal Control Problem of Mayer*) Consider the system

$$\begin{aligned} &\dot{x} = f\,(x, u, t) \\ &f \in C^1(\mathbb{R}^{n+m+1}, \mathbb{R}^n) \\ &u \in C_p^0(\mathbb{R}, \mathbb{R}^m); \end{aligned} \tag{5.17}$$

defined for $t_0 \le t \le t_f$, where t_f is not specified.

$f(x, u, t)$ is a continuously differentiable function of all its arguments, $u(t)$ is a piecewise continuous function of the time.

For this system we specify a complete set of boundary conditions

$$\begin{aligned} &\Psi\left(x_f\right) = 0, \;\; \Psi \in C^1(\mathbb{R}^n, \mathbb{R}^{n_\Psi}) \\ &x\,(0) = x_0 \end{aligned} \tag{5.18}$$

with constraints

$$g\,(x, u, t) = 0, \;\; g \in C^1(\mathbb{R}^{n+m+1}, \mathbb{R}^{n_c}) \tag{5.19}$$

and a performance index J defined as:

$$J = J\left(x_f, t_f\right) \in C^1(\mathbb{R}^{n+1}, \mathbb{R}) \tag{5.20}$$

We want to determine the optimal control law u that minimizes the performance index functional J among all admissible controls.

The preceding expression of J is quite general. An integral performance index

$$J' = \int_{t_0}^{t_f} \phi\,(x, u, t)\, dt$$

can be obtained by (5.20) by defining an additional state variable x_{n+1} such that

$$\dot{x}_{n+1} = f_{n+1}\,(x, u, t) = \phi\,(x, u, t)\,, \;\; \phi \in C^1(\mathbb{R}^{n+m+1}, \mathbb{R}).$$

The constrained minimization problem can be solved using the calculus of variation (see [21]) on the Lagrangian function defined as an *extended* performance index, which includes the constraints Eqs. 5.17–5.19 with the introduction of the Lagrange multipliers λ, μ and ν.

$$\mathscr{L} = J + \int_{t_0}^{t_f} \left[\lambda^T (\dot{x} - f(x, u, t)) + \mu^T g(x, u, t) \right] dt + \nu^T \Psi \left(x_f \right)$$

The necessary condition for the minimization of J under the given constraints is that the first variation of the Lagrangian function is null:

$$\delta \mathscr{L} = 0$$

The Lagrange multipliers λ are time-dependent variables, called *adjoint* variables or *co-state* of the system. The first variation of the Lagrangian is:

$$\begin{aligned} \delta \mathscr{L} = & \frac{\partial J}{\partial x_f} \delta x_f + \frac{\partial J}{\partial t_f} \delta t_f + \lambda^T \delta x \mid_{t_0}^{t_f} \\ & - \int_{t_0}^{t_f} \left[\dot{\lambda}^T \delta x + \lambda^T \frac{\partial f}{\partial x} \delta x + \lambda^T \frac{\partial f}{\partial u} \delta u - \mu^T \left(\frac{\partial g}{\partial x} \delta x + \frac{\partial g}{\partial u} \delta u \right) \right] dt \\ & + \nu^T \frac{\partial \Psi}{\partial x_f} \delta x_f + \int_{t_0}^{t_f} (\dot{x} - f) \, \delta \lambda dt + \int_{t_0}^{t_f} g \delta \mu dt + \Psi \left(x_f \right) \delta \nu = 0 \end{aligned} \tag{5.21}$$

where:

$$\delta x_f = \delta x \mid_{t_f} + \dot{x}_f \delta t_f \quad \Rightarrow \quad \lambda^T \delta x \mid_{t_0}^{tf} = \lambda_f^T \delta x_f - \lambda_f^T \dot{x}_f \delta t_f$$

It follows:

$$\begin{aligned} \delta \mathscr{L} = & \left(\frac{\partial J}{\partial x_f} + \lambda_f^T + \nu^T \frac{\partial \Psi}{\partial x_f} \right) \delta x_f \\ & + \left(\frac{\partial J}{\partial t_f} - \lambda_f^T \dot{x}_f \right) \delta t_f \\ & + \left(\int_{t_0}^{t_f} \left[\dot{\lambda}^T + \lambda^T \frac{\partial f}{\partial x} - \mu^T \frac{\partial g}{\partial x} \right] dt \right) \delta x \\ & + \left(\int_{t_0}^{t_f} \left[\lambda^T \frac{\partial f}{\partial u} - \mu^T \frac{\partial g}{\partial u} \right] dt \right) \delta u \\ & + \left(\int_{t_0}^{t_f} [\dot{x} - f] \, dt \right) \delta \lambda \end{aligned}$$

$$
+\left(\int\limits_{t_0}^{t_f}[g]\,dt\right)\delta\mu
$$

$$
+\left(\Psi\left(x_f\right)\right)\delta\nu
$$

$$
=0 \quad \forall\ \delta x_f,\ \delta t_f,\ \delta x,\ \delta u,\ \delta\lambda,\ \delta\mu,\ \delta\nu
$$

Proposition (Necessary conditions for the Problem of Mayer) *The necessary conditions that the solution of the problem of Mayer must satisfy are:*

- *end-point conditions:*

$$
\begin{cases} \frac{\partial J}{\partial x_f}+\lambda_f+\nu^T\frac{\partial \Psi}{\partial x_f}=0 \\ \frac{\partial J}{\partial t_f}=\lambda_f^T\dot{x}_f \end{cases}
$$

- *characteristic equations + initial conditions:*

$$
\begin{cases} \dot{x}=f\,(x,u,t) \\ \dot{\lambda}=-\lambda^T\frac{\partial f}{\partial x}+\mu^T\frac{\partial g}{\partial x} \end{cases} ;\ x(t_0)=x_0
$$

- *control optimality condition:*

$$
\lambda^T\frac{\partial f}{\partial u}-\mu^T\frac{\partial g}{\partial u}=0
$$

- *constraints:*

$$
\begin{cases} g\,(x,u,t)=0 \\ \psi\left(x_f\right)=0 \end{cases}
$$

Defining the Hamiltonian function as

$$
H=\lambda^T f-\mu^T g \tag{5.22}
$$

the characteristic equations and the optimality condition are expressed as:

$$
\begin{cases} \dot{x}=\frac{\partial H}{\partial \lambda} \\ \dot{\lambda}=-\frac{\partial H}{\partial x} \\ \frac{\partial H}{\partial u}=0 \end{cases} \tag{5.23}
$$

Equation 5.23 are the so-called Euler-Lagrange first order necessary conditions for optimality.

It is possible to derive also additional necessary conditions of the second order, the so-called Legendre-Clebsch and Weistrass necessary conditions to have a local minimum.

It is also possible to give a local description of the *neighboring optimal paths* which can be used to implement a local feedback law for non-linear systems. All these concepts are provided for example in [1, 2].

If the functions f and g do not depend on the time explicitly, a first integral can be found for the system. The Hamiltonian has the following property:

$$\frac{dH}{dt} = \frac{\partial H}{\partial t} + \frac{\partial H}{\partial x}\dot{x} + \frac{\partial H}{\partial \lambda}\dot{\lambda} = \frac{\partial H}{\partial t} + \frac{\partial H}{\partial x}\left(-\frac{\partial H}{\partial \lambda}\right) + \frac{\partial H}{\partial \lambda}\left(\frac{\partial H}{\partial x}\right) = \frac{\partial H}{\partial t} \quad (5.24)$$

so that, if $\frac{\partial f}{\partial t} = 0$ and $\frac{\partial g}{\partial t} = 0$ then the Hamiltonian itself is a first integral $\frac{dH}{dt} = 0$. In this case, being along the solutions $g = 0$ and $\dot{x} = f$, from Eq. 5.22 it follows also that $\lambda^T \dot{x} = const.$

5.4.1 The Maximum Principle

The necessary optimality conditions found via the variational approach apply to smooth constraints. The Pontryagin *Maximum Principle* (which is expressed in many different forms in the literature), is more general than this formulation because it applies also when the control is constrained to be in a generic subset: $u \in \mathbb{U} \subset \mathbb{R}^m$. The set $\mathbb{U} \subset \mathbb{R}^m$ is typically considered as a closed and limited set in $\mathbb{R}^m$ like a ball or an hypercube (see [4]) or, in more weak formulations it is considered as a measurable subset of $\mathbb{R}^m$ (see [22]). Among the wide literature see [4, 5, 22, 23] to visit the many formulations of this principle which vary in terms of requested smoothness of the functions and type of constraints.

Proposition (The Maximum Principle) *Let* $t_f \in [t_0, T]$ *and* $u(t) \in C_p{}^0([t_0, T], \mathbb{U} \subset \mathbb{R}^m)$ *the set of step-wise continuous m-vector functions with value in* $\mathbb{U} \subset \mathbb{R}^m$, *such that the minimum of the following function is attained:*

$$min_{(u,t_f)\in C_p{}^0([t_0,T],\mathbf{R}^m)\times \mathbf{R}} \left(\int_{t_0}^{t_f} l(t, x(t), u(t))dt + \phi(t_f, x(t_f)) \right)$$

with $l(t, x, u) \in C^1(\mathbb{R}^{1+n+m}, \mathbb{R})$ *and* $\phi(t, x) \in C^1(\mathbb{R}^{1+n}, \mathbb{R})$ *continuous and differentiable functions of their arguments. The minimum is attained with the following constraints satisfied:*

$$\dot{x} = f(t, x, u) \quad x(t_0) = x_0, f \in C^1(\mathbb{R}^{n+m+1}, \mathbb{R}^n)$$
$$\psi_k(t_f, x(t_f)) = 0 \; k = 1, \ldots, n_\psi, \; \Psi \in C^1(\mathbb{R}^n, \mathbb{R}^{n_\Psi})$$

Then, being:
$H(x, \lambda) = min_{u\in\mathbb{U}} l(x, u) + \lambda^T f(x, u)$, *the Hamiltonian*
(or also $H(x, \lambda) = sup_{u\in\mathbb{U}} - l(x, u) + \lambda^T f(x, u)$ *where* λ *change sign)*
the necessary conditions for $(u^*, t_f, x^*, \lambda^*)$ *to be optimal are:*

- *Maximum Principle:*

$$u^*(x, \lambda) = argmin_{u \in \mathbb{U}} l(x, u) + \lambda^T f(x, u)$$

(or also $u^*(x, \lambda) = argsup_{u \in \mathbb{U}} - l(x, u) + \lambda^T f(x, u)$*, where* λ *change sign)*

- *Euler-Lagrange equations:*

$$\dot{x} = H_\lambda,\ \dot{\lambda} = -H_x,\ t_0 \leq t \leq t_f$$

- *transversality conditions:*

$$[x - x_0]_{t_0} = 0,$$

$$\left[\lambda - \phi_x + \nu^T \psi_x\right]_{t_f} = 0,$$

$$(or\ also\ using\ the\ sup : \left[\lambda + \phi_x + \nu^T \psi_x\right]_{t_f} = 0)$$

$$\left[H + \phi_t + \nu^T \psi_t\right]_{t_f} = 0,\ if\ t_f\ is\ free$$

$$(or\ also\ using\ the\ sup : \left[H - \phi_t + \nu^T \psi_t\right]_{t_f} = 0,\ if\ t_f\ is\ free)$$

$$[\psi]_{t_f} = 0$$

The transversality conditions define a *two boundary value problem* (TBVP) for the Euler-Lagrange equations. We have 2 n differential equations with n conditions defined at the initial time and n conditions defined at the final time. The final time itself can be known or unknown, in this case we need to introduce an additional terminal condition.

We remark that the Euler-Lagrange equations are of Hamiltonian type (see Appendix C.1). For this kind of equations, techniques have been developed for the solution of dynamical systems close to the integrable ones (see Appendix D) such as the canonical transformations and the Jacobi equation.

We may wonder about the existence of a solution of the optimal control problem, the existence of the optimal control cannot be guaranteed, however, the Hamilton Jacobi Bellman (HJB) weak formulation of the optimal control problem, provides existence and regularity results of the minimum cost function (see [24]).

5.4.2 Open Loop—Closed Loop

The control action can be derived by the Maximum Principle, however this requires the knowledge of the co-state.

The co-state can be found by integrating the TBVP generated by the Euler-Lagrange differential equations and the transversality conditions, the initial co-state is chosen to *shoot* the final state (shooting algorithms).

When such integration can be done in real time, like in linear system where the co-state solution can be expressed in closed form, we can implement a closed loop control.

When the shooting is not possible in real time, for the complexity of the co-state TBVP then, either we accept a suboptimal approach, or the control action is calculated periodically—generally at the ground station—for long windows of time and we have an open loop control.

With respect to the timing between the state information and the control action we can divide the algorithms into two broad classes:

- feedback control algorithms; in which the actions of the actuators are computed on the basis of the real time state information computed from the sensors measurement contextually with the control action. This kind of control is very robust to variation of satellite and environmental parameters, it is fully elaborated automatically on-board the satellite and needs adequate processing capabilities;
- feed-forward open loop control algorithms, used in conjunction with a closed loop control. The actuation plan is elaborated before the start of the control actuation on the basis of the evaluated state. These algorithms are used to improve the performance of the closed loop control. This open loop approach is more sensitive to the parameter deviations and needs a good model of the environment and the satellite;
- very low frequency control loop algorithms. In the case of the orbit control, but also for some attitude control systems, the computation of the control strategy may be implemented in the ground station and a control plan is periodically up-linked to the satellite for the execution. In this case the delay between the ground processing and the actuation can be many tenths of seconds or weeks. This almost open loop approach is sensitive to the parameter deviations and needs a good model of the environment and the satellite. It uses a reduced on board processing capability.

A robust and autonomous AOC design mainly makes use of real time on-board computed closed loop controls. This permits to achieve a better performance at the expense of additional testing and design effort, with respect to a Ground based control that uses TLM and TLC links.

5.4.3 Linear-Quadratic Regulator (LQR)

Given the *linear* dynamical system

$$
\begin{aligned}
\dot{x} &= Ax + Bu \\
x &\in \mathbb{R}^n \\
u &\in \mathbb{R}^m
\end{aligned}
\tag{5.25}
$$

with boundary conditions:

$$\begin{aligned} x(0) &= x_0 \\ x(\infty) &= 0 \end{aligned} \tag{5.26}$$

we look for a control law such that the quadratic cost function J is minimized:

$$J(u) = \frac{1}{2} \int_t^{\infty} \left(x^T Q x + u^T R u \right) dt \tag{5.27}$$

where Q is a symmetric positive semidefinite matrix and R is symmetric positive definite (see Refs. [7, 8]).

We apply *The Maximum Principle* studied in Sect. 5.4.1, the Hamiltonian function for this linear problem is

$$H = \min_u \left(p^T (Ax + Bu) + \frac{1}{2} x^T Q x + \frac{1}{2} u^T R u \right)$$

The optimal control is

$$u_{opt} \mid B^T p + Ru = 0 \Rightarrow \quad u_{opt} = -R^{-1} B^T p$$

Substituting the value for u_{opt} we get:

$$\begin{aligned} H &= p^T Ax - p^T BR^{-1} B^T p + \frac{1}{2} x^T Q x + \frac{1}{2} p^T BR^{-1} B^T p \\ &= p^T Ax - \frac{1}{2} p^T BR^{-1} B^T p + \frac{1}{2} x^T Q x \end{aligned}$$

The equations of the characteristics are then:

$$\begin{aligned} \dot{x} &= Ax + Bu = Ax - BR^{-1} B^T p \\ \dot{p} &= -\frac{\partial H}{\partial x} = -A^T p - Qx \end{aligned}$$

The solution of the characteristics equations for $x(0) = x_0$ and $x(\infty) = 0$ is

$$p = Px$$

so that

$$\begin{cases} \dot{x} = Ax - BR^{-1} B^T Px \\ P\dot{x} = -A^T Px - Qx \end{cases} \Rightarrow \quad PAx - PBR^{-1} B^T Px + A^T Px + Qx = 0 \ \forall \ x$$

Therefore P is the solution of the algebraic Riccati equation:

$$A^T P + PA - PBR^{-1} B^T P + Q = 0 \tag{5.28}$$

In order to study the conditions for the existence of a unique and positive semidefinite solution of Eq. 5.28, we need to introduce the concepts of *stabilizability* and *detectability*.

We define the couple of matrices A, B *stabilizable* if there exists a gain matrix K such that the matrix $A + BK$ is stable (all the eigenvalues have a negative real part), and we define the couple A, Q *detectable* if there exists a gain matrix L such that the matrix $A + LQ$ is stable (see Sect. 5.5.1 or any control theory text-book like [6–8, 10] to learn how to operationally determine these matrix features and how they are linked to the observability and controllability properties).

Sufficient conditions for the existence of a unique, symmetric and positive semi-definite solution of the Riccati equation are that:

1. The couple of matrices A, B is *stabilizable*
2. The couple of matrices A, Q is *detectable*

Through LQR theory it is finally demonstrated that:

Proposition *When the linear time invariant system Eq. 5.25 satisfies the conditions:*

$$\begin{cases} \textit{unbounded state and control} \\ \textit{stabilizable couple}(A, B) \\ \textit{detectable couple}(A, Q) \end{cases} \tag{5.29}$$

the optimal control law, for the performance index Eq. 5.27 and boundary conditions in Eq. 5.26, is a time-independent state feedback:

$$u_{opt} = -R^{-1}BPx \tag{5.30}$$

being the matrix P the unique solution of the Riccati Eq. 5.28.

5.5 Optimal Robust Linear Design

In the case of a linear system with full state measurement, the static feedback given by the LQR technique provides the optimal control. When the full state is not measured but is observable or at least detectable, which requires just the passivity of the system, following the separation principle we could design a state observer in conjunction with the LQR controller. The question about determining the observability of second order systems is studied by many papers, which apply the classic conditions for first order systems to second order ones (see for example [25–27] and for a more modern approach [28]). In the next section we discuss the observability and controllability criteria for the satellite dynamics of Eqs. 3.123, 3.124.

The state observer can be designed knowing the sensor noise characteristic and following a suitable stochastic performance index (see Sect. 5.3).

However, the synthesis of the state observer requires the knowledge of the satellite dynamic parameters, therefore the natural question is what is the level of accuracy that we must provide in our model and what happens if the synthesis is implemented with a model with some inaccuracies.

Let us consider the plant transfer function $P(s) = M(s)^{-1}/s^2$, where $M(s)$ is defined by the Eq. 3.138 in Sect. 3.12.1. As discussed in [26], the error in the plant transfer function $\delta P(j\omega_0)$, due to an error $\delta\omega$ in the elastic frequency estimation, at the nominal elastic frequency ω_0 is: $\frac{|\delta P(j\omega_0)|}{|P(j\omega_0)|} = \frac{1}{\sqrt{1+(\zeta/(\delta\omega/\omega_0))^2}}$, which means that if the damping factor is in the range of 0,01, a relative error of 0,01 in the resonant frequency will produce a maximum error in the frequency response of $1/\sqrt{2}$, which is close to the magnitude of the plant transfer function.

Spacecraft elastic modes have usually very low damping and large uncertainty in the resonant frequency. These problems are related intrinsically to space design, because in space we do not have the friction effects (like air pumping and gravity induced contact forces) which are present in ground tests, whereas ground tests of light flexible appendages are normally affected by errors due to gravity or the 0-gravity set-up.

In Sect. 3.12 we have seen that the knowledge of the spacecraft characteristics is normally assumed in the range of ten per cent and also the damping may be as low as 0,003.

Some attempts have been done to refine the controller parameters after calibration tests in space, this technique may work quite fine but is affected by uncertainties and is weak to the effect of aging, if some instability occurs some time after the calibration we can have effects (like the loss of propellant, service outages) which are highly undesirable.

The answer to these problems are the *robust control theory* and the *adaptive control*. We will address shortly in this book a modern approach to robust synthesis for multi input multi output (MIMO) systems: The H_∞ optimal control design.

5.5.1 Controllability and Observability

The existence of a steady state solution for each of the two Riccati equations Eqs. 5.16, 5.28 is linked to the stabilizability of the couple A, B and the detectability of the couple A, C plus the two auxiliary conditions on the matrices F and Q. Such solutions allow to define the controller and observer gains and to obtain a stable closed loop system. We need to study the stabilizability and the detectability properties of systems because they are sufficient conditions for the existence of a feedback stabilizing controller (see [7]). Therefore whatever synthesis technique we adopt, we need to present the system as stabilizable and detectable.

If we want to define arbitrarily the eigenvalues of the closed loop obtained by our plant and the feedback stabilizing controller, we need to ensure the similar but stronger concepts of controllability of the couple A, B and observability of the couple A, C.

The controllability property of a system is the ability to steer the state from any position to any other position in a finite time with an appropriately selected input. This definition does not consider the amount of energy that the input should have. Limiting the level of the control actuation produces an elongation of the time necessary to reach the final state.

The observability is the ability to determine the history of the state from the history of the output and the input in any finite time interval. The definition does not take into account the presence of noise in the output measurement which allow to determine the state only with a certain delay and in a finite bandwidth and with an error that depends on the noise entering in the observer bandwidth.

The controllability and observability properties are algebraic properties linked to the system matrices (see [10]).

We will expose these conditions directly for second order systems which are the class of systems we have obtained with the linearization technique in Sect. 3.11; setting:

$f = 0,\ c_d = 0,\ u = \delta c_c - \delta \dot{h}_w - [\delta\dot{\phi}, \overline{h_w}]$, in order to express the path from the control torque to the attitude and flexible state, Eqs. 3.123, 3.124 become:

$$\begin{aligned} I_G\delta\ddot{\phi} + S_G\ddot{\eta} &= u \\ S_G^T\delta\ddot{\phi} + \ddot{\eta} + 2\Sigma\Omega\dot{\eta} + \Omega^2\eta &= 0 \end{aligned} \tag{5.31}$$

We see that this model can be represented in the more general form:

$$M\ddot{x} + (C + D)\dot{x} + Kx = Bu \tag{5.32}$$

where: $M = M^T > 0, C = C^T \geq 0, D = -D^T, K = K^T \geq 0$.

In our specific case from Eq. 5.31 we have:

$$\begin{aligned} M &= \begin{pmatrix} I_G, & S_G \\ {S_G}^T, & E_{N_f} \end{pmatrix} \\ C &= \begin{pmatrix} 0_{3x3}, & 0_{3xN_f} \\ 0_{N_fx3}, & 2\Sigma\Omega \end{pmatrix} \\ K &= \begin{pmatrix} 0_{3x3}, & 0_{3xN_f} \\ 0_{N_fx3}, & \Omega^2 \end{pmatrix} \\ D &= 0_{(3+N_f)x(3+N_f)} \\ B &= \begin{pmatrix} E_3 \\ 0_{N_fx3} \end{pmatrix} \end{aligned} \quad , \tag{5.33}$$

The state vector is $x = [\delta\phi^T, \eta^T]^T$. We can associate to these equations a measurement vector y, that can be generically written as:

$$y = Px + Q\dot{x} \tag{5.34}$$

If we measure only the rigid body attitude and rate, which is often the case, then:

$$P = \begin{pmatrix} E_3, & 0_{3xN_f} \\ 0_{3x3}, & 0_{3xN_f} \end{pmatrix}$$
$$Q = \begin{pmatrix} 0_{3x3}, & 0_{3xN_f} \\ E_3, & 0_{3xN_f} \end{pmatrix} \tag{5.35}$$

where $N_f = \dim \eta$.

For the system Eq. 5.32 we have some simple conditions to determine the controllability and the observability derived in [27].

Consider the set Λ composed of the λ such that

$$det(\lambda^2 M + \lambda(C + D) + K) = 0 \tag{5.36}$$

The set Λ is a finite set with $N_f + 3$ pairs of complex conjugate solutions which we call the *spectrum* of the eigenvalues of Eq. 5.32. Each eigenvalue corresponds to an eigen-solution $x = x_\lambda e^{\lambda t}$, where x_λ is the corresponding eigenvector. We will consider $\Lambda = \Lambda^+ \cup \Lambda^-$, where $\Lambda^+ = \lambda | Re(\lambda) \geq 0$ is the set of the unstable eigenvalues and $\Lambda^- = \lambda | Re(\lambda) < 0$ is the set of the stable ones.

The following results can be demonstrated (see [27]):

Proposition (General Conditions for Controllability and Observability) *The system Eq. 5.32 is controllable (stabilizable) if and only if*

$$rank(\lambda^2 M + \lambda(C + D) + K, B) = N_f + 3, \forall \lambda \in \Lambda(\Lambda^+). \tag{5.37}$$

The system Eq. 5.32 is observable (detectable) if and only if

$$rank(\lambda^2 M + \lambda(C + D) + K; P + \lambda Q) = N_f + 3, \forall \lambda \in \Lambda(\Lambda^+). \tag{5.38}$$

We now specify these conditions for the system defined by Eq. 5.31. We use the Modal Matrix defined by Eq. 3.130 in Sect. 3.12.1 to transform this system and simplify the determination of the rank:

$$W = \begin{pmatrix} I_G{}^{-\frac{1}{2}}, & -I_G{}^{-1} S_G \eta_1, & -I_G{}^{-1} S_G \eta_2, & \ldots\ldots, & -I_G{}^{-1} S_G \eta_{N_f} \\ 0_{N_f x3}, & \eta_1, & \eta_2, & \ldots\ldots, & \eta_{N_f} \end{pmatrix} \tag{5.39}$$

with the transformation $Wz = x$, the controllability matrix becomes:

$$
\begin{aligned}
&[\lambda^2 W^T M W + \lambda W^T C W + W^T K W,\ W^T B] \\
&= [\lambda^2 \begin{pmatrix} E_3, & 0_{3xN_f} \\ 0_{N_f x3}, & E_{N_f} \end{pmatrix} + \lambda \begin{pmatrix} E_3, & 0_{3xN_f} \\ 0_{N_f x3}, & 2\Sigma_f \Omega_f \end{pmatrix} + \begin{pmatrix} E_3, & 0_{3xN_f} \\ 0_{N_f x3}, & {\Omega_f}^2 \end{pmatrix},\ W^T B] \\
&W^T B = \begin{pmatrix} I_G^{\frac{1}{2}} \\ -{\eta_1}^T {S_G}^T \\ \ldots\ldots \\ -{\eta_{N_f}}^T {S_G}^T \end{pmatrix} {I_G}^{-1}
\end{aligned}
\tag{5.40}
$$

and the observability matrix:

$$
\begin{aligned}
&\begin{bmatrix} \lambda^2 W^T M W + \lambda W^T C W + W^T K W \\ (P + \lambda Q) W \end{bmatrix} \\
&= \begin{bmatrix} \lambda^2 \begin{pmatrix} E_3, & 0_{3xN_f} \\ 0_{N_f x3}, & E_{N_f} \end{pmatrix} + \lambda \begin{pmatrix} E_3, & 0_{3xN_f} \\ 0_{N_f x3}, & 2\Sigma_f \Omega_f \end{pmatrix} + \begin{pmatrix} E_3, & 0_{3xN_f} \\ 0_{N_f x3}, & {\Omega_f}^2 \end{pmatrix} \\ (P + \lambda Q) W \end{bmatrix} \\
&(P + \lambda Q) W = -{I_G}^{-1} \begin{pmatrix} [\ \ -I_G^{\frac{1}{2}}, S_G \eta_1, S_G \eta_2 \ldots, S_G \eta_{N_f}] \\ \lambda[\ -I_G^{\frac{1}{2}}, S_G \eta_1, S_G \eta_2 \ldots, S_G \eta_{N_f}] \end{pmatrix} \\
&(P + \lambda Q) W = \begin{pmatrix} B^T W \\ \lambda\ B^T W \end{pmatrix}
\end{aligned}
\tag{5.41}
$$

Apart from the sub-matrix $\lambda B^T W$, which has no effect on the rank, the controllability matrix is the transpose of the observability one, therefore the system is controllable if and only if it is also observable and it is stabilizable only if it is also detectable.

We can study the controllability (stabilizability) from the rank of the controllability matrix defined by Eq. 5.40 calculated for $\lambda \in \Lambda(\Lambda^+)$.

The first three eigenvalues of the *spectrum* are null; when $\lambda = 0$ the controllability has full rank because $\Omega_f > 0$. If Λ^+ is formed only by these three eigenvalues we can ensure the stabilizability and detectability.

Proposition (Conditions for Detectability and Stabilizability) *System Eq. 5.32 is detectable and stabilizable provided all the flexible modes have a not null damping factor.*

However, the damping factors are so low that we cannot accept just the stabilizability, we need to improve the convergence time of our controller, therefore we often request the controllability and observability properties, to see them we need to analyze the other eigenvalues of the *spectrum.*

When we set: $\lambda = -\zeta_k \Omega_{fk} \pm j \Omega_{fk} \sqrt{1 - {\zeta_{fk}}^2}$ the kth element of the diagonal matrix $[\lambda^2 E_{N_f} + \lambda 2 \Sigma_f \Omega_f + {\Omega_f}^2]$ becomes null, if we have multiple eigenvalues the diagonal block related to that eigenvalue will become null.

Le us consider first the case of a single multiplicity eigenvalue. We lose the full rank of the controllability matrix unless we find a column in $W^T B$ which, displaced in the $(k+3)$th column restores the full rank. The $(k+3)$th row of $W^T B$ is formed by $({I_G}^{-1} S_G \eta_k)^T$. Considering that I_G is full rank, it is enough that one of the three entries of the vector $S_G \eta_k$ is different from zero. The case where we have multiplicity two or three follows the same logic. When the multiplicity is higher than three, we cannot recover the full rank with the columns of the $N_f \times 3$ matrix $({I_G}^{-1} S_G \eta_k)^T$, $k = 1..N_f$. Therefore we have the following proposition:

Proposition (Conditions for Controllability and Observability) *System Eq. 5.32 is controllable and observable if for each kth free eigen-mode of single multiplicity* $rank(S_G \eta_k) = 1$, *and, for any block of eigen-modes with multiplicity* $N_e < 4$, *we have:*

$$rank(S_G[\eta_k, \eta_{k+1}, .., \eta_{k+N_e-1}]) = N_e.$$

If we have a block of eigenmodes with multiplicity higher than 3 the system Eq. 5.32 is not controllable and observable.

The demonstration follows from Eq. 5.37.
We can derive the additional following considerations:

- detectability and stabilizability can always be provided when $\Sigma_f > 0$. Nevertheless we may have modes that are both uncontrollable and unobservable;
- if some eigen-modes are unobservable and uncontrollable, we can eliminate them from the dynamics using a minimal realization of the *dynamic mass matrix*. These modes will not be excited by the controller, as can be seen using the Kalman decomposition (see [6]), however they may be excited by some other exogenous input (like the force f during the maneuver), in such a case, we cannot do much about it without a direct control and measurement on the flexible modes;
- the controllability/observability of a single multiplicity mode η_k is provided whenever the vector $S_G \eta_k \neq 0$. This means that the free oscillations are coupled with the rigid motion;
- the controllability/observability of a block of eigen-modes with the same frequency can be provided only if their multiplicity is less than 4. A subspace of a block of eigen-modes having the same frequency may be presented as the sum of two subspaces, one controllable/observable and the other unobservable/uncontrollable.

5.5.2 H_∞ Norms

We consider the vector space L_2 of all finite energy signals $u(t) \in R^n$ with Fourier transform $u(j\omega) \in C^n$. We indicate $|| \cdot || = \sqrt{u^H u}$ the norm for vectors $u \in C^n$ and the induced operator norm for the complex matrices operating on such vectors. In the space L_2 we introduce the norm:

$$\|u\|_2 = \sqrt{\int_0^\infty u^T(t)u(t)dt} = \sqrt{\int_{-\infty}^\infty u^H(j\omega)u(j\omega)d\omega/2\pi} \tag{5.42}$$

We now consider a transfer matrix $P(j\omega)|Pu \in L_2\ \forall u \in L_2$ we can consider the operator norm of this transfer matrix induced by the L_2 norm.

$$\|P\|_\infty = sup_{\|u\|_2} \frac{\|Pu\|_2}{\|u\|_2} \tag{5.43}$$

We have that $\|P\|_\infty$ is the square root of the maximum energy that the system can provide in output for an input of unitary energy.

It is possible to demonstrate that each component of a matrix with a bounded H_∞ norm is in the set of the functions analytic and bounded in the right half positive real domain of the complex plane. The domain of these functions will be represented by RH_∞. Any function $F \in RH_\infty$ is proper, that is bounded in the imaginary axis, and stable (in this chapter stability will always be intended as asymptotic stability) because its fourier transform is convergent to zero as $t \to \infty$. We will indicate with RH_∞ the space of the functions and the space of the matrices composed of such functions.

The operator $P(j\omega) \in RH_\infty$ at any given frequency $j\omega$ is a matrix that can be presented in singular values decomposition (SVD): $P = U^H \Sigma V$ with U, V unitary matrices and Σ diagonal matrix with the same number of rows and columns of P and containing its real positive singular values on the main diagonal (see [29]).

For the unitary property it follows that:

$$\frac{\|Pu\|_2}{\|u\|_2} = \frac{\|\Sigma V u\|_2}{\|V u\|_2}, \tag{5.44}$$

so it is easy to demonstrate that:

$$\|P\|_\infty = sup_\omega \overline{\sigma}(P(j\omega)) \tag{5.45}$$

where $\overline{\sigma}(P(j\omega)) = max_i(\Sigma_{i,i}(j\omega))$ is the maximum singular value of the matrix P at the angular frequency ω.

This norm is very useful because it allows to characterize well MIMO system transfer functions, the minimization of this norm applied to certain transfer functions of the closed loop system, with respect to the set of all stabilizing controllers allows to select a robust and performing closed loop system.

There are some properties of the H_∞ norm which can be easily demonstrated and will be useful following this chapter:

- *Given two matrix transfer functions $F_1, F_2 \in RH_\infty$ with the same number of columns, if*:

$$\left\| \begin{matrix} F_1 \\ F_2 \end{matrix} \right\|_\infty \leq \gamma \tag{5.46}$$

then, it follows $\|F_i\|_\infty \leq \gamma, i = 1, 2.$
This can be easily demonstrated partitioning the output in $y_1 = F_1 u$ and $y_2 = F_2 u$, applying the definition of H_∞ norm and considering that $y_1^H y_1 + y_2^H y_2 \geq y_i^H y_i,\ i = 1, 2.$

- *Given two matrix transfer functions* $F_1, F_2 \in RH_\infty$ *with the same number of rows, if*:

$$\|F_1,\ F_2\|_\infty \leq \gamma \tag{5.47}$$

then, it follows $\|F_i\|_\infty \leq \gamma, i = 1, 2.$
This can be easily demonstrated partitioning the input in u_1 and u_2. The output is $y = F_1 u_1 + F_2 u_2$.
The output energy is $y^H y = u_1^H F_1^H F_1 u_1 + u_2^H F_2^H F_2 u_2 + 2 u_1^H F_1^H F_2 u_2$.
For any couple $u = (u_1; u_2)$ the output energy of $u = (u_1; 0)$ is surely less than that generated by at least one of the couples $(u_1; u_2)$ and $(u_1; -u_2)$. The square root of the maximum energy we have as output with only one input channel active is always less than H_∞ norm.
- *From the results of the previous two bullets it follows that if* $\|F\|_\infty \leq \gamma$ *then we have the same for each component of the matrix* $|F_{i,j}| \leq \gamma$.
- *Given three matrix transfer functions* $F, F_1, F_2 \in RH_\infty$ *such that* $F = F_1 F_2$ *the following chain of inequalities is true*:

$$\overline{\sigma}(F_1(j\omega))\overline{\sigma}(F_2(j\omega)) \geq \overline{\sigma}(F(j\omega)) \geq \underline{\sigma}(F_1(j\omega)\overline{\sigma}(F_2(j\omega)) \tag{5.48}$$

where we denote $\underline{\sigma}(F_1(j\omega)$ *the minimum singular value of* F_1 *at* $j\omega$.
This relationship can be obtained from the properties of singular values:

$$\overline{\sigma}(F_1(j\omega)u^H {F_2}^H F_2 u \geq u^H {F_2}^H {F_1}^H F_1 F_2 u \geq \underline{\sigma}(F_1(j\omega)u^H {F_2}^H F_2 u \tag{5.49}$$

dividing each member by $u^H u$ and taking the *sup* of the chain we obtain the property. From this chain of dis-equations, applying the max for any ω it also follows:

$$||F_1||_\infty ||F_2||_\infty \geq ||F||_\infty \tag{5.50}$$

- *Given a square matrix transfer function* $F(j\omega)$, *and its SVD decomposition* $F = U^H \Sigma V$, *we have that* $F^H F = V^H \Sigma^H \Sigma V$, *the second member is the diagonal transformation in eigenvalues of the Hermitian matrix* $F^H F$. *It follows that the singular values are the square roots of the non null eigenvalues of* $F^H F$ *which are the same as those of* $F F^H$.
- *Given a square matrix transfer function* $F(j\omega)$ *such that* $\underline{\sigma}(F(j\omega)) > 0$, *we have the following relationship:*

$$\overline{\sigma}(F(j\omega)^{-1}) = \frac{1}{\underline{\sigma}(F(j\omega))} \tag{5.51}$$

5.5.3 Small Gain Theorem and Robustness

The small gain theorem gives a sufficient condition for the stability of the feedback connections. We provide the result in terms of Laplace transforms, recall that E is the identity matrix.

We consider two transfer matrices $F_1(s), F_2(s) \in RH_\infty$ connected in a feedback loop as in Fig. 5.1, with input u injected after F_1 and output y taken before F_2. For this loop $y = F_1(s)F_2(s)y + u$, then we have the following:

Proposition (The Small Gain Theorem) *The transfer function from u to y, $F_{uy}(s) = (E - F_1(s)F_2(s))^{-1}$ is asymptotically stable if $||F_1F_2||_\infty < 1$, or also if $||F_1||_\infty||F_2||_\infty < 1$.*

The existence of a matrix F_2 with $||F_2||_\infty \leq 1$ which makes the system unstable if $||F_1||_\infty \geq 1$ can also be demonstrated by construction.

The demonstration of the sufficient condition is simple.

Let us consider any point s^+ of the complex plane so that $Re(s^+) > 0$. Then $\forall y \in L_2$ it follows

$$\begin{aligned} &||(E - F_1(s^+)F_2(s^+))y||_2 \geq ||y||_2 - ||F_1(s^+)F_2(s^+))y(s)||_2 \geq \\ &||y||_2 - ||F_1F_2||_\infty||y||_2 \geq ||y||_2 - ||F_1||_\infty||F_2||_\infty||y||_2 > 0 \end{aligned} \tag{5.52}$$

The third inequality derives from the *Maximum Modulus Theorem*, for which the modulus of any analytic complex function F in a certain domain, attains its maximum modulus at the boundary of this domain.

If s^+ was a pole of $(E - F_1(s)F_2(s))^{-1}$, we would have a zero response of $(E - F_1(s^+)F_2(s^+))y$ for some $y \in L_2$, this cannot be possible because $||(E - F_1(s^+)F_2(s^+))y||_2 > 0$. This demonstrates the *Small Gain Theorem.*

A simple application of this theorem allows to state the stability conditions in the presence of plant multiplicative perturbations. Consider a plant with an external block $\Delta(s) \in RH_\infty$ which represents the plant possible perturbations.

$$\begin{aligned} y &= M(s)u + N(s)w \\ z &= H(s)u + L(s)w \end{aligned} \tag{5.53}$$

We assume that $M, N, L, H \in RH_\infty$ and $u = \Delta(s)y$, the transfer function from w to z is

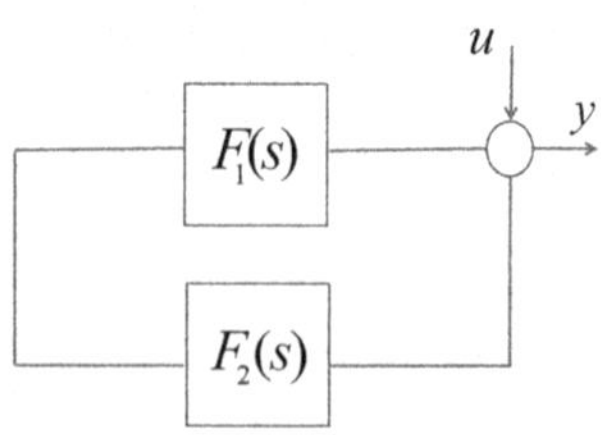

Fig. 5.1 The small gain theorem set-up

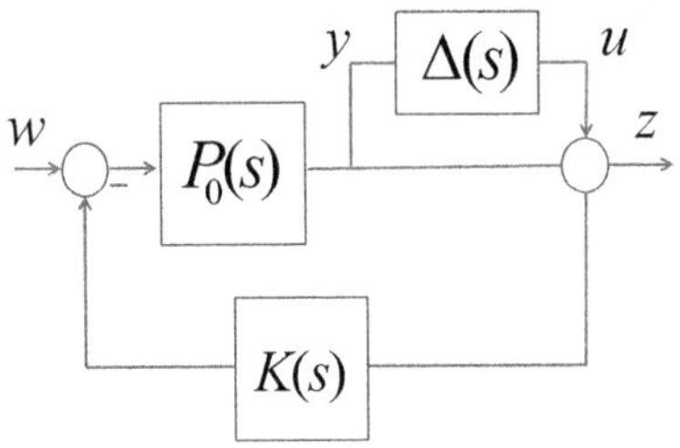

Fig. 5.2 Plant multiplicative perturbation

$$T_{wz} = L + H\Delta(E - M\Delta)^{-1}N \tag{5.54}$$

The *Small Gain Theorem* states that we have asymptotically stable transfer matrix from $w \to z$ if

$$||M\Delta||_\infty \leq 1 \tag{5.55}$$

The set of allowed plant perturbations is

$$||\Delta||_\infty \leq \frac{1}{||M||_\infty} \tag{5.56}$$

We see in Fig. 5.2 a system whose plant P_0 has been stabilized with the controller K designed for the case $\Delta = 0$. We can calculate all the transfer functions:

$$N = L = (E + P_0K)^{-1}P_0, M = -(E + P_0K)^{-1}P_0K, H = E + M$$

We suppose that all these transfer functions are $\in RH_\infty$, this requires that $F = (E + P_0K)^{-1}$ and P_0, K are stable. Now, suppose we want to analyze the stability of this system considering a set of plants $(E + \Delta)P_0$, which is called the *multiplicative plant perturbation*, from the *Small Gain Theorem* the stability is guaranteed if

$$||\Delta||_\infty \leq \frac{1}{||(E + P_0K)^{-1}P_0K||_\infty} \tag{5.57}$$

We can establish similar results for other types of perturbations like the additive plant perturbation (see for example [10]).

5.5.4 H_∞ Optimal Control

The H_∞ methods are used in control theory to synthesize controllers for MIMO systems achieving stabilization with guaranteed performance even in the presence of plant perturbations.

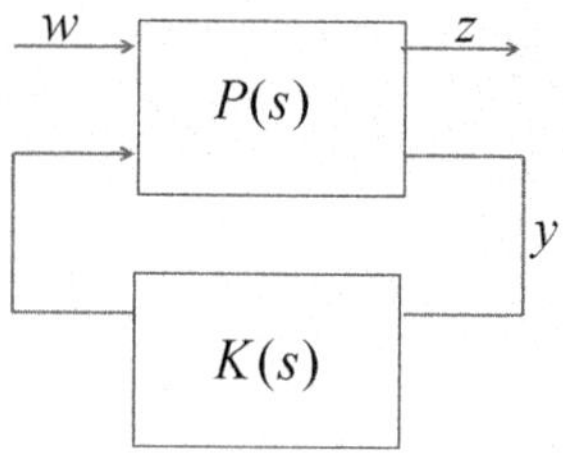

Fig. 5.3 H_∞ set up

The control designer expresses the control problem as the minimization of the H_∞ norm of a closed loop matrix transfer function over the class of the stabilizing controllers and then finds the controller that solves this optimization problem. The H_∞ techniques have the advantage over classical control techniques in that they are readily applicable to problems involving multivariable systems with cross-coupling between channels.

In order to apply the H_∞ algorithms the design problem has to be represented according to the following standard configuration (Fig. 5.3).

The plant P, formed by the process plant connected with some additional blocks which shape the output signals as will be clarified later, has two inputs, the exogenous input w, that includes reference signal and disturbances, and the control variables u. There are two outputs, the error signals z that we want to minimize, and the measured variables y, that we use to feed the controller K, which gives in output the control variable u. Notice that all these signals are vectors, whereas P and K are matrices. The objective of control design is to find a controller K such that the H_∞ norm of the transfer function from w to z is minimized.

The plant can be modeled in time domain as:

$$\begin{aligned} \dot{x} &= Ax + B_1 w + B_2 u \\ z &= C_1 x + D_{11} w + D_{12} u \\ y &= C_2 x + D_{21} w + D_{22} u \end{aligned} \tag{5.58}$$

the H_∞ design problem is to find a proper realizable controller $y = K(s)u$ such that the norm of $||T_{wz}||_\infty$ is minimized. In order to have a solution of $||T_{wz}||_\infty \leq \gamma$ for some γ there are some necessary conditions to be satisfied (see [10]):

- The plant must be stabilizable from the control inputs u and detectable from the measurement output y otherwise no stabilizing controller exist. When unobservable and uncontrollable stable modes of Eq. 5.32 exist, it is preferable to work on a minimal realization of Eq. 5.32 in order to remove these states.
- $rank(D_{12}) = dim(u)$ and $rank(D_{21}) = dim(y)$ in order to ensure that the stabilizing controller is proper and realizable.
- The following rank must be verified to ensure that the controller does not try to implement a zero or pole cancellation on the imaginary axis, which would result in an unstable closed loop solution.

$$
\begin{aligned}
rank \begin{bmatrix} A - j\omega E \ , B_2 \\ C_1 \quad\quad , D_{12} \end{bmatrix} &= dim(A) + dim(u) \\
rank \begin{bmatrix} A - j\omega E \ , B_1 \\ C_2 \quad\quad , D_{21} \end{bmatrix} &= dim(A) + dim(y)
\end{aligned}
\tag{5.59}
$$

Under such conditions there are available algorithms that can provide a controller which realizes $||T_{wz}||_\infty \leq \gamma$ for some γ. In particular there is an optimal controller, based on a classic observer structure, the so-called *central controller* (see [13]) which can be expressed in terms of the solutions of two Riccati equations.

The matrices of these Riccati equations depend on γ, so the algorithm starts with γ sufficiently high and then decreases its value until the Riccati equations admit a solution. The order of the *central controller* is the same of the extended plant, which is often very high, making the *central controller* not very used in practice.

New algorithms have been developed where it is possible to fix some controller properties such as its maximum order or a predefined controller structure dependent on some parameters (see [30, 31]); in this way, we obtain quasi-optimal solutions that are easier to implement and less demanding in terms of on-board computer (OBC) processing time.

The main task of the designer using the H_∞ control optimization routines, is to select suitable inputs w and outputs z so that all the problem requirements, such as robust stability and performance, are taken into account. It is usual to introduce to this end, specific weight functions that are placed after the input w and before the output z. These weight functions allow to shape the loop response.

A typical design approach is the mixed sensitivity problem where the plant P is formed by the primitive plant G augmented in order to realize three classes of outputs z, we refer to Fig. 5.4:

$$
T_{wz} = \begin{bmatrix} W_S S \\ W_U U \\ W_T T \end{bmatrix}
$$

with:

$$
\begin{aligned}
S &= (E + GK)^{-1} \\
U &= -(E + GK)^{-1} K \\
T &= (E + GK)^{-1} GK
\end{aligned}
\tag{5.60}
$$

The sensitivity S brings in output the control error, the control sensitivity U brings in output the control action and the complementary sensitivity brings in output the plant state before the injection of the inputs w, as shown in Fig. 5.4.

All the weight functions must be stable and proper function $\in RH_\infty$, in order to have a solution, the stability is necessary in order to have an overall plant which admits a stabilizing controller.

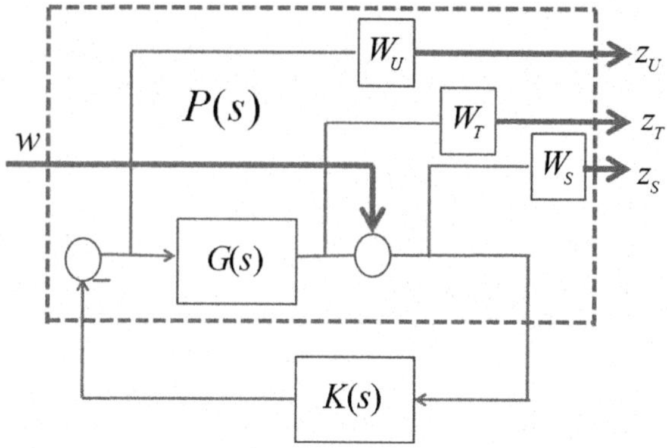

Fig. 5.4 H_∞ set up

After the H_∞ optimization a controller K such that:

$$||T_{wz}||_{H_\infty} < \gamma$$

is finally determined for γ enough high.

From the properties of the H_∞ norm the following *loop shaping* characteristics are provided:

$$\begin{aligned}\overline{\sigma}(S(j\omega)) &\leq \gamma/\underline{\sigma}(W_S(j\omega))\\ \overline{\sigma}(U(j\omega)) &\leq \gamma/\underline{\sigma}(W_U(j\omega))\\ \overline{\sigma}(T(j\omega)) &\leq \gamma/\underline{\sigma}(W_T(j\omega))\end{aligned} \tag{5.61}$$

The weight matrices are also called *templates* of the corresponding system transfer functions.

We note from Eq. 5.60 that, with the assumption of a strictly proper plant and a proper controller, $S(j\infty) = E$, this ensures that $\gamma \geq \underline{\sigma}(W_S(j\infty))$.

If we choose $\underline{\sigma}(W_S(j\infty)) = 1$, γ will never be less than 1.

With this design approach or the many variants of it, we can shape the closed loop response to be robust and performing.We now address the choice of W_S, W_U, W_T in order to implement specific design requirements.

- The sensitivity $S(j\omega)$ defines how much the output z_s (refer to Fig. 5.4) is sensitive to the input w, we desire typically a low sensitivity at low frequency up to a certain ω_S.
  Considering that from Eq. 5.61, $\overline{\sigma}(S(j\omega)) \leq \gamma/\underline{\sigma}(W_S(j\omega))$, we select a $|W_S|$ very high at low frequency.
  We cannot impose a low sensitivity also at high frequency because of the *Bode Integral Theorem*.
  From [32], we have the following

Proposition (Bode Integral Theorem) *In a MIMO system* $\int_0^\infty \ln(||S(j\omega)||)d\omega \geq 0$ *where:* $||\cdot||$ *is the matrix norm,* $S = (E + GK)^{-1}$ *is the sensitivity,* $|GK| \approx O(s^{-2})$*; in SISO systems if GK is stable and proper* $\int_0^\infty \ln(||S(j\omega)||)d\omega = 0$.

The average value of $\ln(||S(j\omega)||)$ being above zero, if we have a low sensitivity $\ln(||S(j\omega)||) < 0$ in the low frequency domain, it follows that we must have a high sensitivity $\ln(||S(j\omega)||) > 0$ in the high frequency domain. Above a certain frequency the $S(j\omega)$ shall approach the identity matrix and $\ln(||S(j\infty)||) \to 0$.

We can select a diagonal W_S and choose for any sensitivity output a specific weight $W_{S_i}, i = 1 \ldots 3$.

The following considerations can be applied to any element in the i-th row of the matrix S, so we will not indicate the index of S and W_S.

In each sensitivity output we can choose $W_S(s) = \frac{as+\omega_S}{s+b\omega_S}$.

Other templates can be chosen but the general considerations are the same. In the hypothesis that γ results close to unity:

- b limit from above the low frequency gain of $S(j\omega)$that can be set to a very low value depending on the application ($b \approx 1.e-3 \ldots 1.e-6$)
- ω_S is the cross-over frequency where $|S(j\omega)|$ is allowed to cross 0-dB
- $1/a$ limit from above is the high frequency gain and the possible undesired peaking of $|S(j\omega)|$. This limit should be set to $\frac{1}{a} \approx 1 \ldots 2$. The high frequency gain of S is related to the stability margins. Choosing $||W_S(j\omega)|| \geq a, \forall\omega$ and in all output channels of S, we can derive from the resulting γ the robustness margins as demonstrated in the following bullet.

- Let us introduce M_S the *module margin*, the minimal distance from the critical point $(-1+0j)$ of all the eigenvalues of the open loop transfer $L(s) = G(s)K(s)$ while s moves along the imaginary axis. We will demonstrate that $\frac{a}{\gamma} \leq M_S$.

  Designing around the critical point $(-1+0j)$ a circle of radius M_S, in the hypothesis that none of these eigenvalues encircles it, it follows that the curve of each eigenvalue in the complex plane has a gain margin $G_M \geq \frac{1}{1-M_S}$, and a phase margin $sin(P_M) \geq M_S/2$, so we can guarantee a phase margin better than $P_M = arcsin(M_S/2)$.
  We can demonstrate that such margins can be applied in all the lines of the loop as margins of the open loop transfer $L(j\omega)$.

Proposition *$det(E + kL(j\omega)) \neq 0, \forall k|\ |k| < M_S$, in addition $\frac{a}{\gamma} \leq M_S$.*

Let us consider $S = (E + L(j\omega))^{-1}$, from

$$\overline{\sigma}(S(j\omega)) \leq \frac{\gamma}{\underline{\sigma}(W_S(j\omega))} \leq \frac{\gamma}{a} \tag{5.62}$$

using Eq. 5.51 it follows that:

$$\underline{\sigma}(E + L(j\omega)) \geq \frac{a}{\gamma} \tag{5.63}$$

The Nyquist criterion in MIMO systems (see [9]) states that $S = (E + L(j\omega))^{-1}$ is stable if and only if the plot of $det(E + L)$ encircles the origin counter-clockwise a number of times equal to the number N_p of the unstable poles of the plant. In our case the plant is stable $N_p = 0$. Let us consider the Spectrum Λ of $L(j\omega)$, we have from the definition of eigenvalue that:

$$\Lambda = \{\lambda_i | \exists x_i, Lx_i = \lambda_i x_i, (E + L)x_i = (1 + \lambda_i)x_i\} \tag{5.64}$$

It follows that $x_i^H(E + L)^H(E + L)x_i = |1 + \lambda_i|^2 x_i^H x_i$, and also that

$$\frac{a}{\gamma} \leq \underline{\sigma}(E + L(j\omega)) \leq |1 + \lambda_i|, \forall i \tag{5.65}$$

So, as we wanted to demonstrate: $\frac{a}{\gamma} \leq M_S$, being M_S the minimal distance of all the eigenvalues of $L(j\omega)$ from the critical point $(-1 + j0)$.

It is well known that $det(E + L(j\omega)) = \prod_i (1 + \lambda_i)$ where the λ_j are considered with their multiplicity.

If the system is stable, and we multiply the open loop transfer L by a factor k, we have $det(E + kL(j\omega)) = \prod_i (1 + k\lambda_i(j\omega))$.

So, as we wanted to demonstrate: *if* $|k| < M_S$ it follows that $det(E + kL(j\omega)) \neq 0$ and the closed loop system remains stable after the application of the multiplicative factor k.

The gain margin G_M of the MIMO is the minimum real amplification factor $k > 1$ which brings an eigenvalue to cross the critical point $(-1 + j0)$. The phase margin of the MIMO is the minimum rotation P_M such that $k = e^{-jP_M}$ brings an eigenvalue to pass across the critical point $(-1 + j0)$. When an eigenvalue of L crosses the critical point the complete system becomes unstable.

- The complementary sensitivity T is such that $T + S = E$ as can be easily verified from the definition given by Eq. 5.60. This means that when we have a low sensitivity the complementary matrix must be close to the unit matrix.
  From Eq. 5.61 $\overline{\sigma}(T(j\omega)) \leq \gamma / \underline{\sigma}(W_T(j\omega))$. We can use this property to guarantee the stability versus multiplicative perturbations of the primitive plant. Reusing the scheme of Fig. 5.2 on the $G(s)$ of Fig. 5.4 any multiplicative perturbation Δ of the plant G such that $\frac{\underline{\sigma}(W_T(j\omega))}{\gamma} \geq \overline{\sigma}(\Delta(j\omega))$ ensures that:

  $$\overline{\sigma}(T(j\omega))\overline{\sigma}(\Delta(j\omega)) \leq 1 \tag{5.66}$$

  which guarantees the stability of the system by Eq. 5.56 where the place of $M(s)$ is taken here by the complementary sensitivity.
  We normally desire that T is low at high frequency because the plant uncertainties are higher in high frequency. This requires that W_T in chosen very high at high frequency (high-pass).

- The matrix transfer function U, is equal to the inverse of the plant when $|L| = |GK| \gg 1$ (see Eq. 5.60) which generally happens at low frequency, instead when the loop gain is low $|L| = |GK| \ll 1$, $U \approx K$ the controller.
From Eq. 5.61
$$\overline{\sigma}(U(j\omega)) \leq \gamma/\underline{\sigma}(W_U(j\omega)) \tag{5.67}$$
so we can use W_U to limit the high frequency gain of the controller which prevents a control actuation too sensible to the high frequency sensor noise and also from actuator high frequency mis-use. This is obtained by setting W_U as a low pass filter (or roll-off filter) with a cut off frequency ω_U. This frequency is placed ahead of the controller bandwidth ω_S.

5.5.5 H_∞ Design with Acceleration Sensitivity

The controller design method, that we illustrate in this section, is an H_∞ method specific for second order systems. We introduce it following the work of Alazard and others (see [11, 33]) in order to provide the basic theory for a design example that will be described in Sect. 6.3 of Chap. 6.

The method implemented in [33] uses the acceleration in a second order system to define the sensitivity: the sensitivity function will be the transfer function between the input three axis angular acceleration (which may be derived by an input disturbing torque) and the output angular acceleration which twice integrated provides the pointing accuracy.

With reference to the system defined by Eq. 5.31, its input-output transfer function can be written in terms of the inverse of the *dynamic mass matrix* defined by Eq. 3.138 followed by a double integrator. This is illustrated in the Fig. 5.5 that shows the set up necessary to implement the mixed acceleration sensitivity method. Figure 5.5 recalls Fig. 5.4. The differences are:

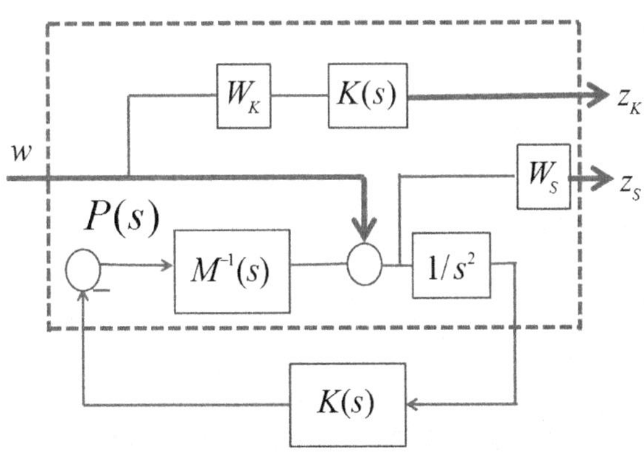

Fig. 5.5 Mixed acceleration sensitivity set up

- we have put the double integrator outside the plant,
- instead of using the direct controller output to limit the controller we duplicate the controller in a dedicated line. This approach is used here because $U = -SK/s^2$ instead of $U = -SK$,
- the complementary sensitivity is not used in order to limit the complexity of the H_∞ problem; the stability will be guaranteed using the *module margin* M_S.
- $W_S = O(\frac{1}{s^2})$ in order to limit the energy transfer to the attitude pointing $\delta\phi = \frac{\delta\ddot{\phi}}{s^2}$. To ensure the existence of a stabilizing controller, we should avoid unstable poles in W_S, therefore it must be realized using the position and rate states of the plant.

The loop transfer function is $L(s) = M(s)^{-1}K(s)/s^2$, the sensitivity is as usual $S = (E + L)^{-1}$

The input w is an acceleration, we can obtain a disturbing acceleration deriving it from the disturbing torque c_d considering, $w = M(s)^{-1}c_d$.

The transfer function $M(s)^{-1}$ from the input torque to the output angular acceleration is a proper matrix transfer function and is the inverse of the *dynamic mass matrix* $M(s)$ determined in Chap. 3, Eq. 3.138.

$$M(s) = I_G - s^2 \sum_{k=1..N_f} \frac{S_G{}^k S_G{}^{kT}}{s^2 + 2\zeta_k \Omega_k s + \Omega_k{}^2} \tag{5.68}$$

The angular response to a disturbing torque c_d is:

$$\delta\phi = \frac{SM^{-1}}{s^2} c_d = (Es^2 + M^{-1}K)^{-1} M^{-1} c_d \tag{5.69}$$

From Eq. 5.69, we see that at low frequency ($\omega \ll \Omega_1$) we have $\delta\phi \approx K^{-1}c_d$. In addition from Eq. 5.61 we have that;

$$||\delta\phi(j\omega)|| \leq \gamma \frac{||M^{-1}(j\omega)||}{\underline{\sigma}(\omega^2 W_S(j\omega))} ||c_d(j\omega)|| \tag{5.70}$$

If we evaluate the transfer from the noise n of the position measurement to the position output $\delta\phi$ we have:

$$\delta\phi = -\frac{SM^{-1}}{s^2} Kn \tag{5.71}$$

This transfer function is the same as that in Eq. 5.69 post-multiplied by the controller K. With the mixed acceleration method we try to keep Ss^{-2} low in a certain region of the bandwidth.

In order to limit the noise effect and to avoid the control saturation we need to limit as well the controller amplitude above a certain frequency. This is obtained by limiting the output z_K (see Fig. 5.5).

We give some indications on how to select appropriate weight matrices.

We choose as W_S a diagonal matrix, each diagonal term satisfies

$|W_{S_i}(j\infty)| \geq 1, i = 1, 2, 3$ so, the *module margin* is simply $M_S = \frac{1}{\gamma}$. After the optimization, having reached a certain value of γ, such that: $\underline{\sigma}(E + L(j\omega)) \geq \frac{1}{\gamma}$ we have:

- the module margin is $\frac{1}{\gamma}$
- the guaranteed gain margin is $20\ Log_{10}(\frac{\gamma}{\gamma-1})$
- the guaranteed phase margin is $(\frac{360}{\pi})2\ arcsin(\frac{1}{2\gamma})$

We select for each axis a sensitivity of the form:

$$W_{S_i}(s) = \frac{s^2 + 1.4\omega_{S_i}s + {\omega_{S_i}}^2}{s^2}, i = 1..3 \tag{5.72}$$

The $W_{S_i}(s)$ must be realized taking the angular and rate states available in the plant output and avoiding to introduce new integrators that would be unstable and not controllable.

$$W_{S_i}(s) = \delta\ddot{\phi}_i + 1.4\omega_{S_i}\delta\dot{\phi}_i + {\omega_{S_i}}^2\delta\phi_i, i = 1..3 \tag{5.73}$$

This template specifies that input frequencies above ω_{S_i} are not rejected in the i-th output. With this choice, each term of the acceleration sensitivity satisfies (by the definitions Eqs. 5.61, 5.72):

$$|s^{-2}S_{i,j}| \leq \frac{\gamma}{|s^2 + 1.4\omega_{S_i}s + {\omega_{S_i}}^2|}, \quad s = j\omega \tag{5.74}$$

We can determine now the influence of the chosen template on the controller static gains. Given an input torque $c_{d,i}, i = 1..3$

$$\delta\phi_i = s^{-2}\sum_{k,j} S_{i,k}I_{G_{k,j}}^{-1}c_{d,j} \tag{5.75}$$

For a static input, we have already seen that at low frequency $\delta\phi = K^{-1}c_d$, therefore a static disturbing torque c_{d0}, will give a static angular displacement: $\delta\phi_0 = K^{-1}(s = 0)c_{d0}$, where:

$$\begin{aligned} K_{i,j}^{-1}(s=0) &= \textstyle\sum_k (s^{-2}S_{i,k})|_{s=0}I_{G_{k,j}}^{-1} \\ |K_{i,j}^{-1}(s=0)| &\leq \textstyle\sum_k |(s^{-2}S_{i,k})|_{s=0}|I_{G_{k,j}}^{-1}| \\ |K_{i,j}^{-1}(s=0)| &\leq \frac{\gamma}{{\omega_{S_i}}^2}\textstyle\sum_k |I_{G_{k,j}}^{-1}| \end{aligned} \tag{5.76}$$

If we assume, only for simplicity, that the inertia matrix is almost diagonal then $\sum_k |I_{G_{k,j}}^{-1}| \approx \frac{1}{I_{G_{j,j}}}$ then, we can derive that:

$$|K_{i,j}^{-1}(s=0)| \leq \frac{\gamma}{I_{G_{j,j}}{\omega_{S_i}}^2} \tag{5.77}$$

The value of ω_{S_i} defines the bandwidth of the sensitivity function for the i-th axis and also the loop static rejection to external disturbances.

In addition to shaping the sensitivity function it is also necessary to roll-off the controller in high frequency to limit the effects of the high frequency noise and actuation.

We use the additional output lines $z_K = K(s)W_K w$, where $W_K(s)$ is a diagonal matrix of roll-off filters one for each output control axis. The form of these filters can be set as:

$$W_{K_i}(s) = \frac{1}{\sqrt{2}I_{G_{i,i}}\omega_{S_i}{}^2} \frac{s + \omega_{R_i}}{s/100 + \omega_{R_i}}, i = 1..3 \tag{5.78}$$

where $I_G{}^{-1}{}_{i,i} \approx 1/I_{G_{i,i}}$ is used to have a low frequency controller limit balanced with the limits imposed by the sensitivity function. Where ω_{R_i} is the roll-off frequency of the controller in the ith axis. We place $\omega_{R_i} > \omega_{S_i}$, for example a decade ahead. When it is necessary to improve the cut-off effect, which may be necessary to manage flexible frequencies close to the controller bandwidth, the strength of the roll-off filter can be improved using the second order template:

$$W_{K_i}(s) = \frac{1}{\sqrt{2}I_{G_{i,i}}\omega_{S_i}{}^2} \frac{s^2 + 1.4\omega_{R_i}s + \omega_{R_i}^2}{(s/100)^2 + 1.4\omega_{R_i}(s/100) + \omega_{R_i}^2}, i = 1..3 \tag{5.79}$$

References

1. A.E. Bryson, Y.C. Ho, *Applied Optimal Control* (Hemisphere Publishing Co., Washington, 1975)
2. G.D. Hull, *Optimal Control Theory for Applications* (Springer, New York, 2003)
3. C. Bruni, G. Di Pillo, *Metodi Variazionali per il controllo ottimo* (Masson, Milano, 1993)
4. B. Chachuat, Non linear and dynamic optimization, from theory to practice, La-Teaching, 2007-001, Ecole Polytechnique Federale de Lausanne
5. H.J. Sussmann, *Nonlinear Controllability and Optimal Control*, Pure and Applied Mathematics (Marcel Dekker Inc., 1990)
6. G. Basile, G. Marro, *Controlled and Conditioned Invariants in Linear System Theory* (Prentice Hall, Englewood Cliffs, 1992)
7. B. Friedland, *Control System Design* (Mc Graw-Hill, Singapore, 1987)
8. A. Isidori, *Sistemi di Controllo*, vol. 2 (Siderea, Roma, 1993)
9. I. Postlethwaite, A.G.J. MacFarlane, *A Complex Variable Approach to the Analysis of Linear Multivariable Feedback Systems*, Lecture Notes in Control and Information Sciences (Springer, Berlin, 1979)
10. S. Skogestad, I. Postlethwaite, *Multivariable Feedback Control-Analysis and Design* (Wiley, Chichester, 2005)
11. D. Alazard, C. Cumer, P. Apkarian, M. Gauvrit, G. Ferreres, *Robustesse et commande optimale*, Cepadues Editions (1999)
12. J.C. Doyle, B.A. Francis, A.R. Tannenbaum, *Feedback Control Theory* (Macmillan Publishing Company, New York, 1992)
13. A. Stoorvogel, *The H-infinity Control Problem: A State Space Approach* (Prentice-Hall, New York, 1992)
14. K. Zhou, *Essentials of Robust Control* (Prentice Hall, New Jersey, 1998)

15. R.E. Kalman, R.S. Bucy, New results in linear filtering and prediction theory. Trans. ASME J. Basic Eng. **86D**(1), 95–108 (1961)
16. B.P. Gibbs, *Advanced Kalman Filtering, Least-Squares and Modeling—A Practical Handbook* (Wiley, Hoboken, 2011)
17. M. Fu, Lack of separation principle for quantized linear quadratic gaussian control. IEEE Trans. Autom. Control **57**(9) (2012)
18. M. Maggiore, K.M. Passino, A separation principle for a class of non uniformly completely observable systems. IEEE Trans. Autom. Control **48**(7) (2003)
19. M. Fu, On separation principle for linear quadratic control with input saturation, in *Fifth World Congress on Intelligent Control and Automation (WCICA)*, vol. 6, pp. 5605–5607, 5–19 June 2004
20. T. Georgiou, A. Lindquist, The separation principle in stochastic control. IEEE Trans. Autom. Control **57**(9) (2012)
21. I.M. Gelfand, S.V. Fomin, *Calculus of Variations* (Dover publications, New York, 2000)
22. F.H. Clarke, Maximum principles without differentiability. Bull. Am. Math. Soc. **81**(1) (1975)
23. F.H. Clarke, M.R. de Pinho, The non smooth maximum principle. J. Control Cybern. **38**(4A) (2009)
24. M.G. Crandall, H. Ishii, P.L. Lions, User's guide to viscosity solutions of second order partial differential equations. Bull. Am. Math. Soc. **27**(1), 1–67 (1992)
25. P.C. Hughes, R.E. Skelton, Controllability and observability of linear matrix second-order systems. ASME J. Appl. Mech. **47**, 415–424 (1980)
26. S.M. Joshi, *Control of Large Flexible Space Structures*. Lecture Notes in Control and Information Sciences, vol. 131 (Springer, Berlin, 1989)
27. A.J. Laub, Controllability and observability criteria for multivariable linear second-order models. IEEE Trans. Autom. Control **AC-29**(2) (1984)
28. J.P. Sharma, R.K. George, Controllability of Matrix second order systems: a trigonometric based approach. Electron. J. Differ. Equ. **2007**(80), 114 (2007)
29. S. Roman, *Advanced Linear Algebra* (Springer, New York, 2007)
30. P. Apkarian, M.N. Dao, D. Noll, Parametric robust structured control design. IEEE Trans. Autom. Control. doi:10.1109/TAC.2015.2396644
31. P. Apkarian, D. Noll, Nonsmooth H_∞ synthesis. IEEE Trans. Autom. Control **51**(1) (2006)
32. S. Boyd, C.A. Desoer, Subharmonic functions and performance bounds on linear time invariant feedback systems. IMA J. Math. Control Inf. **2**, 153–170 (1985)
33. N. Fezans, D. Alazard, N. Imbert, B. Carpentier, H_∞ control design for generalized second order systems based on acceleration sensitivity function, in *16th Mediterranean Conference on Control and Automation*, Ajaccio (France), 25–27 June 2008

Chapter 6
Attitude Control Methods

Abstract The parameters of the equations of a satellite including the flexible modes and the gravitational coupling, are calculated following the theory developed for the flexible satellite. After, we illustrate the attitude stabilization with two different approaches: one nonlinear, providing a global stabilization, and the other using a classic linear technique, which works only in proximity of the equilibrium. We conclude with an example of the H_∞ design presented in the previous chapter using the mixed acceleration sensitivity method on a satellite with large flexible appendages.

6.1 Introduction

The problem of the attitude control design of an artificial satellite lies in the definition of a control law such that the attitude of the *reference trajectory* is obtained and maintained with certain tolerances.

The *reference trajectory* depends on the mission and is derived from the payload pointing requirements, it is generally a free dynamics trajectory or a trajectory which is free under a particular—defined a priori—model of the external actions.

The torques, both predictable and unpredictable, will have to be compensated by the on-board actuation systems like thrusters (with chemical, cold gas or ion propellant), reaction or momentum wheels, control moment gyros, magnetic actuators, orientable solar sails, etc.

The typical environmental disturbing torques acting on artificial satellites are:

- gravity gradient: due to $\frac{1}{r^2}$ gravitational field variation for long, extended bodies (e.g. space shuttle, tethered vehicles, satellite with large appendages);
- aerodynamic drag: effect due to the offset between the *center of mass* (CoM) and the drag *center of pressure* (CP). Only a factor in LEO;
- magnetic torques: induced by residual magnetic moment. Model the spacecraft as a magnetic dipole. Only within magnetosphere;
- solar radiation: torques induced by CoM and solar CP offset. It can be compensated with differential reflectivity or reaction wheels;
- mass expulsion: torques induced by propellant leaks or jettisoned objects;

© Springer International Publishing Switzerland 2016

L. Mazzini, *Flexible Spacecraft Dynamics, Control and Guidance*, Springer Aerospace Technology, DOI 10.1007/978-3-319-25540-8_6

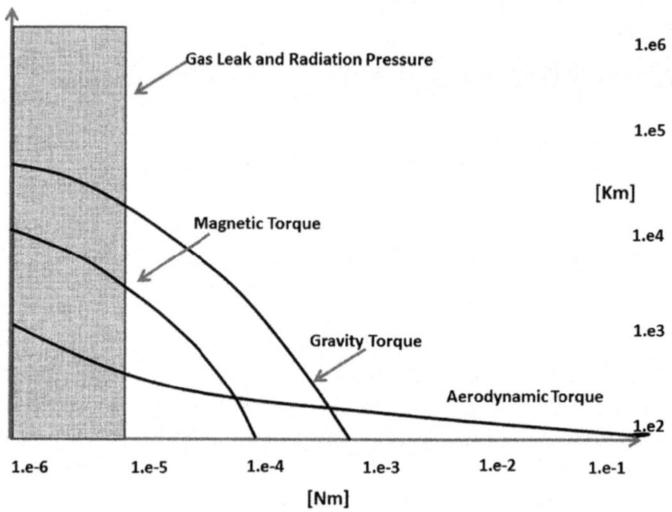

Fig. 6.1 Environmental disturbing torques order of magnitude [Nm] versus the satellite altitude

- Internal: on-board equipment (machinery, wheels, cryo-coolers, pumps etc., sloshing...). No net effect, but internal momentum exchange affects the attitude dynamics;

The Fig. 6.1 provides an order of magnitude of the environmental torques for a small satellite.

The environmental torques are normally quite low but act continuously; in addition to these torques the satellite may be subject to torques produced by the orbital maneuver thrusters which cannot be perfectly balanced with respect to the CoM. These last torques are normally higher than the environmental ones but last for limited time windows.

The errors of the satellite status with respect to the *reference trajectory* are measured by the on-board sensors and the errors, in keeping the correct attitude, feed an appropriate control logic which determines the commands to the actuators (RW, RCT, MGT...).

Since the equations of motion are highly nonlinear, the problem is often linearized, with the method illustrated in Sect. 3.11. The linearization allows to study the small motions of the dynamics around the *reference trajectory*, and gives the possibility to use the tools of linear control theory to provide attitude stabilization.

The techniques of optimal control that have been introduced in Chap. 5 are often utilized because they allow to ensure an optimal performance with a minimal use of actuators.

However, linearizing the equations is not always possible and in space applications the use of nonlinear techniques is frequent. This is due to:

- The satellite needs to acquire a given attitude trajectory from a very large set of possible unknown states;

- The satellite's attitude stability must be robust under very large perturbations of the state such that the use of linear techniques is not feasible.

In this chapter, we develop the equations for a satellite including the flexible modes and the gravitational coupling. We will calculate the mass properties following the theory developed in Chap. 3. After, we will illustrate the attitude stabilization reached with two different approaches: one nonlinear, providing a global stabilization, and the other using a classic linear technique, which works only in proximity of an equilibrium. We conclude with an example of H_∞ design using the mixed acceleration sensitivity method on the linearized equations of a satellite with large flexible appendages.

We consider the system described by (**2**), (**3**) of Eq. 3.94, we assume the existence of an equilibrium point $(\theta_0, \dot{\theta}_0, \epsilon_0, \dot{\varepsilon}_0)$, which is our *reference trajectory*; we aim to find a stabilizing control law to keep the state of the system close to this equilibrium point under the action of external disturbances. As the system is *nonlinear* we can follow two different ways:

1. study the stability of the nonlinear system using the *Lyapunov function method.*
2. synthesize an optimal control law for the linearized system in the neighborhood of an equilibrium point $(\theta_0, \dot{\theta}_0, \epsilon_0, \dot{\varepsilon}_0)$, the local stability will be guaranteed by the *Grossman-Hartman Theorem.*

Before passing to the concrete examples we review these two theorems of the *dynamical system theory.*

6.1.1 The Grossman-Hartman Theorem

When we implement a linear feedback control law to stabilize an equilibrium point, we basically create a new dynamic system which is locally stable in a neighborhood of this point. The nonlinear dynamics is not only locally stable but each local trajectory can be mapped into a trajectory of the linearized system and has the same rate of convergence to the equilibrium point. This property can be demonstrated when the equilibrium point is hyperbolic that is, all the eigenvalues of the linearized system have non zero real part. This is stated by the Grossman-Hartman theorem (see [1]).

Theorem (Grossman-Hartman). *Consider the dynamic system defined by the following set of differential equations* $\dot{x} = f(x)$ *with* $f \in C^1(\mathbb{R}^n)$ *and admitting an equilibrium point in* x_e. *The linearized system equations in* x_e *are*

$$\dot{x} = Ax \quad with \quad A_{ij} = \left[\frac{\partial f_i}{\partial x_j}\right]_{x=x_e}$$

where A *is the Jacobian matrix of* f *in* x_e. *If* x_e *is a hyperbolic equilibrium point, there is a homeomorphism*[1] H *from a neighborhood* S *of* 0 *to a neighborhood* R *of*

[1] A homeomorphism is a continuous map with a continuous inverse.

x_e, such that

$$x(t) = H\left(e^{At}x_0\right)$$

is a solution to $\dot{x} = f(x)$ whenever $e^{At}x_0 \in S$.

The use of this theorem is implicit any time we use a linear synthesis technique. The problem of defining the neighborhood of validity of the linear synthesis cannot be easily solved by analytical means, therefore usually simulations are performed to cover the region of interest.

6.1.2 Nonlinear Systems: Lyapunov-La Salle Theorem

When we want to stabilize a system of differential equations in a large neighborhood of the equilibrium point, the method of Lyapunov is the most powerful tool available to the control engineer. This method is based on the Lyapunov-La Salle theorem which states the sufficient conditions for the asymptotic stability of *nonlinear* systems (the cited theorem can be found in [2], the fundamental work of Lyapunov can be found in [3]).

Theorem (Lyapunov-La Salle.) *Consider the dynamic system $x_t = \Phi_t(x_0)$ defined by the following set of differential equations $\dot{x} = f(x)$ with $f \in C^1(\mathbb{R}^n)$ and $x \in \mathbb{R}^n$ the state of the nonlinear dynamic system.*

Given

$$\begin{aligned}
&V(x) \in C^1(\mathbb{R}^n),\\
&\Omega_L = \{x | V(x) < L\},\ \text{with } \Omega_L \text{ bounded open set in } \mathbb{R}^n,\ \text{such that};\\
&\dot{V} = \partial_x V f(x) \le 0, \forall x \in \Omega_L
\end{aligned}$$

and $R = \left\{x | \dot{V} = 0, x \in \Omega_L\right\}$.

If $M = \cup_\alpha M_\alpha$ with $M_\alpha \subseteq R$ the collection of all the <u>invariant sets</u> in R, i.e. $\phi_t(M_\alpha) \subset M_\alpha$ (Equilibrium points and cycles),

then: $\forall x_0 \in \Omega_L$, $\phi_{t\to\infty}(x_0) \to M$.

If M is an isolated equilibrium point, all the solutions in Ω_L asymptotically approach it (local asymptotic stability).

If $V(x) \to +\infty$ for $|x| \to \infty$ there is global asymptotic stability.

The method of Lyapunov resides in constructing a function of the state which is not increasing along the trajectories and to characterize the invariant sets which are contained in the set of null time derivative of such a function.

The theorem stated here is a quite general version based on the so called *Barbalat Lemma*. In the simpler version and by far the most used in control applications the set of the null time derivatives of the selected function coincides with an isolated equilibrium point which is the only attractive point of the system and the function

$V(x)$ is positive definite. The control design consists in determining a feedback law which establishes the desired properties of the dynamic system. Energy is very often a good starting point as Lyapunov function.

6.2 The Attitude Control of a Satellite with Flexible Appendages

6.2.1 The Equations of the Dynamics

We consider the equations of the dynamics of the flexible body of Fig. 6.2: the central body is assumed rigid, while the antenna mounted on the Earth facing panel ($+Z$ direction) is flexible. The satellite kinematics is such that all displacements happen in the orbital plane. The central body rotational motion is subject to the action of a control torque c_c.

Equation 3.94 translated to the CoM-**BRF** (with $p = 0, L = 0, P = 0$) can be applied; however, we will not use the specific symbols to indicate the Center of mass and the associated mass properties to simplify the notation. The orbital dynamics is assumed free from any control action, we have:

$$\begin{cases} \ddot{x} = -\mu_e \frac{k}{|x|^2} - \frac{3}{2}\mu_e k \frac{tr(\widetilde{I})}{m|x|^4} - 3\mu_e \frac{\widetilde{I}k}{m|x|^4} + \frac{15}{2}\mu_e k \frac{k^T \widetilde{I} k}{m|x|^4} & \text{translational dynamics} \\ \widetilde{I}\dot{\omega} + \widetilde{S}\ddot{\varepsilon} + \omega \times \widetilde{I}\omega + \omega \times S\dot{\varepsilon} + < \widetilde{J}\dot{\varepsilon} > \omega = 3\frac{\mu_e}{|x|^3} k \times \widetilde{I}k + c_c & \text{rotational dynamics} \\ \widetilde{S}^T \dot{\omega} + T\ddot{\varepsilon} + C\dot{\varepsilon} + K\varepsilon - \frac{1}{2}\omega^T \widetilde{J}\omega = \frac{\mu_e}{2} \frac{tr(\widetilde{J})}{|x|^3} - \frac{3}{2}\frac{\mu_e}{|x|^3} k^T \widetilde{J}k & \text{flexible dynamics} \end{cases} \tag{6.1}$$

where the state is (x, θ, ε) and c_c is the control torque generated for example with a *momentum management system* based on reaction wheels. If the momentum of such system is h_w, as shown by (3.74), then $c_c = -\dot{h}_w - \omega \times h_w$.

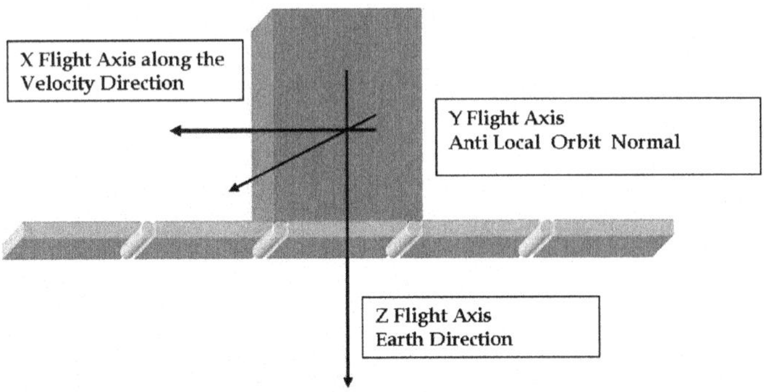

Fig. 6.2 Flexible satellite: rigid central body + flexible antenna

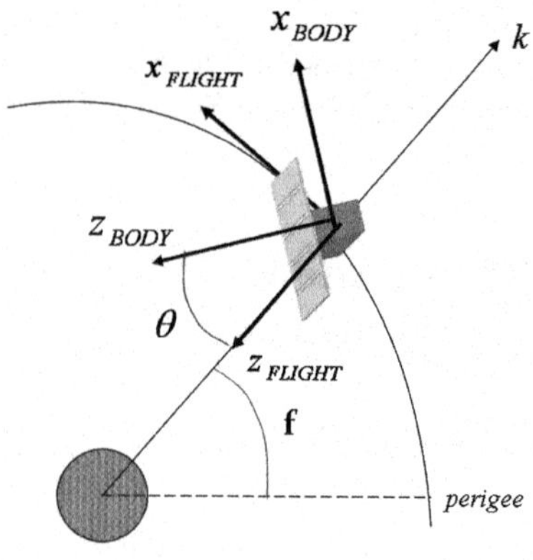

Fig. 6.3 Orbit anomaly f and attitude angle θ

In Eq. 6.1 ω represents the body angular rate in the inertial reference frame and k is the *anti-Earth* unit vector (Fig. 6.3).

We assume the following simplifying hypotheses:

- translational dynamics is independent on rotational and flexible dynamics:

$$\frac{I}{m\,|x|^2} \ll 1, \text{ typically} \approx 10^{-12}$$

- Plane motion: $\omega = e_2\left(\dot{\theta} - \omega_0\right)$

  where $k = [\sin\theta, 0, -\cos\theta]^T$ is the unit vector of the anti-Nadir direction (radius) in body coordinates, e_2 is the unit vector Y_{ORF} of the flight to the orbit plane, $\omega_0 = \dot{f}$ is the angular rate of the flight reference frame.

For the specific structure we have introduced (see Sect. 6.2.3):

$$I = \begin{bmatrix} I_x & 0 & 0 \\ 0 & I_y & 0 \\ 0 & 0 & I_z \end{bmatrix};\ J^k = \begin{bmatrix} J_y^k & 0 & J_{xz}^k \\ 0 & J_y^k & 0 \\ J_{xz}^k & 0 & 0 \end{bmatrix};\ K^{kl} = \begin{bmatrix} T^{kl} & 0 & 0 \\ 0 & T^{kl} & 0 \\ 0 & 0 & 0 \end{bmatrix}$$

$$S_x^k = S_z^k = 0;$$

$$J_y^k = \int 2z\phi_z^k \rho dV\,;\ J_{xz}^k = \int -x\phi_z^k \rho dV\,;\ S_y^k = \int -x\phi_z^k \rho dV\,;\ T^{kl} = \int \phi_z^k \phi_z^l \rho dV$$

$$\widetilde{I}_y = I_y + J_y^T\varepsilon + \varepsilon^T T\varepsilon\,;\ \widetilde{I}_z = I_z\,;\ \widetilde{I}_x = I_x + J_y^T\varepsilon + \varepsilon^T T\varepsilon$$

$$\widetilde{J}_{xz} = J_{xz}\,;\ \widetilde{J}_y = J_y + 2T\varepsilon$$

The equations of the rotational and flexible dynamics are:

$$\widetilde{I}_y\left(\ddot{\theta}-\dot{\omega}_0\right)+S_y\ddot{\varepsilon}+\widetilde{J}_y^T\varepsilon\left(\dot{\theta}-\omega_0\right)=\tfrac{3}{2}\tfrac{\mu_e}{|x|^3}\left(\widetilde{I}_z-\widetilde{I}_x\right)\sin 2\theta+3\tfrac{\mu_e}{|x|^3}\widetilde{J}_{xz}^T\varepsilon\cos 2\theta+c_c$$

$$S_y^T\left(\ddot{\theta}-\dot{\omega}_0\right)+T\ddot{\varepsilon}+C\dot{\varepsilon}+K\varepsilon-\tfrac{1}{2}\widetilde{J}_y\left(\dot{\theta}-\omega_0\right)^2=\tfrac{\mu_e}{|x|^3}\widetilde{J}_y-\tfrac{3}{2}\tfrac{\mu_e}{|x|^3}\widetilde{J}_y\sin^2\theta+\tfrac{3}{2}\tfrac{\mu_e}{|x|^3}\widetilde{J}_{xz}\sin 2\theta \tag{6.2}$$

6.2.2 *Orbit Equations*

The motion of the CoM is described by the equation of the Keplerian orbit:

$$\begin{aligned} r &= |x| = \tfrac{a\left(1-e^2\right)}{1+e\cos f} \\ h &= r^2\dot{f} = \sqrt{\mu_e a\left(1-e^2\right)}, \end{aligned} \tag{6.3}$$

being f the orbit *true anomaly*.

The position x of the CoM and the orbital angular speed ω_0 are functions of time through the orbit anomaly:

$$\begin{aligned} \omega_0 &= \dot{f} = \tfrac{\sqrt{\mu_e a\left(1-e^2\right)}}{a^2\left(1-e^2\right)^2}\left(1+e\cos f\right)^2 = \sqrt{\tfrac{\mu_e}{a^3}}\tfrac{\left(1+e\cos f\right)^2}{\left(1-e^2\right)^{-\frac{3}{2}}} \\ \dot{\omega}_0 &= \ddot{f} = -\tfrac{2\mu_e e}{a^3\left(1-e^2\right)^3}\sin f\left(1+e\cos f\right)^3 \\ \tfrac{\mu_e}{|x|^3} &= \tfrac{\mu_e\left(1+e\cos f\right)^3}{a^3\left(1-e^2\right)^3} \end{aligned} \tag{6.4}$$

where, introducing the *mean anomaly* M, and expanding in the Taylor series of the eccentricity, one has:

$$\begin{cases} f = M+\left(2e-\frac{e^3}{4}\right)\sin M+\frac{5}{4}e^2\sin 2M+\frac{13}{12}e^3\sin 3M+O\left(e^4\right) \\ \cos f = \cos M+e\left(\cos 2M-1\right)+e^2\left(\frac{9}{8}\cos 3M-\frac{9}{8}\cos M\right)+e^3\left(\frac{4}{3}\cos 4M-\frac{4}{3}\cos 2M\right)+O\left(e^4\right) \\ \sin f = \sin M+e\sin 2M+e^2\left(\frac{9}{8}\sin 3M-\frac{7}{8}\sin M\right)+e^3\left(\frac{4}{3}\sin 4M-\frac{7}{6}\sin 2M\right)+O\left(e^4\right) \\ \dot{M} = \sqrt{\frac{\mu_e}{a^3}} \end{cases}$$

6.2.2.1 Circular Orbit

For a *circular* orbital motion

$$\begin{aligned} \omega_0 &= \sqrt{\tfrac{\mu_e}{|x|^3}} = constant \\ \dot{\omega}_0 &= 0 \end{aligned} \tag{6.5}$$

therefore:

$$\begin{aligned}
&\widetilde{I}_y\ddot{\theta} + S_y\ddot{\varepsilon} + \widetilde{J}_y^T\dot{\varepsilon}\left(\dot{\theta}-\omega_0\right) + \tfrac{3}{2}\omega_0^2\left(\widetilde{I}_x-\widetilde{I}_z\right)\sin 2\theta - 3\omega_0^2\widetilde{J}_{xz}^T\varepsilon\cos 2\theta = c_c \\
&S_y^T\ddot{\theta} + T\ddot{\varepsilon} + C\dot{\varepsilon} + K\varepsilon = \tfrac{1}{2}\widetilde{J}_y\left(\dot{\theta}-\omega_0\right)^2 + \omega_0^2\widetilde{J}_y - \tfrac{3}{2}\omega_0^2\widetilde{J}_y\sin^2\theta + \tfrac{3}{2}\omega_0^2 J_{xz}\sin 2\theta
\end{aligned} \tag{6.6}$$

6.2.3 Computation of Mass/Geometry Properties

The satellite body presented in Fig. 6.2 is composed by a central rigid body plus a flexible antenna. The antenna is symmetric with respect to the YZ body plane and it is made of rigid panels of length L connected by flexible joints (see in Fig. 6.4 the geometry of the antenna). We name h the distance between the satellite CoM and the central section of the antenna; moreover we assume $x = \frac{m_A}{m_s}$ as the ratio between the mass of the antenna and the total mass and Ω as the first elastic frequency of the antenna.

One half of the antenna is represented with N panels and $N+1$ nodes located in the joints. E.g. for $N = 3$ the coordinates of the nodes are:

$$x_p = \left[\frac{L}{2}, \frac{3L}{2}, \frac{5L}{2}, \frac{7L}{2}\right]$$

The shape functions for the representation of the elastic modes are defined on every antenna element as follows:

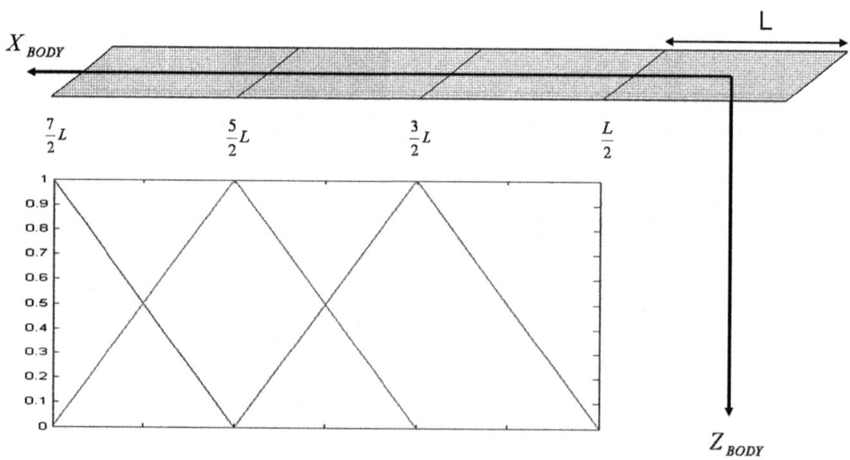

Fig. 6.4 Antenna nodes and shape functions

$$Z_p = \begin{bmatrix} -\frac{\frac{L}{2}-x}{L} & \frac{\frac{5L}{2}-x}{L} & 0 \\ 0 & -\frac{\frac{3L}{2}-x}{L} & \frac{\frac{7L}{2}-x}{L} \\ 0 & 0 & -\frac{\frac{5L}{2}-x}{L} \end{bmatrix}$$

where the (i, j) element contains the definition of the i-th shape function on the j-th antenna element.

The antenna elastic displacement along the z direction is expressed in nodal components as: $z(x) = Z_p^T(x)\,\varepsilon$, where ε is a column vector containing the nodal displacement.

From Eq. 3.3 we compute the mass, the inertia and modal participation factors and all the mass parameters for the antenna and for the whole satellite.

For the half-antenna (case $N = 3$):

$$\rho = \frac{m_A}{6L} \qquad \text{linear density}$$

$$m_A = \int_0^{3L} \rho dX = \frac{m_A}{2} \qquad \text{mass}$$

$$p_{Az} = \int_0^{3L} \rho h dX = \frac{m_A h}{2} \qquad \text{static momentum}$$

$$I_{Ay} = \int_0^{3L} (X^2 + h^2) dX = \frac{m_A}{8}\left(19L^2 + 4h^2\right) \text{ moment of inertia}$$

$$L_{Az} = \begin{bmatrix} \int_0^{3L} Z_{p1}\rho dX & \int_0^{3L} Z_{p2}\rho dX & \int_0^{3L} Z_{p3}\rho dX \end{bmatrix} \qquad \text{translational modal participation factors}$$
$$= \begin{bmatrix} \frac{m_A}{6} & \frac{m_A}{6} & \frac{m_A}{12} \end{bmatrix}$$

$$S_{Ay} = \begin{bmatrix} \int_0^{3L} Z_{p1} X \rho dX & \int_0^{3L} Z_{p2} X \rho dX & \int_0^{3L} Z_{p3} X \rho dX \end{bmatrix}$$
$$= \begin{bmatrix} \frac{-Lm_A}{4} & \frac{-5Lm_A}{12} & \frac{-19Lm_A}{72} \end{bmatrix} \qquad \text{rotational modal participation factors}$$

$$T_A = \int_0^{3L} Z_p^i Z_p^j \rho dX = \begin{bmatrix} \frac{m_A}{9} & \frac{m_A}{36} & 0 \\ \frac{m_A}{36} & \frac{m_A}{9} & \frac{m_A}{36} \\ 0 & \frac{m_A}{36} & \frac{m_A}{18} \end{bmatrix} \qquad \text{modal mass matrix}$$

$$J_{Ax} = J_{Ay} = \int_0^{3L} (2hZ_p^k)\rho dX = \begin{bmatrix} \frac{m_A h}{3} & \frac{m_A h}{3} & \frac{m_A h}{6} \end{bmatrix}^T$$

$$J_{Az} = \int_0^{3L} (2hZ_p^k - hZ_p^k - hZ_p^k)\rho dX = 0$$

moment of inertia first variation

$$J_{Ayz} = J_{Axy} = 0$$

$$J_{Axz} = -\int_0^{3L} (XZ_p^k)\rho dX = \begin{bmatrix} -\frac{m_A L}{4} & \frac{5m_A L}{12} & \frac{19m_A L}{72} \end{bmatrix}^T$$

$$K_{Az} = K_{Axy} = K_{Axz} = K_{Ayz} = 0$$

$$K_{Ax} = K_{Ay} = \int_0^{3L} Z_p^i Z_p^j \rho dX$$

moment of inertia second variation

$$K_{Ay} = \begin{bmatrix} \frac{m_A}{9} & \frac{m_A}{36} & 0 \\ \frac{m_A}{36} & \frac{m_A}{9} & \frac{m_A}{36} \\ 0 & \frac{m_A}{36} & \frac{m_A}{18} \end{bmatrix}$$

The matrix P is null because there are no orthogonal modes.
For the whole satellite:

m_A total antenna mass

$m_s = m_A + m_b$ total satellite mass

$p = p_b + m_A h = 0$ total satellite static momentum choosing the origin of the body frame such that $p_b = -hm_A$

$I_y = 2I_{Ay} + I_b = \frac{19m_A L^2}{4} + m_A h^2 + I_b$ total satellite moment of inertia

$L_z = [L_{Az}, \ L_{Az}] = [\frac{m_A}{6} \ \frac{m_A}{6} \ \frac{m_A}{12} \ \frac{m_A}{6} \ \frac{m_A}{6} \ \frac{m_A}{12}]$ total translational modal participation factors

$$S_y = [S_{Ay}, \ -S_{Ay}]$$

$= [-\frac{Lm_A}{4} \ -\frac{5Lm_A}{12} \ -\frac{19Lm_A}{72} \ \frac{Lm_A}{4} \ \frac{5Lm_A}{12} \ \frac{19Lm_A}{72}]$ total rotational modal participation factors

$$T = \begin{bmatrix} T_A & 0 \\ 0 & T_A \end{bmatrix} = \begin{bmatrix} \frac{m_A}{9} & \frac{m_A}{36} & 0 & 0 & 0 & 0 \\ \frac{m_A}{36} & \frac{m_A}{9} & \frac{m_A}{36} & 0 & 0 & 0 \\ 0 & \frac{m_A}{36} & \frac{m_A}{18} & 0 & 0 & 0 \\ 0 & 0 & 0 & \frac{m_A}{9} & \frac{m_A}{36} & 0 \\ 0 & 0 & 0 & \frac{m_A}{36} & \frac{m_A}{9} & \frac{m_A}{36} \\ 0 & 0 & 0 & 0 & \frac{m_A}{36} & \frac{m_A}{18} \end{bmatrix}$$

total mass matrix

$$J^k = \begin{bmatrix} J_y^k & 0 & J_{xz}^k \\ 0 & J_y^k & 0 \\ J_{xz}^k & 0 & 0 \end{bmatrix}$$

$$J_y = \left[J_{Ay}\ ;\ J_{Ay}\right] = \left[\tfrac{m_A h}{3}\ \tfrac{m_A h}{3}\ \tfrac{m_A h}{6}\ \tfrac{m_A h}{3}\ \tfrac{m_A h}{3}\ \tfrac{m_A h}{6}\right]^T$$ total moment of inertia first variation

$$J_x = J_y$$

$$J_{xz} = \left[J_{Axz}\ ;\ -J_{Axz}\right] = \left[-\tfrac{Lm_A}{4}\ -\tfrac{5Lm_A}{12}\ -\tfrac{19Lm_A}{72}\ \tfrac{Lm_A}{4}\ \tfrac{5Lm_A}{12}\ \tfrac{19Lm_A}{72}\right]^T$$

$$K^{kl} = \begin{bmatrix} T^{kl} & 0 & 0 \\ 0 & T^{kl} & 0 \\ 0 & 0 & 0 \end{bmatrix}$$

total moment of inertia second variation

$$K_y = T$$

$$K_x = T$$

$$K_z = K_{xy} = K_{xz} = K_{yz} = 0$$

Following Sect. 3.9, the expressions of the equations of motion can be further simplified with the introduction of a coordinate transformation translating the origin of the body reference frame to the point:

$$x^g = x^0 + M_{E\leftrightarrow B}\frac{\left(p + L^k \varepsilon_k\right)}{m}$$

such that $p_j = 0$ and $L_j = 0$.

In the new reference frame the deformation shape functions are also translated:

$$\phi\,(x) = Z_p\,(x) - \frac{L_z}{m_s}$$

The mass properties are then recomputed using the new origin and shape functions.

$$m_s = \int_{V_0} \rho_o\, dV_o = m_s$$

$$p_G = \int_{V_0} (r - d)\, \rho_o\, dV_o = 0$$

$$L_G^k = \int_{V_0} (\phi^k - L^k/m_s)\rho_o\, dV_o = 0$$

$$S_G^q = S^q - [d, L^q] = S^q$$

$$I_G = I - m_s(d^T d I_{3x3} - dd^T) = I$$

$$\boxed{T_G^{kq} = T^{kq} - \frac{L^{kT} L^q}{m_s}}$$

$$J_G^k = J^k - 2d^T L^k I_{3x3} + dL^{kT} + L^k d^T = J^k$$

$$K_G^{kq} = K^{kq} - \frac{L^{kT} L^q I_{3x3}}{m_s} + \frac{L^k L^{qT}}{m_s} \rightarrow \boxed{K_{Gx}^{kq} = K_{Gy}^{kq} = T_G^{kq}}$$

$$P_G^{kq} = P^{kq} - [L^k, L^q] = 0$$

being $d = \frac{p}{m_s} = 0$, since the static momentum p is already null in the original coordinates. To compute the stiffness and damping matrices K and C, we introduce the elastic potential energy. Each nodal spring has an elastic energy given by the elastic constant multiplied by the square of the angular displacement divided by two. Summing up and setting the boundary condition $\varepsilon_0 = 0$, we can calculate the total elastic potential energy as

$$\mathscr{V}_I = \sum_{i=1}^{N} \left(\frac{K_e}{2} \delta\theta_i{}^2 \right) = \sum_{i=1}^{N} \left(\frac{K_e}{2} \left(\frac{\varepsilon_{i+1} - \varepsilon_i}{L} - \frac{\varepsilon_i - \varepsilon_{i-1}}{L} \right)^2 \right)$$

and the stiffness matrices for the half antenna and for the whole structure:

$$K_A = \left[\frac{\partial^2 \mathscr{V}_I}{\partial \varepsilon_i \partial \varepsilon_j} \right] \; K = \begin{bmatrix} K_A & 0 \\ 0 & K_A \end{bmatrix}$$

For $N = 3$ we have:

half-antenna elastic potential Energy $\mathscr{V}_I = \frac{K_e}{2L^2} \left(6\varepsilon_1^2 - 8\varepsilon_1\varepsilon_2 + 2\varepsilon_1\varepsilon_3 + 5\varepsilon_2^2 - 4\varepsilon_2\varepsilon_3 + \varepsilon_3^2 \right)$

half-antenna stiffness matrix $\begin{bmatrix} 6\frac{K_e}{L^2} & -4\frac{K_e}{L^2} & \frac{K_e}{L^2} \\ -4\frac{K_e}{L^2} & 5\frac{K_e}{L^2} & -2\frac{K_e}{L^2} \\ \frac{K_e}{L^2} & -2\frac{K_e}{L^2} & \frac{K_e}{L^2} \end{bmatrix}$

We introduce the reference angular frequency $\Omega = \sqrt{\frac{K_e}{m_A L^2}}$, computing the cantilevered angular frequencies of the antenna it turns out: $\Omega_1 = 0,5133\Omega$, $\Omega_2 = 22,006\Omega$, $\Omega_3 = 158,864\Omega$, for the right and the left wing.

We calculate the damping matrix of the total antenna as proportional to the stiffness matrix:

$$\boxed{C = qK}$$

We choose q such that in the first mode we get a damping factor of $\zeta = 0.01$.

To find q we consider the general equations of the MDOF (multi degrees of freedom) linear structural dynamics

$$T\ddot{x} + qK\dot{x} + Kx = 0$$

In diagonalized coordinates $\tilde{x}$:

$$\begin{aligned}\tilde{x} &= Ux \\ U^T T U &= I \\ U^T K U &= diag\left(\Omega_i^2\right)\end{aligned}$$

$$\ddot{\tilde{x}}_i + q\Omega_i^2\dot{\tilde{x}}_i + \Omega_i^2\tilde{x}_i = 0$$

$$q\Omega_1^2 = 2\zeta\Omega_1 \rightarrow q = 2\frac{\zeta}{\Omega_1}$$

6.2.4 Nonlinear System Stabilization: Application of Lyapunov-La Salle Theorem

We are looking for a Lyapunov function $\mathcal{V}\,(\theta, \omega, \varepsilon, \dot{\varepsilon}) \in C^1\,(\mathbb{R}^n)$ such that, in some open set of the state space, it is $\dot{\mathcal{V}} \leq 0$ and $\mathcal{V} < L$, for some L.

In the equations of the rotational and flexible dynamics Eq. 6.6 it is possible to recognize the derivatives of the gravitational potential $\mathfrak{V}$:

$$\mathfrak{V} = -\frac{\omega_0^2}{2}tr\left(\widetilde{I}\right) + \frac{3}{2}\omega_0^2 k^T \widetilde{I} k \tag{6.7}$$

Therefore:

$$\begin{cases}\widetilde{I}\dot{\omega} + \widetilde{S}\ddot{\varepsilon} + \omega\times\widetilde{I}\omega + \omega\times\left\langle\widetilde{S}\dot{\varepsilon}\right\rangle + \left\langle\widetilde{J}\dot{\varepsilon}\right\rangle\omega + \frac{\partial\mathfrak{V}}{\partial\theta} = c_c \\ \widetilde{S}^T\dot{\omega} + T\ddot{\varepsilon} + C\dot{\varepsilon} + K\varepsilon - \frac{1}{2}\omega^T\widetilde{J}\omega + \frac{\partial\mathfrak{V}}{\partial\varepsilon} = 0\end{cases} \tag{6.8}$$

Substituting the expressions derived from the hypotheses made for the flexible structure, we have:

$$\begin{cases}\omega = e_2\left(\dot{\theta} - \omega_0\right) \\ \widetilde{I} = \begin{pmatrix} I_x & 0 & 0 \\ 0 & I_y & 0 \\ 0 & 0 & I_z\end{pmatrix} + \begin{pmatrix} J_y^T\varepsilon & 0 & J_{xz}^T\varepsilon \\ 0 & J_y^T\varepsilon & 0 \\ J_{xz}^T\varepsilon & 0 & 0\end{pmatrix} + \begin{pmatrix}\varepsilon^T T\varepsilon & 0 & 0 \\ 0 & \varepsilon^T T\varepsilon & 0 \\ 0 & 0 & 0\end{pmatrix} \\ k = [\sin\theta, 0, -\cos\theta]^T \\ \widetilde{J}_y = J_y + 2T\varepsilon\end{cases}$$

The function $\mathfrak{V}$ becomes:

$$
\begin{aligned}
\mathfrak{V} &= -\frac{\omega_0^2}{2} tr\left(\widetilde{I}\right) + \frac{3}{2}\omega_0^2 \left(\widetilde{I}_x \sin^2\theta + \widetilde{I}_z \cos^2\theta - 2\widetilde{I}_{xz} \sin\theta\cos\theta\right) \\
&= -\frac{\omega_0^2}{2} tr\left(\widetilde{I}\right) + \frac{3}{2}\omega_0^2 \left[I_z + (\widetilde{I}_x - I_z)\sin^2\theta - \widetilde{I}_{xz}\sin 2\theta\right] \\
&= -\omega_0^2 \left(J_y^T \varepsilon + \varepsilon^T T \varepsilon\right) + \frac{3}{2}\omega_0^2 \left[(\widetilde{I}_x - I_z)\sin^2\theta - \widetilde{I}_{xz}\sin 2\theta\right] + \underbrace{\frac{\omega_0^2}{2}\left(3I_z - I_x - I_y\right)}_{\text{const}}
\end{aligned}
$$

The equations of the dynamics can be written as:

$$
\begin{cases}
\widetilde{I}_y \ddot{\theta} + S_y \ddot{\varepsilon} + \widetilde{J}_y^T \dot{\varepsilon}\left(\dot{\theta} - \omega_0\right) + \frac{\partial \mathfrak{V}}{\partial \theta} = c_c \\
S_y^T \ddot{\theta} + T\ddot{\varepsilon} + C\dot{\varepsilon} + K\varepsilon - \frac{1}{2}\widetilde{J}_y \left(\dot{\theta} - \omega_0\right)^2 + \frac{\partial \mathfrak{V}}{\partial \varepsilon} = 0
\end{cases}
\tag{6.9}
$$

Multiplying the first equation by $\dot{\theta}$, the second one by $\dot{\varepsilon}^T$ and then summing the two equations, we get:

$$
\begin{aligned}
&\widetilde{I}_y \ddot{\theta}\dot{\theta} + \widetilde{J}_y^T \dot{\varepsilon}\dot{\theta}^2 - \widetilde{J}_y^T \dot{\varepsilon}\dot{\theta}\omega_0 - \tfrac{1}{2}\widetilde{J}_y^T \dot{\varepsilon}\left(\dot{\theta}^2 - 2\omega_0\dot{\theta} + \omega_0^2\right) \\
&+ \tfrac{d}{dt}\left(\dot{\theta} S_y \dot{\varepsilon} + \tfrac{1}{2}\dot{\varepsilon}^T T\dot{\varepsilon} + \tfrac{1}{2}\varepsilon^T K\varepsilon + \mathfrak{V}\right) = -\dot{\varepsilon}^T C\dot{\varepsilon} + c_c\dot{\theta}
\end{aligned}
\tag{6.10}
$$

That is:

$$
\frac{d}{dt}\phi = -\dot{\varepsilon}^T C\dot{\varepsilon} + c_c\dot{\theta} \tag{6.11}
$$

with

$$
\phi = \widetilde{I}_y \frac{\dot{\theta}^2}{2} - \frac{1}{2}\widetilde{I}_y \omega_0^2 + \dot{\theta} S_y \dot{\varepsilon} + \frac{1}{2}\dot{\varepsilon}^T T\dot{\varepsilon} + \frac{1}{2}\varepsilon^T K\varepsilon + \mathfrak{V} \tag{6.12}
$$

Introducing the state feedback $c_c = -K_P\theta - K_D\dot{\theta}$ it is possible to identify a Lyapunov function $\mathscr{V}$:

$$
\mathscr{V} = \phi + K_p \frac{\theta^2}{2}, \tag{6.13}
$$

such that

$$
\frac{d}{dt}\mathscr{V} = -\dot{\varepsilon}^T C\dot{\varepsilon} - K_D\dot{\theta}^2 \leq 0 \quad \text{for } K_D > 0 \tag{6.14}
$$

To satisfy Lyapunov-La Salle theorem hypotheses for asymptotic stability, we need to prove that $\mathscr{V}$ is bounded below, i.e. that the Hessian matrix of $\mathscr{V}$ is positive definite.

The manifold where

$$
\dot{\mathscr{V}} = 0
$$

is

$$R = \left\{(\theta, \varepsilon) \,|\dot{\theta} = 0, \dot{\varepsilon} = 0\right\};$$

therefore the maximal invariant set in R is the collection of the equilibrium points $(\theta,\ \varepsilon)$ such that

$$\begin{aligned} &\tfrac{\partial \mathfrak{V}}{\partial \theta} + K_P \theta = 0 \\ &K\varepsilon - \tfrac{1}{2}\widetilde{J}_y \omega_o^2 + \tfrac{\partial \mathfrak{V}}{\partial \varepsilon} = 0, \end{aligned}$$

namely,

$$\begin{aligned} &\tfrac{3}{2}\omega_0^2\left(\widetilde{I}_x - I_z\right)\sin 2\theta - 3\omega_0^2 J_{xz}^T \varepsilon \cos 2\theta + K_P \theta = 0 \\ &K\varepsilon - \tfrac{3}{2}\omega_0^2 \widetilde{J}_y \cos\theta^2 - \tfrac{3}{2}\omega_0^2 J_{xz} \sin 2\theta = 0 \end{aligned}$$

An equilibrium point lies in a neighborhood of $(0, 0)$. Neglecting the terms of order ω_0^4, it can be computed approximately as:

$$\begin{aligned} \hat{\theta} &= 0 + O\left(\omega_0^4\right) \\ \hat{\varepsilon} &= \tfrac{3}{2}\omega_0^2 K^{-1} J_y + O\left(\omega_0^4\right) \end{aligned} \tag{6.15}$$

To prove that $\mathscr{V}$ is bounded below it is sufficient that the Hessian matrix is positive-definite in a neighborhood of the equilibrium point. In fact, the Lyapunov function can be expressed as the sum of kinetic energy, that is always positive definite, plus the component of $\mathscr{V}$ in R:

$$\mathscr{V} = \mathscr{T} + \mathscr{V}|_R \,.$$

$$H = \begin{bmatrix} \frac{\partial^2 \mathscr{V}|_R}{\partial \theta^2} & \frac{\partial^2 \mathscr{V}|_R}{\partial \theta \partial \varepsilon} \\ \frac{\partial^2 \mathscr{V}|_R}{\partial \theta \partial \varepsilon} & \frac{\partial^2 \mathscr{V}|_R}{\partial \varepsilon^2} \end{bmatrix} > 0$$

Neglecting terms of order ω_0^4:

$$H = \begin{bmatrix} K_P + 3\omega_0^2\left(\widetilde{I}_x - I_z\right) & -3\omega_0^2 J_{xz}^T \\ -3\omega_0^2 J_{xz}^T & K - 3\omega_0^2 T \end{bmatrix} > 0$$

If U is such that $U^T\left(K - 3\omega_0^2 T\right)U = \Omega^2 - 3\omega_0^2$, the transformation $W = \begin{bmatrix} 1 & 0^T \\ 0 & U \end{bmatrix}$ brings H into:

$$\begin{bmatrix} K_p + 3\omega_0^2\left(\widetilde{I}_x - I_z\right) & -3\omega_0^2 J_{xz}^T U \\ -3\omega_0^2 U^T J_{xz} & \Omega^2 - 3\omega_0^2 \end{bmatrix}$$

A necessary and sufficient condition for the matrix H to be definite-positive is the Routh-Hurwitz rule: *all the leading minors must be positive* (see any linear algebra text, i.e. [4]). In order to calculate the leading minors we can use the fact that $\Omega^2 - 3\omega_0^2$ is diagonal. The value of the leading minors m_k results to be:

$$m_1 = K_p + 3\omega_0^2 \left(\widetilde{I}_x - I_z\right)$$
$$m_k = (m_1 - \sum_{i<k} \frac{9\omega_0^4 (U^T J_{xz})|_i{}^2}{\Omega_i{}^2 - 3\omega_0^2}) \prod_{i<k} (\Omega_i{}^2 - 3\omega_0^2) \tag{6.16}$$

In order to have all the minors positive we must have:

$$K_P + 3\omega_0^2 \left(\widetilde{I}_x - I_z\right) > \sum_i \frac{9\omega_0^4 (U^T J_{xz})|_i{}^2}{\Omega_i{}^2 - 3\omega_0^2}$$
$$\Omega_i{}^2 > 3\omega_0^2, \forall i \tag{6.17}$$

These conditions show the role played by the term $\widetilde{I}_x - I_z$; in the rigid body case, $\Omega_i \rightarrow \infty, \forall i$, therefore the conditions of Eq. 6.17 are satisfied if

$$K_P + 3\omega_0^2 \left(\widetilde{I}_x - I_z\right) > 0. \tag{6.18}$$

In the rigid case the dynamics is stable even without an active control ($K_P = 0$) if the satellite Z-axis, the one pointing the Earth, has an inertia smaller with respect to the inertia of the axis aligned to the orbital velocity.

The gravity gradient exerts the torque of a stabilizing spring of stiffness $3\omega_0^2 \left(\widetilde{I}_x - I_z\right)$. In the flexible case, the stability is provided if both Eq. 6.17 are satisfied. Therefore, when $K_P = 0$, $\widetilde{I}_x$ must be higher than I_z of a quantity of order ω_0^2 and, the elastic frequencies must be relatively higher than the orbital ones: $\Omega_i{}^2 > 3\omega_0^2, \forall i$.

If $K_P \gg 3\omega_0^2 \left(\widetilde{I}_x - I_z\right)$ the gravity gradient term is not relevant, and stability is provided if the cantilevered frequencies are sufficiently higher than the orbital one.

6.2.5 Linearized Equations and LQR Control Application

The Eq. 6.6 can be re-written in non-dimensional variables multiplying the first one by $\frac{1}{m_A L^2 \Omega^2}$ and the second one by $\frac{1}{\Omega^2 m_A L}$, where:

$$m_A = \text{ antenna mass}$$
$$L = \text{ panel length}$$
$$\Omega = \sqrt{\frac{K_e}{m_A L^2}} = \Omega_1 / 0,513$$

Then linearizing around ($\theta = 0$; $\varepsilon = 0$) brings to

$$\begin{aligned} I_y\ddot{\theta} + S_y\ddot{\varepsilon} - \omega_0 J_y^T\dot{\varepsilon} - 3\omega_0^2\left(I_z - I_x\right)\theta - 3\omega_0^2 J_{xz}^T\varepsilon &= \quad c_c \\ S_y^T\ddot{\theta} + T\ddot{\varepsilon} + C\dot{\varepsilon} + K\varepsilon + J_y\omega_0\dot{\theta} - 3\omega_0^2 J_{xz}\theta \quad &= \tfrac{3}{2}J_y\omega_0^2 \end{aligned}$$

with the non-dimensional variables:

$$\begin{array}{lll} I = \frac{I}{m_A L^2} & T = \frac{T}{m_A} & D = \frac{I_z - I_x}{m_A L^2} \\ J = \frac{J}{m_A L} & K = \frac{K}{m_A \Omega^2} & w = \frac{c_c}{m_A L^2 \Omega^2} \\ S = \frac{S}{m_A L} & C = \frac{C}{m_A \Omega} & \varepsilon = \frac{\varepsilon}{L} \\ K_y = \frac{K_y}{m_A} & \omega_0 = \frac{\omega_0}{\Omega} & t = t\Omega \end{array}$$

The linearized second-order equations for the rotational and flexible dynamics of the structure sketched in Fig. 6.2 are:

$$M\ddot{\xi} + P\dot{\xi} + Q\xi = bw \tag{6.19}$$

with

$$\xi = \begin{pmatrix} \theta - \hat{\theta} \\ \varepsilon - \hat{\varepsilon} \end{pmatrix}, \text{ with } \left(\hat{\theta} = 0, \hat{\varepsilon} = \frac{3}{2}\omega_0^2 K^{-1} J_y\right) \text{ the equilibrium point}$$

$$M = \begin{bmatrix} I_y & S_y \\ S_y^T & T \end{bmatrix}$$

$$P = \begin{bmatrix} 0 & -\omega_0 J_y^T \\ \omega_0 J_y & C \end{bmatrix}$$

$$Q = \begin{bmatrix} -3\omega_0^2 D & -3\omega_0^2 J_{xz}^T \\ -3\omega_0^2 J_{xz} & K \end{bmatrix}$$

$$b = \begin{bmatrix} 1 & 0_{2Nx1} \end{bmatrix}$$

Writing Eq. 6.19 as first order equations

$$\boxed{\dot{X} = AX + Bw}$$

with state:

$$X = \left(\theta - \hat{\theta}, \varepsilon - \hat{\varepsilon}, \dot{\theta}, \dot{\varepsilon}\right)^T \tag{6.20}$$

and state matrix A:

$$A_1 = \begin{bmatrix} 0 & 0_{1\times 2N} & 1 & 0_{1\times 2N} \\ 0_{2N\times 1} & 0_{2N\times 2N} & 0_{2N\times 1} & I_{2N\times 2N} \end{bmatrix}$$

$$A_2 = M^{-1}\left[Q \quad P\right]$$

$$A = \begin{bmatrix} A_1 \\ A_2 \end{bmatrix}$$

$$B = \begin{bmatrix} 0_{(2N+1)\times 1} \\ M^{-1}b \end{bmatrix}$$

where:

$w = \frac{c_c}{m_A L^2 \Omega^2}$ non-dimensional control torque

$\varepsilon = \frac{\varepsilon}{L}$ non-dimensional antenna displacement

θ rotation angle of Z-axis from Nadir

Having introduced the non-dimensional variables we can cover with the same model a wide class of physical systems; nevertheless, some of them need to be fixed in order to give a numerical value to all the defined matrices.

We then fix the following:

$$\begin{array}{lll} h/L & = 1/4 & \text{CoM-antenna distance} \\ I_b & = m_b L^2 & \text{satellite body momentum of inertia} \\ m_b & = 0.95 m_s = 0.95(m_A + m_b) & \text{satellite body mass} \\ I_y/(m_A L^2) & = (h/L)^2 + I_b/(m_A L^2) + 19/4 = 24.75 & \text{satellite momentum of inertia} \\ \frac{\omega_0}{\Omega} & = 0 & \end{array} \tag{6.21}$$

This formulation of the problem cancels the gravity gradient term, which has been studied in the previous section. We have to consider that in practical cases the ratio $\omega_0/\Omega \approx 10^{-3}$, and this coupling is not relevant when studying an active controller such that $K_P \gg 3\omega_0^2\left(\widetilde{I}_x - I_z\right)$.

The dynamic system defined by Eq. 6.2.5 results not controllable.

We may realize it thinking of the geometrical properties of the eigen-modes. For the symmetry of the problem, the cantilevered eigen-modes of the two wings of the right and left antenna have the same frequencies and a mirrored shape. Therefore, considering the symmetry of the central body, the state subspace of the free eigen-modes can be presented with a system of 3 blocks of symmetric and antisymmetric iso-frequency modal pairs. In such condition we can operate a space linear transformation of the flexible degrees of freedom into three symmetric and three antisymmetric modal amplitudes: $\epsilon_k = \epsilon_{sym,k} + \epsilon_{asy,k}$.

The three symmetric eigen-modes are not controllable using a torque applied on the rigid part of the satellite, while the antisymmetric modes are controllable.

The simplest way to see this, is to excite the symmetric modes and realize that they do not produce any rotation θ. The contrary is also true, any rotation of the spacecraft will not excite the symmetric modes of the antenna but only the antisymmetric ones.

Therefore, before applying the LQR we need to produce a minimal realization of the system which will cancel the three symmetric modes and the 6 associated states.

With this cancellation we assume that the symmetric states are simply null, because they are irrelevant. Should they be excited by some other input, like a reaction control thruster pulse, the symmetric modes could not be actively damped.

The reduced state contains only the antisymmetric part of the flexible amplitudes

$$X = \left(\theta, \epsilon_{asy,1}, \epsilon_{asy,2}, \epsilon_{asy,3}, \dot{\theta}, \dot{\epsilon}_{asy,1}, \dot{\epsilon}_{asy,2}, \dot{\epsilon}_{asy,3}\right)^T .$$

We assume in the LQR synthesis that this state is fully measured by adequate sensors or fully estimated by a very well matched observer.

With such system we study the acquisition of a target with a slew of $\theta = 0.1$ rad, we want to be fast in the acquisition but also we want to have a quick stabilization of the antenna at about $\theta = 0.001$ rad, $\varepsilon = 0.003$ (the antenna antisymmetric tip deflection is equivalent in terms of average pointing to a rotation three times smaller) and we do not want to saturate the control torque.

To apply the LQR method as illustrated in Sect. 5.4.3, the weight matrices Q, R must be defined. In general the appropriate weights are calculated on the basis of the closed loop poles and the torque saturation value. We set a fixed Q matrix

$$Q = diag(2500, 1, 1, 2500, 1, 1, 1, 900) \tag{6.22}$$

and we study the optimal solution for the following values of R:

$$R = \left(10^3, 10^4, 10^5, 10^6\right) \tag{6.23}$$

The solution obtained for the dynamics of θ, $\epsilon_{asy,3}$, w are reported in the Figs. 6.5, 6.6, 6.7.

From these figures we can see that by augmenting the weight on the control we lower the bandwidth of the system reducing the antenna excitation and the necessary control torque.

The requested performances are reached in a time about double the first zero crossing of the satellite rotation.

In order to make a qualitative comparison among the different controllers we evaluate from the simulation the following parameters: τ_{set} the non-dimensional time where θ first crosses the zero, ϵ_{max} is the maximum of $\epsilon_{asy,3}$, w_{max} is the maximum required non-dimensional control torque, $\tau_{min} = \sqrt{\frac{4\theta_0 I_y}{m_A L^2 w_{max}}}$ is the

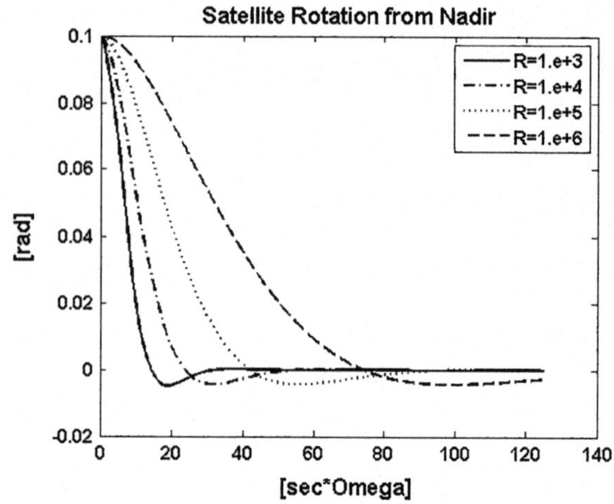

Fig. 6.5 Satellite rotation angle θ

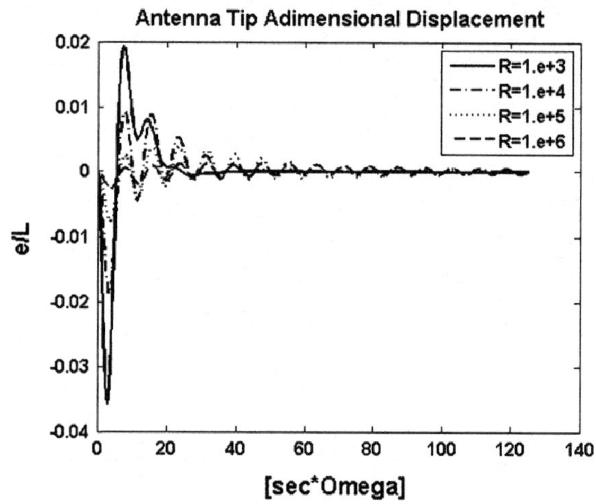

Fig. 6.6 Antenna tip $\epsilon_{asy,3}/L$

non-dimensional time resulting from a minimum time rotation on a rigid spacecraft (using the same maximum control torque and total inertia). The minimum time solution can be used for comparison of the presented solutions.

We analyze the results using the following table:

Case	$R = 10^3$	$R = 10^4$	$R = 10^5$	$R = 10^6$
τ_{set}	14.5	25	42	74
ϵ_{max}	0.035	0.019	0.008	0.002
w_{max}	0.15	0.05	0.015	0.005
τ_{min}	5	8 1	13	26

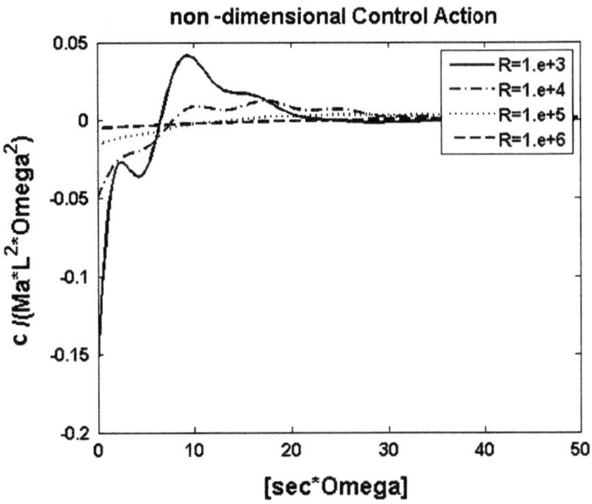

Fig. 6.7 Control torque w

We see that the response time is about 3 times longer than the minimum time-rigid satellite-solution. The minimum time solution however, if applied to the flexible satellite would produce the modal excitation about two times ϵ_{max} and only passively damped (which means that after $\tau = 100.$ we have a reduction of only 0.36 times of the oscillation amplitude due to the damping factor $\zeta = 0.01$).

The operating performances $\theta < 0.001, \epsilon < 0.003$ are reached in about 2 times τ_{set}. To implement the desired dynamics we need to develop a torque $c_c = w_{max} m_A L^2 \Omega^2$. Considering for example $m_A = 40\,\text{Kg}$, $\Omega = 1\,\text{rad/s}$, $L = 0.5\,\text{m}$, we see that the corresponding torques which we should apply are: $c_c = 1.5\,\text{Nm}$, $c_c = 0.5\,\text{Nm}$, $c_c = 0.15\,\text{Nm}$, $c_c = 0.05\,\text{Nm}$. These torque values size the actuation system, for a high-speed solution we should embark CMGs providing control torques up to 10 Nm, for a slower solution we could use Reaction Wheels which can provide torques up to about 0.2 Nm (see Chap. 9).

The use of reaction control thrusters (RCT see Chap. 8) should be avoided because they can produce the desired torques but they would also produce a linear acceleration exciting the symmetric antenna modes.

We conclude this section mentioning that the best performances can be achieved via a torque profile applied open loop and suitably shaped to minimize the flexible mode excitation added with a closed loop stabilizing controller.

This synthesis is the result of the LQR algorithm that assumes the antenna deflection is measured or well estimated (at least the anti-symmetric part). When the full state cannot be directly measured we have to rely upon the use of state observers which pose the question related to their robustness (stability with respect to parameter variations).

6.3 The Robust Control of Flexible Modes

We illustrate now a design exercise of a flexible satellite (see Fig. 6.8) with a very large flexible reflector under study for a future ESA project (see the references [5–7]).

The design approach is the one already described in Sect. 5.5.5 (Fig. 6.9).

The dynamic mass matrix $M(s)$ is formed using Eq. 3.137 of Chap. 3.

We need the rigid body mass properties and both the rotational and translational participation factors. The translational factors are necessary to bring the rotational participation factors to the center of mass CoM following the transformations of Eq. 3.9. The rigid mass properties are:

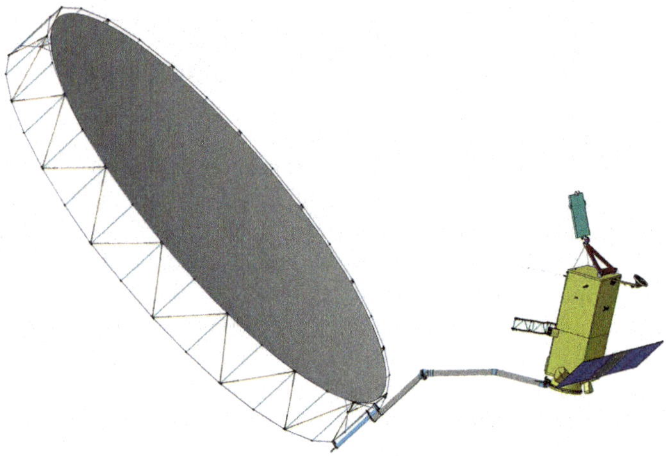

Fig. 6.8 Large antenna radar satellite

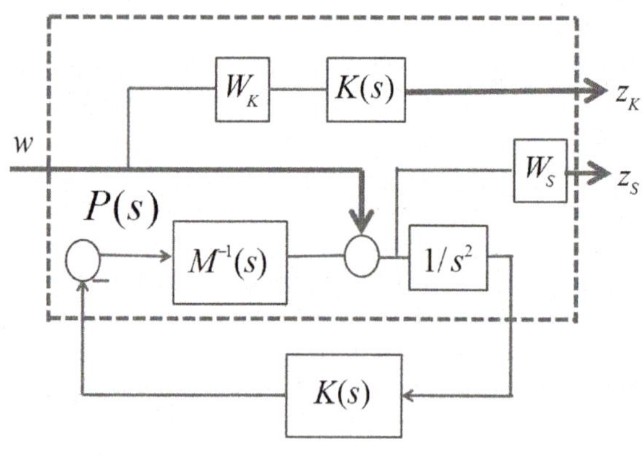

Fig. 6.9 Mixed acceleration sensitivity set up

m_s	1024.	–	–	Kg
X_G, Y_G, Z_G	0.014	0.6258	1.6119	m
I_{xx}, I_{yy}, I_{zz}	9289.1	2895.1	8731.3	Kg/m^2
I_{yz}, I_{xz}, I_{xy}	114.2	−7.0	9.8	Kg/m^2

The Table in Fig. 6.10 provides the frequencies and the modal parameters of this design exercise. The modal participation factors are provided with respect to a frame having the origin at the root of the appendage (located in $X_{i/f}$, $Y_{i/f}$, $Z_{i/f}$ in the **BRF** triad) and axes parallel to **BRF**.

The plant is supposed subject to a disturbing torque of 0, 1 Nm for a duration of 30 s applied after 200 s of simulation; this disturbance can be produced during an orbital maneuver because the small thrusters' unbalance with respect to the CoM, our goal is to keep the rigid attitude dynamics below $|\theta| < 0.1$ deg under such conditions.

We consider as sensors a gyroscope with a flat PSD white noise and $\sigma_{rate} = 6 \times 10^{-6}$ rad/s, and two star trackers, each one having a white noise flat PSD with $\sigma_{angle} = 20$ arcsec around the boresight and $\sigma_{angle} = 2$ arcsec in the orthogonal directions. The star trackers are mounted so that their boresights are 90 deg one from the other, so that the star tracker measurement after the pre-processing can be considered equivalent to a $\sigma_{angle} = 2$ arcsec in all axes.

The wheels are mounted in a pyramidal configuration so that along each body axis they provide a guaranteed torque of 0.56 Nm. The design must avoid wheel saturation during the maneuver.

	ω [rad/s]	ζ	X i/f [m]	Y i/f [m]	Z i/f [m]	Sx $m\sqrt{kg}$	Sy $m\sqrt{kg}$	Sz $m\sqrt{kg}$	Lx $\sqrt{kg}$	Ly $\sqrt{kg}$	Lz $\sqrt{kg}$
Antenna	2.574	0.01	0.685	0.300	0.422	-2.697	18.847	-84.348	5.793	0.194	-0.178
	3.011	0.01	0.685	0.300	0.422	-84.864	-4.823	1.995	-0.203	1.495	-6.059
	6.7535	0.01	0.685	0.300	0.422	0.144	30.549	27.389	-3.273	-0.261	0.149
	12.795	0.01	0.685	0.300	0.422	-27.470	-2.967	3.218	-0.141	-5.362	-5.084
	20.987	0.01	0.685	0.300	0.422	1.015	5.381	15.455	-5.311	-0.619	0.543
	47.881	0.01	0.685	0.300	0.422	-1.303	-0.279	0.337	0.077	-1.266	-0.102
	48.133	0.01	0.685	0.300	0.422	-1.170	-0.251	0.314	0.070	-1.170	-0.079
Solar Array	0.629	0.01	0.002	-0.756	0.483	-0.036	-4.761	10.211	0	0.034	0.016
	1.177	0.01	0.002	-0.756	0.483	3.902	-0.031	0.066	0	-4.644	-2.165
	2.386	0.01	0.002	-0.756	0.483	-1.758	0.027	-0.058	0	0.417	0.194
	3.875	0.01	0.002	-0.756	0.483	0	-11.034	-5.145	-1.815	-0.000	0.000
	4.435	0.01	0.002	-0.756	0.483	-0.856	0.003	-0.006	0	2.003	0.934
Tolerance	±10 %	Min.	±5 %	±5 %	±5 %	±10 %	±10 %	±10 %	±10 %	±10 %	±10 %

Fig. 6.10 Flexible modes properties

With reference to Sect. 5.5.5 will use the following templates:

$$W_{S_i}(s) = \frac{s^2 + 1.4\omega_{S_i} s + \omega_{S_i}{}^2}{s^2}, i = 1..3 \tag{6.24}$$

$$W_{K_i}(s) = \frac{1}{\sqrt{2} I_{G_{i,i}} \omega_{S_i}{}^2} \frac{s + \omega_{R_i}}{s/100 + \omega_{R_i}}, i = 1..3 \tag{6.25}$$

or the second order version:

$$W_{K_i}(s) = \frac{1}{\sqrt{2} I_{G_{i,i}} \omega_{S_i}{}^2} \frac{s^2 + 1.4\omega_{R_i} s + \omega_{R_i}^2}{(s/100)^2 + 1.4\omega_{R_i}(s/100) + \omega_{R_i}^2}, i = 1..3 \tag{6.26}$$

The controller is chosen diagonal, composed of three generic sixth order stable and proper filters placed in the roll, pitch, yaw axes. We use for this design example the MATLAB *hinfstruct* routine, which allows to minimize the H_∞ norm of generalized state-space (GENSS) models that consider tunable parameters, in this way the designer has a lot of flexibility in choosing the structure of the desired controller (see [8]).

First, we want to study the effect of the roll-off filter using (as first trial) $\omega_{S_i} = 0.1$ rad/s and $\omega_{R_i} = 5\omega_{S_i}$ in all axes. We can see, from the Bode plots of the three diagonal open loop transfer functions (Figs. 6.11, 6.12, 6.13), that the effect of augmenting the order of the roll off filter is a sharper filtering of the elastic modes, while the open loop 0 dB crossing frequency remains the same.

In Fig. 6.14 we can see the effect of using a first order or a second order roll off on the control torque.

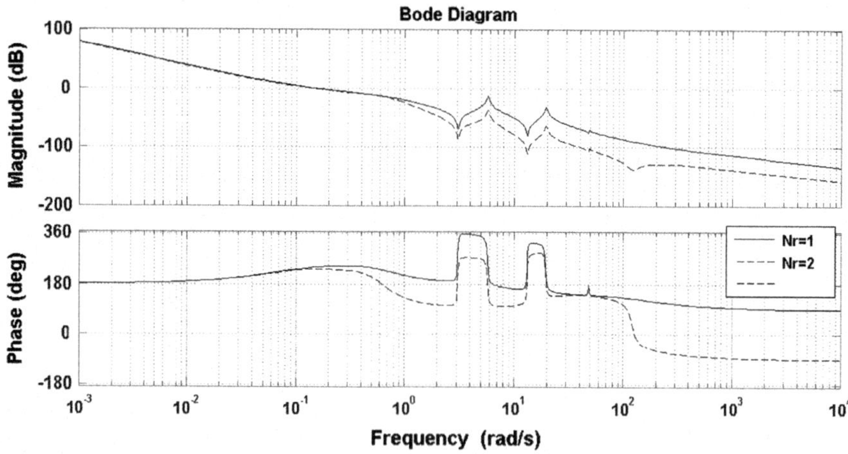

Fig. 6.11 Bode plot of X-axis open loop chain

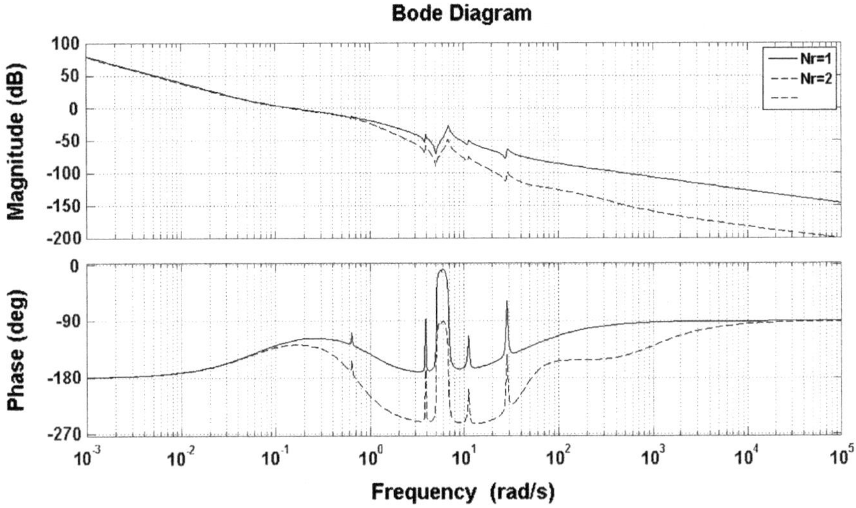

Fig. 6.12 Bode plot of Y-axis open loop chain

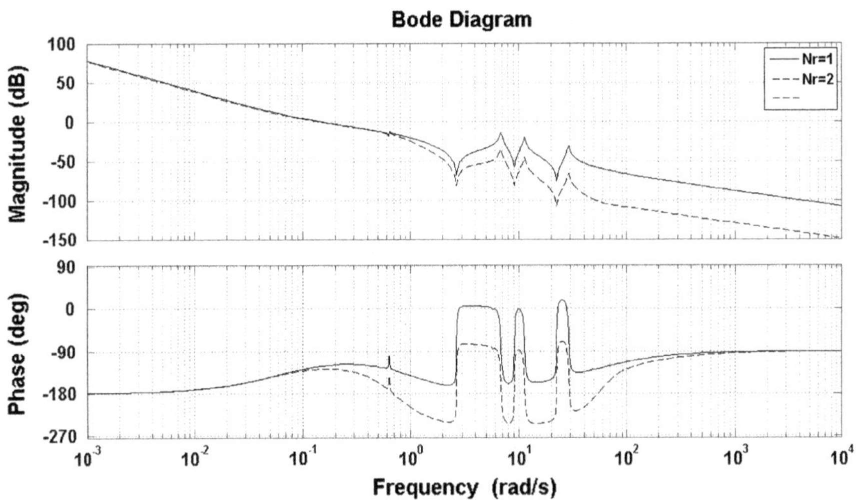

Fig. 6.13 Bode plot of Z-axis open loop chain

The noise effect on the control torque changes drastically while it could be seen that the bandwidth is not very much affected.

From these considerations we choose a second order roll-off for all the following examples.

We now modify ω_{S_i} choosing three design values $\omega_{S_i} = (0.05, 0.1, 0.2)$ rad/s.

The results are summarized in Fig. 6.15.

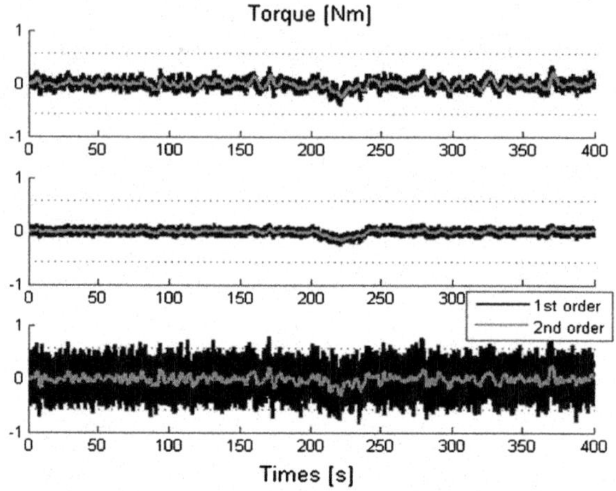

Fig. 6.14 Roll off filter effect on the control torque

	$\omega_S = 0.05$	$\omega_S = 0.1$	$\omega_S = 0.2$
$\sigma(T_{ctrl})$	[0.032 0.029 0.039]	[0.064 0.035 0.075]	[0.178 0.064 0.182]
$\max(T_{ctrl})$	[0.15 0.12 0.15]	[0.25 0.18 0.3]	[0.69 0.22 0.65]
$\max(\theta_{err})$	[0.2 0.6 0.2]	[0.07 0.22 0.07]	[0.04 0.08 0.04]
Υ	1.2974	1.3210	1.3137

Fig. 6.15 Performances using different templates

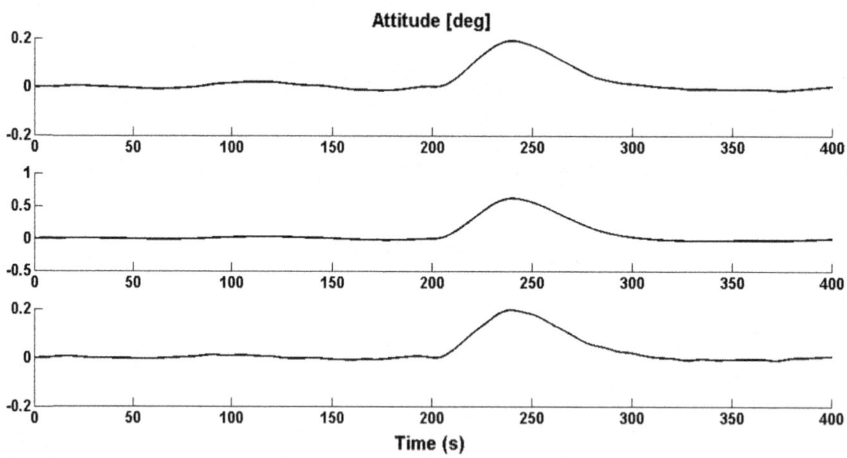

Fig. 6.16 Attitude dynamics with sensor noise and disturbance torques: low gain

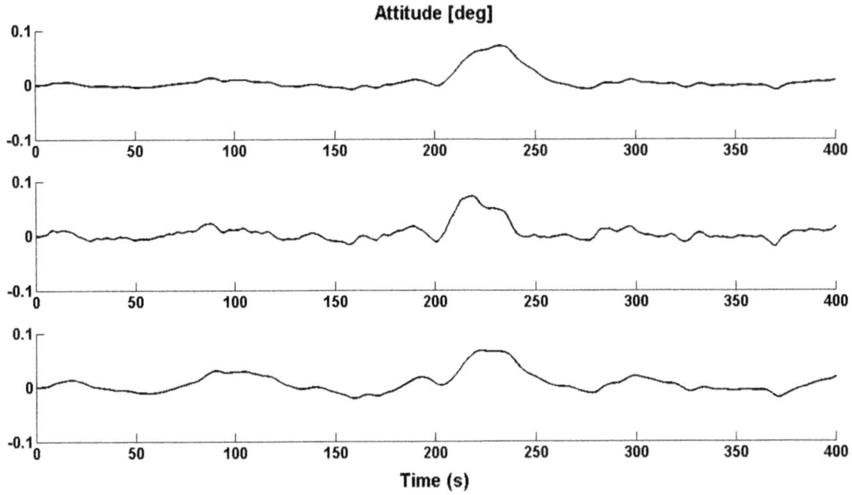

Fig. 6.17 Attitude dynamics with sensor noise and disturbance torques

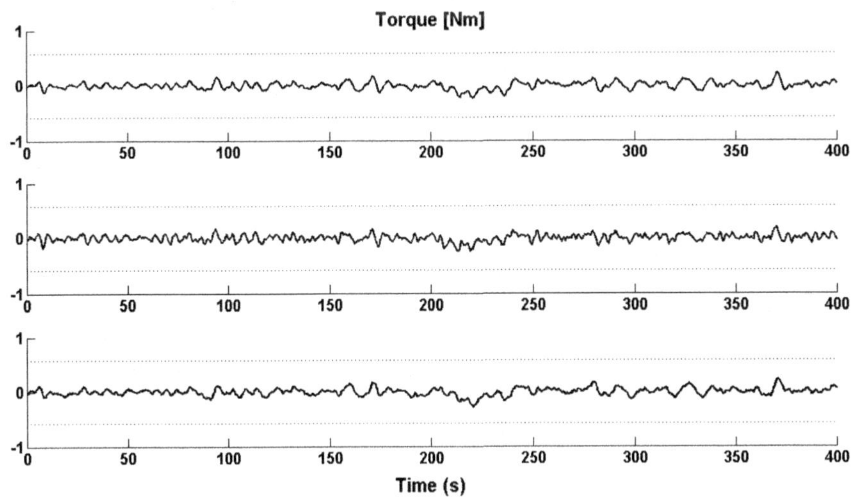

Fig. 6.18 Control torques with sensor noise and disturbance torques

All the design cases provide reasonable robustness margins, however the first template gives a very poor performance in attitude (see Fig. 6.16) while the third gives a max control torque that overcomes the saturation thresholds in the roll and yaw axes. The controller with $\omega_{S_i} = 0.1$ provides a good attitude performance in roll and yaw but is not satisfactory in pitch.

Thus we choose $\omega_{S_1} = 0.1,\ \omega_{S_2} = 0.2,\ \omega_{S_3} = 0.1$ which attains all the control objectives.

The following Figs. 6.17, 6.18 provide the performance in time simulations of this controller.

The value of $\gamma = 1.3092$ obtained in the last controller synthesis, allows to have a guaranteed gain margin $G_M \geq 12.53\,\text{dB}$ and a guaranteed phase margin $P_M \geq 44.9\,\text{deg}$ which are considered satisfactory. Typically after having chosen a controller the performances and the stability are checked using montecarlo methods and changing the plant parameters (see [7]) within the design tolerances.

References

1. J.M. Ginoux, *Differential Geometry Applied to Dynamical Systems* (World Scientific, Singapore, 2009)
2. J.J.E. Slotine, *Applied Nonlinear Control* (Prentice-Hall, Englewood Cliffs, 1991)
3. A.M. Lyapunov, *The General Problem of the Stability of Motion* (Taylor and Francis, London, 1992)
4. A. Jeffrey, *Matrix Operations for Engineers and Scientists* (Springer, Dordrecht, 2010)
5. N. Fezans, D. Alazard, N. Imbert, B. Carpentier, H_∞ control design for generalized second order systems based on acceleration sensitivity function, in *16th Mediterranean Conference on Control and Automation*, Ajaccio (France), 25–27 June (2008)
6. S. Bennani, F. Ankersen, M. Arcioni, M. Casasco, L. Massotti, P. Silvestrin, Robust attitude control design for the BIOMASS satellite (Earth explorer core mission candidate), in *18th IFAC World Congress*, Milano (Italy), Aug. 28–Sept. 2 (2011)
7. P. Gasbarri, R. Monti, G. Campolo, C. Toglia, Control-oriented modelization of a satellite with large flexible appendages and use of worst-case analysis to verify robustness to model uncertainties of attitude control. Acta Astronaut. **81**, 214–226 (2012)
8. P. Apkarian, M. Ngoc Dao, D. Noll, Parametric robust structured control design. IEEE Trans. Autom. Control, doi:10.1109/TAC.2015.2396644

Chapter 7
Orbital Dynamics and Guidance

Abstract The satellite orbital state must be transferred from the injection orbit to the operational one with a minimum consumption of propellant. In order to define the guidance law to be implemented the optimal control theory must be applied to the orbital equations. After this phase, the satellite must be maintained in the operational orbit with a certain accuracy and an acceptable expenditure of propellant. An analysis of the perturbations of the nominal orbit is necessary to this scope. In this chapter, we review all these subjects and then we illustrate the optimal control theory applied to the orbital transfer for the two propulsion models of infinite thrust and finite thrust showing the relations between the two methods and various applications.

7.1 Introduction

The satellite orbital state must be transferred from the injection orbit to the operational one with a minimum consumption of propellant. After this phase, the satellite must be maintained in the operational orbit with a certain accuracy and an acceptable expenditure of propellant. In order to define the guidance to be implemented by the propulsion system we apply the optimal control theory to the orbital equations. The guidance plan is normally calculated in the ground station and the thruster firing orders are transmitted to the AOCS for the execution. Autonomous satellites are able to define the transfer strategy directly on board. In this chapter we illustrate the optimal control theory applied to the orbital transfer for the two basic conditions of infinite thrust and finite thrust showing the relations between the two.

The equations of the orbital motion Eq. 3.95, can be written in **ECI**:

$$\ddot{x^g} = -\mu_e \frac{x^g}{|x^g|^3} + F/m \tag{7.1}$$

In this chapter we will use the symbol x in place of x^g and v in place of v^g to indicate the center of mass (CoM) position and velocity in **ECI**.

© Springer International Publishing Switzerland 2016

L. Mazzini, *Flexible Spacecraft Dynamics, Control and Guidance*,
Springer Aerospace Technology, DOI 10.1007/978-3-319-25540-8_7

F/m is the resultant in **ECI** of all the applied external forces excluding the Keplerian term of the central gravitational field divided by the satellite mass, it includes also the guidance forces produced by the on-board logic (AOC) using the propulsion system with the aim to maintain or change the orbital status.

We will study Eq. 7.1 with the specific objective of steering the orbit to a desired status and maintain this status against the perturbing actions due to the environment and the attitude control when mass ejection systems (like RCTs) are used for this purpose.

In order to apply the *Maximum Principle* (see Sect. 5.4.1) to Eq. 7.1, it is convenient to introduce some additional general concepts of orbital dynamics like canonical variables, invariants, Lagrange planetary equations, singularity free variables. Useful expressions equivalent to Eq. 7.1 can be produced in appropriate variables that can be found using the Hamiltonian method. Basic information on the Hamiltonian method can be found in Appendix C.1.

After the canonical variables, another useful set of non-singular coordinates will be introduced. This set of coordinates is well suited for low-eccentricity and low-inclination orbits, and it will be used to study the orbit keeping maneuvers, that is, the strategies for maintaining the satellite in its operational orbit; the specific orbit keeping strategy is dependent on the specific nominal orbit and its perturbations in terms of amplitude and frequencies. We will give a general description of these perturbations; in order to study in more detail the orbit keeping of specific applications it is possible to consult papers and monographic books like [1] dedicated to geostationary satellites.

In this chapter, after having introduced the orbit keeping techniques, two sections will follow: the optimal orbit control with infinite thrust and the optimal orbit control with finite thrust. These two last sections will be developed using the optimal control theory studied in Chap. 5.

Satellites are often injected by launchers into orbits that are different from their operational one. For this reason satellites may be equipped with important propulsion systems that allow them to reach their operational conditions. As shown in Chap. 8, propulsion systems can be divided into two broad classes: (1) those using a high force level which allows them to reach the final orbit in a few burns, each one lasting a small percentage of the orbital period, (2) those using a low thrust system which needs to be active for a very large percentage of the transfer duration. The mathematical treatment of the *Maximum Principle* will be different for these two cases.

7.2 The Kepler Problem

The invariants of motion in a central gravitational field have already been studied in Sect. 3.10.2. We have derived two integrals: the angular of momentum 3D vector $\mathscr{I}_{ROT}$ defined by Eq. 3.97 and the energy scalar $\mathscr{E}$ defined by Eq. 3.99.

These integrals contain also the components due to attitude and flexibility which by arguments already introduced in Sect. 3.10.1 are negligible for bodies of the dimensions of artificial satellites when treating the translational motion.

When the mass is constant, during un-propelled coasting phases ($F = 0$), we can consider the mass as unitary and the integrals are called angular momentum and energy per unit mass.

In this chapter we will call $\mathscr{I}_{ROT} = \mathscr{I}$ in order to simplify the notation: under these conditions the invariants, for unitary mass, become:

$$\mathscr{I} = x \times v \tag{7.2}$$

$$\mathscr{E} = \frac{v^T v}{2} - \frac{\mu_e}{|x|} \tag{7.3}$$

An additional integral of motion, during the un-propelled phases, can be found by cross multiplying Eq. 7.1 by $\mathscr{I}$. Following [2] we get a third 3D vector integral of motion, which is called the Runge-Lenz vector or eccentricity vector. The equalities:

$$\frac{d(x/|x|)}{dt} = -\frac{\mathscr{I} \times x}{|x|^3} = \frac{\mathscr{I} \times \dot{v}}{\mu_e} \tag{7.4}$$

can be demonstrated easily by developing the time derivative and using Eq. 7.1 and the definition of $\mathscr{I}$ in Eq. 7.2.

Then, integrating in time the first and the last member of Eq. 7.4, for the constancy of $\mathscr{I}$, it results that the following Runge-Lenz vector is constant along the motion:

$$L = \frac{v \times \mathscr{I}}{\mu_e} - \frac{x}{|x|} \tag{7.5}$$

In addition, if we scalar multiply both sides of Eq. 7.5 by x, using the identity

$$x^T (v \times \mathscr{I}) = \mathscr{I}^T (x \times v) = |\mathscr{I}|^2,$$

we get the general conic section equation in terms of the integrals of motion. Being f the angle between the vectors L and x, the result is:

$$|x| = \frac{|\mathscr{I}|^2/\mu_e}{1 + |L|\cos(f)} = \frac{p}{1 + e\cos(f)} \tag{7.6}$$

where the symbols p and e find their definition from Eq. 7.6 and will be treated in more detail in the next section.

7.3 Classic Orbit Elements

Classic orbit elements represent the most intuitive and widely used set of coordinates describing the orbit trajectory. Such coordinates are in fact easily correlated with the invariants of motion and the geometrical characteristics of the trajectory.

The geometrical properties of the conic section can be related to the integrals of motion by direct comparison of the right and left members of Eq. 7.6:

$$p = \frac{|\mathscr{I}|^2}{\mu_e} \tag{7.7}$$

$$e = |L| \tag{7.8}$$

where $p = a\left(1 - e^2\right)$ is the semi latus rectum, a is the semi-major axis and e is the eccentricity of the conic section.

The constant direction of the angular momentum as it is often called, defines the invariant plane in which the conic section lies in the 3D space. The plane can be fully characterized by two angles with respect to the **ECI** reference frame. Let us consider the x, y, z components of $\mathscr{I}$ in the **ECI** frame:

$$I = \cos^{-1}\left(\frac{\mathscr{I}_z}{|\mathscr{I}|}\right) \tag{7.9}$$

$$h = \tan^{-1}\left(-\frac{\mathscr{I}_x}{\mathscr{I}_y}\right) \tag{7.10}$$

The inclination I, is the angle between the angular momentum direction and the **ECI** Z-axis; the right ascension of the ascending node h (also indicated in the literature as Ω), is the angle between the **ECI** X-axis and the intersection of **ECI** XY plane with the orbital plane at the so-called ascending node, where the satellite crosses the equatorial plane from south to north. The right ascension is measured positively moving from the **ECI** X-axis towards the **ECI** Y-axis.

The constant direction of the Runge-Lenz Vector defines the peri-apsis of the trajectory, i.e. the orientation of the conic section inside the orbit plane.

The argument of perigee, g (also indicated in the literature as ω), is the angle in the orbit plane from the ascending node to the periapsis direction.

$$g = \tan^{-1}\left(\frac{L_z|\mathscr{I}|}{L_y\mathscr{I}_x - L_x\mathscr{I}_y}\right) \tag{7.11}$$

The previous 5 elements (a or p, e, I, h, g) are the integrals of motion of the Keplerian restricted two-body problem during un-propelled coasting arc. The body position along the trajectory at time t is defined by a sixth component, typically the True Anomaly f, representing the angular position along the orbit measured from the periapsis direction.

The orbit time evolution, for elliptical or circular orbits, is regulated by Kepler equation:

$$M(t) = M_0 + n\,(t - t_0)$$

being $M = E - e\sin(E)$ the mean anomaly, E such that $|x| = a(1 - e\cos(E))$ is the eccentric anomaly and $n = \sqrt{\frac{\mu_e}{a^3}}$ is the mean angular rate.

The definition of classical orbit elements becomes singular for orbits with very low inclination and/or eccentricity, when the *periapsis* or the *ascending node* are not defined. In such cases it is possible to adopt alternative sets of coordinates, as the following nonsingular elements, often used for circular and/or equatorial orbits (e.g. geostationary satellites):

$$\begin{aligned}
a &= -\tfrac{\mu_e}{2\mathscr{E}}\\
e_x &= e\cos\,(g+h) = -L_x\\
e_y &= e\sin\,(g+h) = L_y\\
I_x &= \sin I\cos h = -\mathscr{I}_y/|\mathscr{I}|\\
I_y &= \sin I\sin h = \mathscr{I}_x/|\mathscr{I}|\\
\lambda &= h + g + f - \alpha_G(t)
\end{aligned} \tag{7.12}$$

The couples e_x, e_y and I_x, I_y are called eccentricity and inclination vectors. The rotating anomaly $\alpha_G(t) = \omega_{Earth}t + \alpha_G(0)$ represents the right ascension of the Greenwich meridian in the **ECI** frame, where ω_{Earth} is the Earth rotation rate.

7.3.1 Kepler Motion in Hill Canonical Variables

The equations of motion of the restricted two-body problem can be written in different sets of variables that may be useful for specific purposes. The *Hill* canonical variables for example allow to express very simply the maneuver equations Eq. 7.1. The derivation of the *Hill* variables using the Hamiltonian formulation of the Kepler problem can be found in Appendix C.1. Hill variables set (p, G, H, r, u, h) is defined as follows:

Generalized momenta	Generalized coordinates
$p = \dot{r}$	r = range = $\|x\|$
$G = \sqrt{\mu_e a(1-e^2)}$	$u = g + f$ = argument of latitude
$H = G\cos I$	$h = \Omega$ = right ascension of the ascending node

where g and f represent the argument of perigee and the true anomaly respectively, G is the modulus of the angular momentum vector, while H corresponds to the component of angular momentum along the direction of Earth's spin axis (Fig. 7.1).

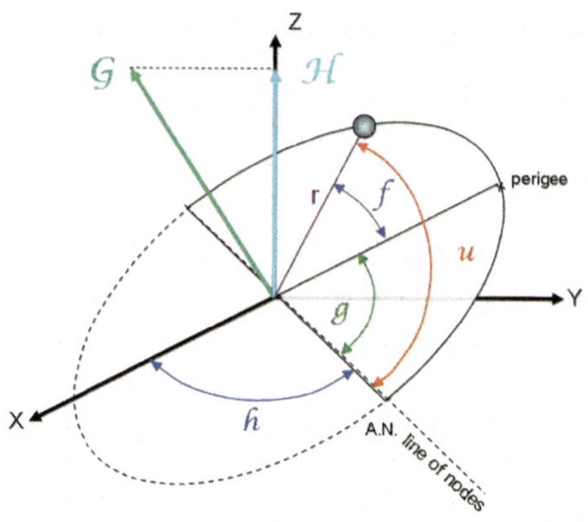

Fig. 7.1 Representation of hill variables

In Appendix C.1 we have derived the equations of Kepler motion in the Hill variables:

$$
\begin{cases}
\dot{r} = \frac{\partial F}{\partial p} & = p \\
\dot{u} = \frac{\partial F}{\partial G} & = \frac{G}{r^2} \\
\dot{h} = \frac{\partial F}{\partial H} & = 0 \\
\dot{p} = -\frac{\partial F}{\partial r} & = \frac{G^2}{r^3} - \frac{\mu_e}{r^2} \\
\dot{G} = -\frac{\partial F}{\partial u} & = 0 \\
\dot{H} = -\frac{\partial F}{\partial h} & = 0
\end{cases}
\tag{7.13}
$$

The Hamiltonian formulation of the Kepler problem is also widely used in perturbation theory as it allows to find closed-form solutions for some perturbed two-body problems (see Appendix D), these solutions are useful in computing desired orbit trajectories and orbit keeping techniques. We will not study in detail the perturbation theory of Eq. 7.1 which can be found in many classic works such as [2–4].

7.4 Satellite Orbital Maneuvering

In order to steer the satellite orbit to a new orbit it is necessary to use the Eq. 7.1 with $F \neq 0$. However Eq. 7.1 expressed in Cartesian coordinates is not very attractive to study the orbital transfers. It is more useful to develop these equations in terms of invariant coordinates like those defined by Eq. 7.12 or the Hill variables of Fig. 7.1. In the literature these equations are called for historical reasons *Lagrange planetary equations* but we will often refer to them as maneuver equations.

Let us define the vector

$$A = F/m \tag{7.14}$$

as the satellite acceleration produced by the propulsion system in **ORF** components, namely the radial, ortho-radial and normal components A_r, A_t, A_n of the acceleration.

The equations describing the change in Hill variables as a consequence of an external force per unit of mass (acceleration), such as small impulsive velocity change provided by the reaction control thrusters (RCT), are obtained by adding to Eq. 7.13 the derivatives of each Hill parameter with respect to the velocity variation $\delta V_h,\ h = r, t, n$, evaluated in $\delta V_h = 0$ and multiplied by the external accelerations $A_h,\ h = r, t, n$.

With reference to the satellite **ORF**, let us call the radial, ortho-radial (quasi tangential) and normal velocity variations as δV_r, δV_t and δV_n. It is easy to evaluate the variation of the Hill parameters under a finite instantaneous change of these quantities. From a simple geometrical construction, denoting the Hill parameters before the maneuver with the subscript 0 and those just after without any subscript, we have:

$$\begin{aligned}
r &= r_0\\
p &= p_0 + \delta V_r\\
G &= \sqrt{G_0^2 + 2r\delta V_t G_0 + r^2(\delta {V_t}^2 + \delta {V_n}^2)}\\
H &= H_0 + r\cos(I_0)\delta V_t - r\sin(I_0)\cos(u_0)\delta V_n\\
u &= \arcsin(\sin(\beta)/\sqrt{1 - H^2/G^2})\\
h &= \alpha - \arctan((H/G)\tan(u))
\end{aligned}$$

We name here α, β as the azimuth and elevation spherical coordinates of the satellite in **ECI**.

From spherical trigonometry (i.e. see [4]) it results that $\sin(u) = \sin(\beta)/\sin(I)$ and $\tan(\alpha - h) = \cos(I)\tan(u)$.

By deriving the above expressions with respect to $\delta V_r, \delta V_t, \delta V_n$ and then evaluating the resulting expressions in $\delta V_r = 0$, $\delta V_t = 0$ and $\delta V_n = 0$, we obtain:

$$\begin{aligned}
&\frac{\partial G}{\partial V_r} = 0,\ \frac{\partial G}{\partial V_t} = r,\ \frac{\partial G}{\partial V_n} = 0\\
&\frac{\partial H}{\partial V_r} = 0,\ \frac{\partial H}{\partial V_t} = -r\cos(I),\ \frac{\partial H}{\partial V_n} = -r\sin(I)\cos(u)\\
&\frac{\partial p}{\partial V_r} = 1,\ \frac{\partial p}{\partial V_t} = 0,\ \frac{\partial p}{\partial V_n} = 0\\
&\frac{\partial u}{\partial V_r} = 0,\ \frac{\partial u}{\partial V_t} = 0,\ \frac{\partial u}{\partial V_n} = -\frac{r}{G}\frac{\cos(I)\sin(u)}{\sin(I)}\\
&\frac{\partial h}{\partial V_r} = 0,\ \frac{\partial h}{\partial V_t} = 0,\ \frac{\partial h}{\partial V_n} = \frac{r}{G}\frac{\sin(u)}{\sin(I)}\\
&\frac{\partial r}{\partial V_r} = 0,\ \frac{\partial r}{\partial V_t} = 0,\ \frac{\partial r}{\partial V_n} = 0
\end{aligned}$$

The maneuver equations due to external accelerations are:

$$\begin{aligned}
\dot{r} &= p \\
\dot{p} &= G^2/r^3 - \mu_e/r^2 + A_r \\
\dot{G} &= rA_t \\
\dot{u} &= G/r^2 - r\cos(I)\sin(u)A_n/(G\sin(I)) \\
\dot{H} &= r\cos(I)A_t - r\sin(I)\cos(u)A_n \\
\dot{h} &= r\sin(u)/(G\sin(I))A_n
\end{aligned} \tag{7.15}$$

The same equations for the non-singular variables set Eq. 7.12 can be derived from Eq. 7.15 using the relationships:

$$\begin{aligned}
a &= \frac{\mu_e}{-p^2+2\frac{\mu_e}{r}-\frac{G^2}{r}} \\
e_x &= \left(\frac{G^2}{\mu_e r}-1\right)\cos(h+u)+\frac{Gp}{\mu_e}\sin(h+u) \\
e_y &= \left(\frac{G^2}{\mu_e r}-1\right)\sin(h+u)-\frac{Gp}{\mu_e}\cos(h+u) \\
I_x &= \sqrt{1-\frac{H^2}{G}}\cos h \\
I_x &= \sqrt{1-\frac{H^2}{G}}\sin h \\
\lambda &= h+u-\alpha_G
\end{aligned} \tag{7.16}$$

We obtain, without demonstration:

$$\dot{a} = \left(\frac{2a\left(1+e_x\cos\theta+e_y\sin\theta\right)}{V_c\sqrt{1-e^2}}\right)A_t + \left(\frac{2a\left(e_x\sin\theta-e_y\cos\theta\right)}{V_c\sqrt{1-e^2}}\right)A_r$$

$$\dot{e_x} = \left(\frac{\sqrt{1-e^2}\left(2\cos\theta+e_x+e_x\cos\theta^2+e_y\cos\theta\sin\theta\right)}{V_c\left(1+e_x\cos\theta+e_y\sin\theta\right)}\right)A_t + \left(\frac{\sqrt{1-e^2}}{V_c}\sin\theta\right)A_r$$

$$+\left(\frac{e_y\sqrt{1-e^2}\left(I_y\cos\theta-I_x\sin\theta\right)}{V_c\left(1+\cos I\right)\left(1+e_x\cos\theta+e_y\sin\theta\right)}\right)A_n$$

$$\dot{e_y} = \left(\frac{\sqrt{1-e^2}\left(2\sin\theta+e_y+e_y\sin\theta^2+e_x\cos\theta\sin\theta\right)}{V_c\left(1+e_x\cos\theta+e_y\sin\theta\right)}\right)A_t + \left(-\frac{\sqrt{1-e^2}}{V_c}\cos\theta\right)A_r$$

$$+\left(-\frac{e_x\sqrt{1-e^2}\left(I_y\cos\theta-I_x\sin\theta\right)}{V_c\left(1+\cos I\right)\left(1+e_x\cos\theta+e_y\sin\theta\right)}\right)A_n$$

$$\dot{I_x} = \left(-\frac{\sqrt{1-e^2}\left(I_y^2\cos\theta+I_x^2\cos I\cos\theta-I_xI_y\sin\theta+I_xI_y\cos I\sin\theta\right)}{V_c\sin I^2\left(1+e_x\cos\theta+e_y\sin\theta\right)}\right)A_n$$

$$\dot{I}_y = \left(-\frac{\sqrt{1-e^2}\left(I_x^2 \sin\theta + I_y^2 \cos I \sin\theta - I_x I_y \cos\theta + I_x I_y \cos I \cos\theta\right)}{V_c \sin I^2 \left(1 + e_x \cos\theta + e_y \sin\theta\right)} \right) A_n$$

$$\dot{\lambda} = \left(\frac{\sqrt{1-e^2}\left(I_x \sin\theta - I_y \cos\theta\right)}{V_c\left(1+\cos I\right)\left(1 + e_x \cos\theta + e_y \sin\theta\right)} \right) A_n$$
$$+ \left(\frac{V_c}{a}\left(1 + \frac{3}{2}e^2 + \left(e_x \cos\theta + e_y \sin\theta\right)\left(2 + 3e^2 + e_x \cos\theta + e_y \sin\theta\right)\right) - \omega_{Earth} \right) \tag{7.17}$$

where

$$\begin{aligned} I &= \sin^{-1}(\sqrt{I_x^2 + I_y^2}), \\ e &= \sqrt{e_x^2 + e_y^2}, \\ V_c &= \sqrt{\tfrac{\mu_e}{a}} = na \quad \text{is the circular velocity}, \\ \theta &= u + h = f + g + h, \quad \text{and} \\ \lambda &= \theta - \alpha_G. \end{aligned} \tag{7.18}$$

Even if Eq. 7.17 are more used in the literature than Eq. 7.15, we prefer in the analytical derivations to use Eq. 7.15 because they are much simpler.

When considering orbits with very low eccentricity and inclination the following form of maneuver equations, found by 0-th order truncation of e and I terms, can be used:

$$\begin{pmatrix} \dot{a} \\ \dot{e}_x \\ \dot{e}_y \\ \dot{I}_x \\ \dot{I}_y \\ \dot{\lambda} \end{pmatrix} = \begin{bmatrix} \frac{2a}{Vc} & 0 & 0 \\ \frac{2}{Vc}\cos\theta & \frac{1}{Vc}\sin\theta & 0 \\ \frac{2}{Vc}\sin\theta & -\frac{1}{Vc}\cos\theta & 0 \\ 0 & 0 & -\frac{1}{Vc}\cos\theta \\ 0 & 0 & -\frac{1}{Vc}\sin\theta \\ 0 & 0 & 0 \end{bmatrix} \begin{pmatrix} A_t \\ A_r \\ A_n \end{pmatrix} + \begin{pmatrix} 0 \\ 0 \\ 0 \\ 0 \\ 0 \\ \frac{Vc}{a} - \omega_{Earth} \end{pmatrix} \tag{7.19}$$

7.5 Orbit Keeping

It has been pointed out that the environmental forces and the attitude control thrusters (RCT) firings acting on a spacecraft modify its orbital elements w.r.t. the nominal ones. This deviation can result, after some time, in the degradation of the mission performances, for example the antenna coverage or the revisit times. *orbit keeping* is the practice of periodically performing orbit maneuvers of small intensity in order to maintain the S/C in proximity of the nominal or *reference orbit*. Such maneuvers are typically performed by chemical or electrical propulsion. It is important to underline that the orbit keeping must be implemented when the satellite is operational, therefore the satellite attitude cannot be changed to orient the thruster in a desired optimal direction that is normally the approach followed for the optimal orbit steering which is performed before the operational phase.

The strategy to perform the corrections depends on

- the free evolution of the spacecraft around the nominal trajectory, depending on the project orbit and the body characteristics;
- the propulsion system features (basically thrust and thrusters' pointing directions in **BRF**);
- the required *control box*, i.e. the maximum allowed deviation from the project orbit;
- if commanded from ground, the ground station visibility;
- specific operational needs (maneuver duration, eclipse constraints, operations schedule compatibility, etc.).

The first step for the study of every orbit keeping strategy is to model and understand the free motion of the body around the assigned orbit, distinguishing short-medium-long term[1] and secular components.

Then the maneuver plan, i.e. the sequence of small firings defined in the directions and the anomalies and the periodicity of this sequence, is optimized in order to minimize the associated DV budget, given the allowed control box and the relevant constraints. The maneuver plan must be updated to take into account long term variations like seasonal effects.

The minimum DV budget is obtained by controlling only the secular drifts of the orbital parameters. This solution is feasible only if the control box is large enough to contain all periodic terms plus some allocation for the free evolution of the secular terms. The control box amplitude is generally well-suited to contain small amplitude short term components of the error without the need for correcting them. Nevertheless some medium and long term components could not fit inside the dead-band, so that the associated deviation requires dedicated maneuvers to eliminate or mitigate its effects. The following section provides a discussion about the types of maneuvers adopted for orbit keeping in the case of infinite thrust propulsion systems (like chemical propulsion system). In the finite thrust case similar types of maneuvers can be applied, but this time the maneuvers are implemented along finite firing arcs centered around the optimal firing anomalies defined by the infinite thrust strategy. Therefore, an additional effect of the geometrical inefficiency due to the finite arc firing must be taken into account (the effect due to the θ-trigonometric functions in Eq. 7.19).

[1]Here we denote with *short term* the components whose periodicity is of the order of the orbit period (hours or days), with *medium term* the components with periodicity of weeks or months and with *long term* the components with periodicity of years.

7.5.1 Orbit Keeping Maneuvers

As introduced previously, the orbit keeping strategy is the plan of the orbit maneuvers needed to maintain all orbit parameters inside an assigned tolerance. The frequency and intensity of the station keeping maneuvers depend on the tolerance of the orbit parameters deviation from the project orbit ones and on the intensity of the disturbing forces acting on the specific satellite, therefore on its orbit, mass, surface properties. In the next Chapter we will see that the external environmental accelerations produce secular drifts of the orbital parameters. These drifts must be counteracted by the action of small control burns.

The application of impulsive burns corresponds to the instantaneous variations of the radial, ortho-radial and normal velocities in **ORF** frame (ΔV_r, ΔV_t, ΔV_n). The maneuvers plan can be identified using the Eqs. 7.17 or 7.15. The effect of small burns on near-circular near-equatorial orbits can be evaluated using Eq. 7.19.

From Eq. 7.19 we can verify that with A_t we can control all the in-plane parameters (a, e_x, e_y, λ), while with A_n we can control the out-of-plane ones (I_x, I_y). Normally A_r is not used intentionally because it is less effective in controlling the eccentricity vector (see the factor 2 in the first column of Eq. 7.19) and does not provide any effect on the semi-major axis.

The actions of A_t and A_n are uncoupled. Therefore, a possible approach for orbit keeping, when the **BRF** and the **ORF** frames are aligned, is to place one thruster along the X-**BRF** direction producing A_t and one thruster along the Y-**BRF** direction producing A_n and to operate the first for the *in-plane* parameters control and the second for the *out-of-plane* parameters control.

These thrusters are positioned to point the spacecraft CoM in order not to generate torques. This approach, which allows to have the optimal efficiency, is not always possible due to the effect of the thruster plume and the uncertainty in the CoM position: work around solutions include the use of more thrusters and squinted thrusters.

Orbit keeping maneuvers are preferably not performed in phases when the satellite is maneuvering its attitude because they would have a low efficiency.

When the **BRF** and the **ORF** frames are not aligned—e.g. satellites in continuous yaw steering or in inertial pointing—the maneuver are performed with thrusters properly squinted and fired only at certain anomalies of the orbit in order to have the maximum possible maneuver efficiency.

In-plane maneuvers are performed through tangential boosts ΔV_t, inside the orbit plane. These affect only the parameters that describe the orbit geometry inside the plane, i.e. a, e_x, e_y and λ.

Out-of-plane maneuvers are performed through normal burns ΔV_n. They affect the orbit plane orientation and, consequently, also the parameters whose definition relies on the orientation of the line of nodes: in practice, the out-of-plane maneuvers impact all orbit elements except a and e. This fact can be easily seen from Eq. 7.15 considering that A_n does not change the dynamics of r, p, G which directly depend on a,e and the true anomaly.

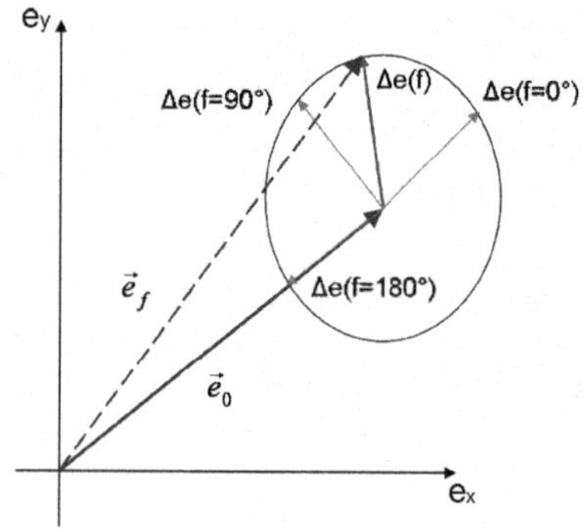

Fig. 7.2 Zero order effect of an ortho-radial Δ_V on eccentricity vector

Hereafter the effects of each kind of burn are summarized and described for zero and 1st order in eccentricity; from these effects, in the case of quasi-circular operational orbits, knowing the specific perturbations discussed in Sect. 7.6, it is possible to define a suitable station keeping strategy.

7.5.1.1 In-Plane Maneuvers

In-plane maneuvers can be applied to correct a, e_x, e_y, but also λ through the indirect effect of a bias on the semi-major axis a over a finite time interval (see Eqs. 7.17, 7.19).

The application of an ortho-radial (quasi-tangential) thrust produces on the different orbital elements, to the 0th order in eccentricity, the following effects:

- semimajor axis: a variation equal to $\frac{2a}{V_c}$ regardless of where the burn is provided along the orbit;
- eccentricity vector: a vector variation of magnitude $\frac{2}{V_c}$ and direction defined by true anomaly f with respect to the original direction of eccentricity vector in the e_x, e_y plane. As depicted in Fig. 7.2, a maneuver performed at perigee or the apogee ($f = 0 - 180$ deg) produces a variation of modulus only, while a maneuver at $f = 90 - 270$ deg produces a pure rotation of the line of apsides.

To the first order in the eccentricity (see Eq. 7.17), the maneuver produces an additional small variation of semi-major axis, maximum in modulus at perigee and apogee (with opposite signs) and null at 90 deg from the apsides, and an eccentricity vector additional variation in the tangential direction at the firing anomaly, whose amplitude is maximum at 90 deg from the apsides and zero at perigee/apogee.

The application of a radial burn provides, to the 0th order, only a variation of the eccentricity vector of half the magnitude with respect to an ortho-radial burn, in the tangential direction of the firing anomaly in the e_x, e_y plane. Radial burns are rarely used in practical cases due to their low efficiency.

Combinations of ortho-radial maneuvers can be used to achieve any modification of the orbit size, shape and in-plane orientation. Considering only 0th order terms:

- A semimajor axis pure correction of magnitude Δa is obtained from two ortho-radial burns of magnitude $\frac{1}{4V_c}\frac{\Delta a}{a}$ with 180 deg separation in the true anomaly;
- An eccentricity vector pure correction is obtained from the application of two burns of equal magnitude and opposite sign with 180 deg separation in true anomaly. The ΔV magnitude for each burn shall be equal to $\frac{1}{2}V_c\Delta\,|e|$ and the firing anomaly shall be $\theta = \pm\arctan\frac{\Delta e_y}{\Delta e_x}$;
- Any correction of both eccentricity vector and semi-major axis can be achieved, in general, by three burns with appropriate splitting of ΔV and true anomaly.

Finally, ortho-radial maneuvers are used to achieve corrections of on-orbit phasing λ. This parameter cannot be corrected instantaneously with an in-plane maneuver, but a change can be obtained by setting a bias on a during an adequate time interval, that is by modifying the *natural* rate of λ:

$$\Delta\dot{\lambda} = -\frac{3}{2}n\frac{\delta a}{a} = -\frac{3}{a}\Delta V_t$$

A change in the spacecraft longitude can therefore be obtained, neglecting perturbations, by providing two ortho-radial firings of intensity ΔV_t in opposite directions, separated by a time interval Δt such that:

$$\Delta\lambda = -\frac{3}{a}\Delta V_t \Delta t$$

7.5.1.2 Out-of-Plane Maneuvers

The application of normal burns produces to 0th order in eccentricity:

- On the inclination vector, a variation of magnitude $\frac{1}{V_c}$ and direction depending on the firing anomaly and the inclination value itself. From Eq. 7.15 we can see that a normal velocity impulse given at nodes, $u = (\theta - h) = 0, 180$ deg, produces only variations in the inclination modulus while, if given at 90 deg from the node, it produces only a rotation of h. For small inclinations the inclination vector variation is oriented with the angle θ in the I_x, I_y plane, i.e. it forms an angle $u = g + f$ with respect to the original inclination vector (see Fig. 7.3);
- On the true longitude, we have a variation with maximum magnitude at 90 deg from the nodes and vanishing for small inclinations. This variation of course does not represent a change of the satellite position and is due to the rotation of the ascending node direction in the inertial reference frame.

From Eq. 7.15 we see that A_n has effect only on u, H, h, therefore normal plane maneuvers are used for the correction of the inclination vector modulus and/or direction. Their effect on the in-plane state due to the change of $u = g + f$ is limited to a small rotation of the perigee in eccentric orbits.

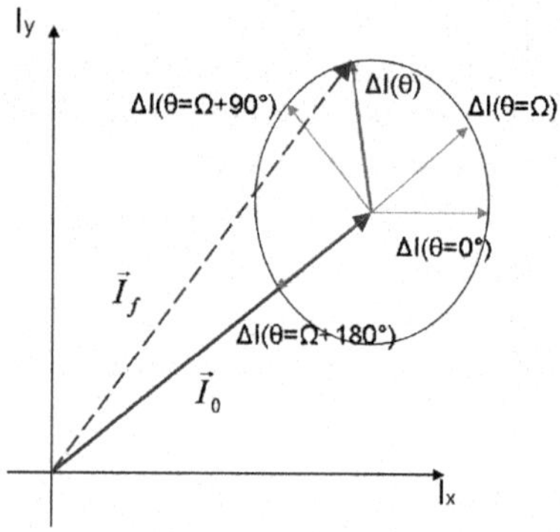

Fig. 7.3 Zero order effect of a normal Δ_V on inclination vector for near equatorial orbits

7.6 Orbital Perturbations

Operational orbits are subject to many perturbations with respect to the Keplerian motion, whose intensity depends on the orbit parameters and the S/C characteristics. The perturbed motion deviates from the Keplerian one on a periodical and secular basis, so that the orbit degrades until, eventually, the mission goals can no longer be fulfilled.

The principal perturbing effects acting on an Earth satellite are briefly introduced here.

7.6.1 Earth Gravitational Potential

As the Earth is not a perfect sphere with uniform density, the *geopotential*, i.e. the spatial distribution of gravitational potential, is not the same in every direction. The effect of Earth's oblateness on a body is represented as an additional acceleration due to further terms perturbing the ideal, spherical gravity field.

The perturbed gravity potential (Eq. 3.91) is recalled hereafter:

$$U = -\frac{\mu_e}{r}\left\{1 - \sum_{k=2}^{\infty} J_k \left(\frac{R_e}{r}\right)^k P_k(\sin(\mathfrak{L})) + \sum_{k=2}^{\infty}\sum_{j=1}^{k} \left(\frac{R_e}{r}\right)^k J_{k,j} P_k^j(\sin(\mathfrak{L}))\cos\left[m\left(\mathfrak{l} - \mathfrak{l}_{k,j}\right)\right]\right\}$$

being $\mathfrak{L}$ and $\mathfrak{l}$ the body latitude and longitude in the **ECEF** frame (see Footnote 5 of Chap. 3) respectively.

The P_k represent the Legendre functions of degree k and P_k^j the associated Legendre function of degree k and order j:

$$\begin{aligned} P_k(x) &= \frac{1}{2^k k!}\frac{d^k}{dx^k}\left(x^2-1\right)^k \\ P_k^j(x) &= \left(1-x^2\right)^{j/2}\frac{d^j}{dx^j}P_k(x) \end{aligned} \tag{7.20}$$

The complete geopotential is then given by the sum of the point-mass potential plus a series of terms expressed in the orthogonal basis of Legendre's polinomials, and indicated by the indexes k and j.

The terms containing P_k are the *zonal harmonics*: they give a description of the planet shape and mass distribution in North-South direction, being function of latitude only.

The terms containing P_k^j with $j = k$ are called *sectorial harmonics*.

The remaining terms in P_k^j with $j \neq k$ are called *tesseral harmonics* and describe Earth's shape and mass distribution through a chessboard-like representation.

The approximate solution of such the orbital perturbation problem under the full Earth geopotential (refer to Appendix D), produces the following structure on the generic orbit element p:

$$p(t) = p(0) + \underbrace{D\,t}_{\text{secular drift}} + \underbrace{\sum_{jk} D\,s_{jk}\sin\left(ju'' + kf\right) + \sum_{jk} D\,c_{jk}\cos\left(ju'' + kf\right)}_{\text{periodic motion}}$$

Depending on the specific orbit of interest, some terms are more important than others in the representation of the gravitational field.

For LEO orbits the action of the first zonal terms is typically stronger than sectorial and tesseral terms and it provides the principal secular drift of orbit elements. An analytical solution of J_2 zonal problem is developed in Appendix D using the Hamiltonian formalism. For example, the evolution of the right ascension of the ascending node h is described by

$$\begin{aligned} h(t) &= h_0'' + n_0\left\{-3\gamma_2'\theta + \tfrac{3}{8}\gamma_2'^2\left[\left(-5+12\eta+9\eta^2\right)\theta + \left(-35-36\eta-5\eta^2\right)\theta^3\right]\right\}t \\ &+ \left\{-\tfrac{1}{8}\gamma_2' e''^2\theta\left[11 + 80\theta^2\left(1-5\theta^2\right)^{-1} + 200\theta^4\left(1-5\theta^2\right)^{-2}\right]\right\}\sin 2g'' \\ &- \tfrac{1}{2}\gamma_2'\theta\left[6\left(f'-l'+e''\sin f'\right) - 3\sin\left(2g'+2f'\right) - 3e''\sin\left(2g'+f'\right) - e''\sin\left(2g'+3f'\right)\right] \end{aligned}$$

The symbols are explained in Appendix D but, just considering that f, f', l' are anomaly variables rotating at the orbital frequency, we see that the solution is made of a constant term, plus a secular term drifting with the time t, plus a long term varying as $\sin 2g''$ and some short rotating terms which contains the anomalies oscillating at the orbital frequency.

For orbits whose period is resonant with Earth's rotation, sectorial and tesseral harmonics can represent the main source of secular perturbation. Many studies exist finalized to give approximations of the long term dynamics due to the effect of resonances. A quite extensive and recent study for MEO and GEO orbits is presented in [5]. For a GEO orbit the main perturbing term is the sectorial harmonic of degree and order 2.

In the case of a geostationary orbit, the main harmonic term J_{22} acting on the spacecraft produces a secular effect on the semi-major axis, and consequently on the *mean longitude drift*, defined as $D = -\frac{3}{2}n_{geo}\frac{a-a_{geo}}{a_{geo}}$.

$$\dot{D}_{J_{22}} = 18n_{geo}^2 J_{22}\left(\frac{Re}{a_{geo}}\right)^2 \sin\left(2(\lambda_s - \lambda_{22})\right)$$

where[2]:

λ_s is target on-station longitude of the spacecraft
R_e is the Earth equatorial mean radius
$J_{22} = 1.81e - 6$
$\lambda_{22} = -14.95$ deg

The resulting secular evolution of semi-major axis is then:

$$a(t) = a_0 - \frac{2}{3}\frac{a_{geo}}{n_{geo}}\dot{D}_{J_{22}}t \tag{7.21}$$

As a consequence of the constant derivative of semi-major axis in time, the mean longitude λ evolves on a parabolic path with acceleration $\dot{D}$:

$$\lambda(t) = \lambda_0 + D_0 t + \frac{1}{2}\dot{D}_{J_{22}}t^2 \tag{7.22}$$

7.6.2 Atmospheric Drag

The air drag due to high atmosphere represents the main source of deviation from the nominal trajectory for LEO satellites.

Drag acts on spacecrafts in the opposite direction with respect to the body velocity relative to the atmosphere, dissipating the orbital energy of the satellite. The direction of the perturbing acceleration is along the anti-velocity and its modulus is given by:

$$a_D = \frac{1}{2m}C_D A_e \rho V^2 \tag{7.23}$$

---

[2]The provided expression of $\dot{D}$ is no longer valid in proximity of λ_{22}. The complete expression for this case can be found in [6].

with:

C_D = drag coefficient
A_e = mean frontal surface
ρ = air density
m = satellite mass
V = modulus of the velocity relative to the atmosphere

The drag coefficient C_D depends on the form, attitude and kind of surface of the body. As a first approximation, such coefficient is assumed equal to 1 if the mean free path of atmosphere molecules is small w.r.t. the satellite dimension, otherwise it takes a value between 2 and 3.

The computation of air drag is highly complicated by the unsteadiness of the air density, that varies with altitude, latitude and solar activity (day/night cycles, 27-days cycles due to UV radiations, 11-years cycles due to solar flux).

The effect of the air drag on elliptical orbits is a progressive decrease of apogee altitude due to the intense force acting at perigee, where both velocity and air density are higher. As a consequence LEO orbits with no drag control are subject to circularization and then to a semi-major axis decay becoming faster as the altitude reduces. The effect of atmosphere drag is purely secular, a perturbation analysis of this effect can be found in [2].

7.6.3 Sun and Moon Gravity

For satellites in high altitude orbits the effect of Sun and Moon gravitational attraction is significant. Following [4], the equation of motion of the satellite with respect to the Earth, can be written adding a perturbing acceleration equal to the acceleration due to the force expressed by the third body on the satellite minus the acceleration expressed by the third body on the Earth. The last term is a correction because the **ECI** Earth's frame is not inertial.

Considering the perturbing acceleration of a third body of mass M_P, $\mu_p = GM_p$ with G the Universal Gravitational Constant, the equations of motion become:

$$\frac{d^2x}{dt^2} = -\frac{\mu_e}{|x|^3}x + a_p \tag{7.24}$$

where

$$a_p = \left[\frac{\mu_p}{|R|^3}R - \frac{\mu_p}{|\rho|^3}\rho\right] \tag{7.25}$$

is the perturbing acceleration and R, ρ represent respectively the position vectors of the Earth and the satellite from the third body, i.e. $R + x = \rho$.

For near-Earth satellites the assumption $\frac{x}{R} \ll 1$ is normally applicable, so that the perturbing acceleration takes the simplified form:

$$a_p = -\mu_p(I - 3RR^T/|R|^2)\frac{x}{R^3} \tag{7.26}$$

The integration of the equation of motion can be implemented by transferring the expression of the acceleration in **ORF** coordinates and integrating Eq. 7.17. An approximate solution can be obtained using perturbation techniques like the averaging (see Sect. 7.9.4 for an introduction to the averaging technique and [6] to have an explicit expression of the orbital perturbation).

The long period effect of the Sun perturbation results in a precession of the orbit pole around the pole of the ecliptic, while the lunar attraction realizes a precession of the orbit pole around the axis normal to the Moon's orbit. For a spacecraft in geostationary orbit, assuming the Sun right ascension on the equatorial plane null, the Sun and Moon's gravity affect the inclination vector's secular and long-term evolution as follows (see [6]):

$$\begin{aligned}
I_x(t) &= I_{x0} - \tfrac{3}{8}K_m \sin(2I_m)\cos(I_s)\sin(\Omega_m)(t - t_0) + \\
&\tfrac{3}{8n_s}K_s \sin(I_s)(1 - \cos(2L_s(t)) + \text{ medium/short period terms} \\
I_y(t) &= I_{y0} + (\tfrac{3}{8}(K_m\cos(I_m)^2 + K_s)\sin(2I_s) + \tfrac{3}{8}K_m \sin(2I_m)\cos(2I_s)\cos(\Omega_m))(t - t_0) \\
&- \tfrac{3}{16n_s}K_s \sin(2I_s)\sin(2L_s(t) + \text{ medium/short period terms}
\end{aligned} \tag{7.27}$$

with:

$K_s = \frac{n_s^2}{n_{geo}}\frac{M_s}{M_s+M_e} = 2.691e - 3\,\text{deg/day}$
$K_m = \frac{n_m^2}{n_{geo}}\frac{M_m}{M_m+M_e} = 5.844e - 3\,\text{deg/day}$
$L_s =$ Sun equatorial right ascension
$M_s,\ M_m,\ M_e =$ Sun, Moon Earth masses
$n_s,\ n_m =$ Sun, Moon apparent mean motion
$I_s = 23, 44$ deg inclination of the ecliptic on the equatorial plane
$I_m = 5, 14$ deg $\pm\, 0, 15$ deg Moon inclination on the ecliptic plane
$\Omega_m = 12.11 - 0.053\Delta T_{1950}$: RAAN of the Moon orbit on the ecliptic plane, in degrees where ΔT_{1950} is the number of days elapsed since 01/01/1950

The mean inclination vector is subject to a secular drift in a direction close to I_y and slowly oscillating around this direction with a period of 18.6 years (moon orbit plane precession). The secular drift magnitude varies from a minimum of 0.75 deg/year to a maximum of 0.95 deg/year, depending on the value of Ω_m, i.e. whether the lunar node is in conjunction or opposition with respect to the Sun's apparent node (X-**ECI** direction).

The principal periodic perturbation is a six month oscillation of about ±0.04 deg amplitude along both I_x and I_y directions produced by the Sun's attraction. Other medium and short period terms have been neglected as their overall amplitude is of the order of 0.01 deg, that is one tenth of the typical inclination vector control threshold for a spacecraft in GEO.

7.6.4 Solar Radiation Pressure

Solar radiation produces a force on the satellite's surface that cannot be neglected when there are large surfaces, e.g. for 3-axis stabilized spacecrafts with large solar panels. If σ is the reflectivity coefficient of the surface, a part of the total radiation $1-\sigma$ is absorbed and a part σ of the total flux is reflected by the satellite's surface. The reflected energy can be either specularly reflected (the part $\sigma\beta$) or diffusively reflected ($\sigma(1-\beta)$). The normal and tangential force components produced on each element of surface hit by the Sun are (see [7]):

$$dF_N = \frac{I}{c}\, dA\,[(1+\beta\sigma)\cos^2(\phi) + \frac{2\sigma(1-\beta)}{3}\cos(\phi)] \tag{7.28}$$

$$dF_T = \frac{I}{c}\, dA\,(1-\beta\sigma)\sin(\phi)\cos(\phi) \tag{7.29}$$

where:

c = light speed,
I = intensity of solar radiation,
σ = surface reflectivity coefficient,
$\sigma\beta$ = surface specular reflection coefficient,
$\sigma(1-\beta)$ = surface diffusive reflection coefficient,
ϕ = Sun incidence angle.

Due to the complexity of surfaces, their orientation and reflection properties, the resulting force acting on the whole satellite is a function of the shape, position and attitude of the body.

Considering $\sigma = 0$ the flux is totally absorbed, which is what happens approximately for solar arrays' surfaces, the overall force lies along the Sun-satellite direction, with a magnitude:

$$F = -\frac{I}{c} A_e \hat{r}_{sun} \tag{7.30}$$

where $A_e = A\cos(\phi)$ is the satellite's effective area projected along the Sun's direction, $\frac{I}{c} \approx 4.6\ 10^{-6} \frac{N}{m^2}$ is the solar pressure per unit surface and $\hat{r}_{sun}$ is the unit vector of the Sun's direction.

For a spacecraft in geostationary orbit, the main effect of solar radiation pressure perturbation is a long-term drift of the eccentricity vector:

$$\begin{aligned} e_x(t) &= e_{x0} + e_N\,(\cos(L_s(t)) - \cos(L_{s0})) \\ e_y(t) &= e_{y0} + e_N\,(\sin(L_s(t)) - \sin(L_{s0})) \end{aligned} \tag{7.31}$$

with $e_N = \frac{3}{2}\frac{I}{c}\frac{A_e}{m\, n_{geo} a_{geo} n_{Sun}} \approx 1.115\ 10^{-2} \frac{I}{c}\frac{A_e}{m}$, being n_{geo}, a_{geo} the mean motion and semi-major axis of the GEO orbit, and n_{Sun} the apparent mean motion of the Sun.

The mean eccentricity vector evolves on a 1-year-long circular path of radius e_N in the e_x, e_y plane, moving in the direction orthogonal to the Sun during the year. The center C of the eccentricity circle depends upon the initial components of the eccentricity vector and the initial right ascension of the Sun in **ECI**:

$$\begin{aligned} C_x &= e_{x0} - e_N \cos(L_{s0}) \\ C_y &= e_{y0} - e_N \sin(L_{s0}) \end{aligned} \tag{7.32}$$

7.7 Orbital Transfers in Non-dimensional Hill Coordinates

In order to deal with the optimal transfer problems in the infinite and finite thrust assumptions we introduce here the equations to maneuver the satellite orbit that will be applied later (the subject is also dealt in the papers [8, 9]).

We have seen in the previous Sect. 7.4 that the choice of the Hill coordinates can simplify considerably the expression of the maneuver equations. In this section we will apply some additional transformations to the Hill coordinates in order to reach two objectives: to have non-dimensional coordinates which put into evidence the scaling rules between different transfer problems and to have a set of coordinates which are integral of motion.

Being A the vector in the **ORF** triad (as defined in Chap. 2) containing the radial, ortho-radial and normal components A_r, A_t, A_n of the acceleration, $|\dot{m}|$ the flow rate, m the mass of the satellite, v_e the exhaust velocity and $\mathfrak{l}$ a unit vector of the local flight triad, which express the direction of the thrust, we have from the *rocket equation* Eq. 8.3 discussed in Chap. 8:

$$A_\alpha = \mathfrak{l}_\alpha \frac{|\dot{m}| v_e}{m}, \quad \frac{dm}{dt} = -|\dot{m}|, \alpha = r, t, n \tag{7.33}$$

From this equation we can introduce the characteristic velocity V that characterizes the evolution of the satellite mass:

$$V = \int_{t_0}^{t} \sqrt{A_\alpha A_\alpha} dt \tag{7.34}$$

$$\frac{m_0}{m} = e^{V/v_e} \tag{7.35}$$

Let us consider the maneuver equations in the Hill parameters (Eq. 7.15). At this point we introduce a new independent variable, the anomaly w defined by the differential equation

$$\dot{w} = G/r^2 \Rightarrow dt = r^2 dw/G \tag{7.36}$$

The vector A multiplied by $\frac{dt}{dw}$ provides the velocity variation per variation of anomaly w:

$$A_\alpha^w = \mathfrak{l}_\alpha \frac{|\dot{m}|v_e r^2}{mG} = A_\alpha \frac{dt}{dw} \tag{7.37}$$

Now, we introduce two new variables Y_1 and Y_2 in place of r and p and we calculate their derivatives with respect to w. The aim of introducing these new variables is to obtain an expression for the in-plane free dynamics similar to that of an oscillator. Such a feature will later allows us to easily introduce as states the invariant amplitudes of this oscillator.

We pose: $Y_1 = 1/r$, $\frac{dY_1}{dw} = Y_2$, deriving these two equations in w and using the maneuver equations Eq. 7.15, we obtain:

$$\frac{dY_2}{dw} = -Y_1 + \mu_e/G^2 + Y_2/(GY_1)A_t^w - A_r^w/G. \tag{7.38}$$

Then, we put in non-dimensional form all the variables using a_0 the semi-major axis of the initial orbit, by posing $Y_1^{ad.} = Y_1 a_0$, $Y_2^{ad.} = Y_2 a_0$, $A_\alpha^{ad.} = A_\alpha^w/\sqrt{\mu_e/a_0}$, $V^{ad.} = V/\sqrt{\mu_e/a_0}$, $G^{ad.} = G/\sqrt{\mu_e a_0}$ and $H^{ad.} = H/\sqrt{\mu_e a_0}$.

The resulting equations in non-dimensional variables are formally identical to the dimensional ones posing $\mu_e = 1$.

The non-dimensional vector $A^{ad.}$ is:

$$A_\alpha^{ad.} = \mathfrak{l}_\alpha \frac{|\dot{m}|v_e r^2}{mG\sqrt{\mu_e/a_0}} = \mathfrak{l}_\alpha e^{V/v_e} f \frac{X_3^{1/2}}{Y_1^2} \tag{7.39}$$

where f is the positive non-dimensional scalar control thrust:[3]

$$f = \frac{|\dot{m}|v_e}{m_0 \frac{\mu_e}{a_0^2}} \tag{7.40}$$

m_0 is the initial mass, $X_3 = 1/G^{ad.^2}$ and the multiplicative term in the expression of the acceleration is $\frac{X_3^{1/2}}{Y_1^2} = \frac{d\tau}{dw}$, being $\tau = t\sqrt{\mu_e/a_0^3}$ the non-dimensional time.

From now on, the non-dimensional variables will always be used without the superscript $ad.$ unless the opposite is explicitly declared.

We are now ready to define a new non-dimensional invariant state vector X based on the non-dimensional Hill parameters and the Keplerian elements, as follows:

[3]The symbol f is used also to mean the external thrust projected in **BRF** and the true anomaly, the context should unequivocally identify the meaning of the symbol in each location.

$$\begin{aligned} X_1 &= (Y_1 - X_3)\cos(u) - Y_2 \sin(u) = eX_3 \cos(g) \\ X_2 &= (Y_1 - X_3)\sin(u) + Y_2 \cos(u) = eX_3 \sin(g) \\ X_3 &= 1/G^2 = (a_0/a)/(1-e^2) \\ X_4 &= H/G = \cos(I) \\ X_5 &= h \\ X_6 &= u - w = g + f - w \end{aligned} \tag{7.41}$$

In the optimal transfer problem between keplerian orbits we can always rotate the two orbits in such a way to bring the first orbit to any plane and line of apses direction: we can assume that initial conditions are $g_0 = 0,\ w_0 = 0,\ u_0 = 0,\ I_0 = 90\ deg,\ h_0 = 0$.

The initial orbit in these non-dimensional variables is parametrized only by the eccentricity e_0, so that $X_{0,1} = e_0,\ X_{0,2} = 0,\ X_{0,3} = 1/(1-e_0^2),\ X_{0,4} = 0,\ X_{0,5} = 0,\ X_{0,6} = 0$, all the possible final orbits are defined by a choice of the first five state variables such that $\sqrt{X_{f,1}^2 + X_{f,2}^2} < X_{f,3}$.

Each state variable is constant if the external accelerations are null. With such positions the state variables X_1 and X_2 can be considered as amplitudes of the non-dimensional inverse range and its rate, $Y_1 = X_1 \cos(u) + X_2 \sin(u) + X_3,\ Y_2 = -X_1 \sin(u) + X_2 \cos(u)$.

With the variables defined by Eq. 7.41, the maneuver equations are written in the independent variable w as:

$$\frac{dX}{dw} = B(X, u = w + X_6)A \tag{7.42}$$

The matrix $B(X, u = w + X_6)$ depends on the first five entries of the first argument X, the dependence on the sixth entry takes place through the 2π-periodic argument $u = w + X_6$.

The matrix $B(X, u = w + X_6)$ is:

$$\begin{pmatrix} \sin(u)\sqrt{X_3} & \frac{2\cos(u) X_3^{\frac{3}{2}}}{Y_1} + \frac{\sin(u)\sqrt{X_3}\, Y_2}{Y_1} & \frac{\sin(u) X_2 \sqrt{X_3} X_4}{\sqrt{1-X_4^2}\, Y_1} \\ -\cos(u)\sqrt{X_3} & \frac{2\sin(u) X_3^{\frac{3}{2}}}{Y_1} - \frac{\cos(u)\sqrt{X_3}\, Y_2}{Y_1} & -\frac{\sin(u) X_1 \sqrt{X_3} X_4}{\sqrt{1-X_4^2}\, Y_1} \\ 0 & -\frac{2 X_3^{\frac{3}{2}}}{Y_1} & 0 \\ 0 & 0 & -\frac{\cos(u)\sqrt{1-X_4^2}\sqrt{X_3}}{Y_1} \\ 0 & 0 & \frac{\sin(u)\sqrt{X_3}}{\sqrt{1-X_4^2}\, Y_1} \\ 0 & 0 & -\frac{\sin(u)\sqrt{X_3} X_4}{\sqrt{1-X_4^2}\, Y_1} \end{pmatrix}$$

In general orbital control problems the quantity to be minimized is the non dimensional characteristic velocity:

$$V_f = \int_{w_0}^{w_f} |A|\, dw \tag{7.43}$$

In many reports V_f is also indicated as DV to mean that this is the contribution of that particular maneuver to the total satellite DV budget.

In the infinite thrust optimization the characteristic velocity DV is minimized without constraints on the non-dimensional thrust $f = \frac{|\dot{m}|v_e}{m_0 \frac{\mu_e}{a_0^2}}$, in the finite thrust case the same cost function is minimized for $f \leq \varepsilon$ a small constant parameter that is determined by the low thrust propulsion system.

The two assumptions lead to a quite different setting for the *Maximum Principle*.

The equations that we have written can be utilized to optimally steer the orbital state between two elliptical orbits. They are homogeneous in the A, this means that during coasting phases the state remains constant.

7.8 Optimal Orbital Transfer: Infinite Thrust

7.8.1 Introduction

In the problems of optimal orbital transfer, the infinite thrust assumption is very often considered in order to derive a simple solution for a given transfer problem. This assumption is made when the propulsion system provides enough large thrust values. This can be the case of mono-propellant and bi-propellant chemical propulsion systems (see Chap. 8). When the thrust value is sufficiently high, the transfer program providing the minimum propellant consumption has a finite number of impulsive firings. We will study this type of transfers and later we will analyze the finite thrust case.

The necessary conditions for the optimality of an orbital transfer are expressed as a system of differential equations in state, satellite position and velocity, and co-state, plus additional boundary conditions coming from the problem at hand and from the so-called transversality conditions (see Chap. 5).

In the infinite thrust assumption the state and co-state equations are integrable in closed form as found in many papers, such as Lawden [10] for the planar problems and [11, 12] for the general case.

However, we do have explicit solutions only in very simple cases due to the wide variety and complexity of the extremal portrait. In general for the same problem we may have many local extrema or we may not have an extremal at all. Apart from the *open-angle* condition, there is no global description of the solutions.

All the classic literature is based on the primer method introduced by [10], the primer method defines costate equations for the infinite thrust maneuver derived with a limit process from the equations of the finite thrust case. The Hamiltonian for the infinite thrust case is not defined in this method. In this chapter we will follow the approach proposed in [8] which uses a Hamiltonian setting specific for the infinite thrust case. We use non-dimensional and invariant state variables and the cost as independent variable. In this frame an analytic Hamiltonian is derived for the time-open, infinite thrust case.

Using this technique the time-open/angle-open problem and a general TBVP (two boundary value problem) will be presented.

7.8.2 The Cost as Independent Variable

Starting from Eq. 7.42, we introduce a new independent variable V, the accumulated non-dimensional velocity increment, or the (non-dimensional) characteristic velocity. This variable is related to w by the equation:

$$\frac{dV}{dw} = e^{V/v_e} f \frac{\sqrt{X_3}}{Y_1^2} \tag{7.44}$$

Using V as independent variable (7.42) becomes:

$$\frac{dX}{dV} = B(X, u = w + X_6)\iota \tag{7.45}$$

The transformation defined by Eq. 7.44 will produce a $V(w)$ stepwise continuous not decreasing function for any given control thrust $f(w)$. We note that, by the use of $f(w)$ in its functional domain, we can produce any $w(V)$ stepwise continuous not decreasing function, it is then possible to consider w and ι as control variables to be optimized in Eq. 7.45, without passing through the optimization in f.

Such position implies that our problem is reduced to a minimal time problem (minimal final V= V_f), with the dynamic constraint of the state evolution equations:

$$\frac{dX}{dV} = B(X, u = w + X_6)\iota, \tag{7.46}$$

where the control variables are $w(V)$ any stepwise continuous not decreasing function and the thrust direction, that is ι such that $|\iota| = 1$. This position of the optimal control problem is new in the literature, because the anomaly is no longer an independent variable but a control variable and there is no integration in the coasting phase, only the firing arcs are integrated.

7.8.3 The Maximum Principle for the Infinite Thrust Case

The optimal transfer problem can be stated as follows:

Problem (*Infinite Thrust Optimal Transfer*) Given the state X evolving as,

$$\frac{dX}{dV} = B(X, u = w + X_6)\iota \tag{7.47}$$

being the controls: $w(V)$ stepwise continuous not decreasing function and $\mathfrak{l}(V)$ such that $|\mathfrak{l}| = 1$, fixing $X(0)$ and some of the entries of the vector $X(V_f)$, *we want to obtain the control actions so that the final* $V = V_f$ is minimal.

This setting is appropriate for any time-open, minimum fuel problem.

Following the *Maximum Principle* presented in Sect. 5.4.1, we introduce the Hamiltonian H and the co-state $L \in R^6$.

$$H = \sup_{w\in[0,2\pi],|\mathfrak{l}|=1} L^T B(X, w + X_6)\mathfrak{l} - 1 \tag{7.48}$$

By maximizing in $\mathfrak{l}$, we get,

$$H = \sup_{w\in[0,2\pi]} \sqrt{L^T B(X, w + X_6) B(X, w + X_6)^T L} - 1 \tag{7.49}$$

Which is equivalent to:

$$H = \sup_{w\in[0,2\pi]} \sqrt{L^T B(X, w) B(X, w)^T L} - 1 \tag{7.50}$$

So, the Hamiltonian does not depend on X_6. The Hamiltonian by direct inspection, does not depend on X_5 either. These facts imply that the co-states L_5 and L_6 are constant for any given problem. The firing direction is:

$$\mathfrak{l} = \frac{B(X,w)^T L}{\sqrt{L^T B(X,w) B(X,w)^T L}} \tag{7.51}$$

calculated at the w which realizes the maximum value of the Hamiltonian. The transversality conditions for the problem at hand are: $L_6|_f = 0$ because the final X_6 is not constrained, and $H|_f = 0$ because the problem is time-open. Due to the fact that both L_6 and H are constant along any extremal trajectory, they are null at any point of the extremal including the initial conditions.

$H = 0$ along the extremal trajectory implies a normalization of the co-state so that $\sqrt{L^T B(X, w) B(X, w)^T L} = 1$ at any point of the optimal trajectory. With this choice we can introduce a new equivalent Hamiltonian, that has the same state and co-state derivatives along the optimal solutions and provides the same extremal flow:

$$H = \sup_{w\in[0,2\pi]} \frac{L^T B(X,w) B(X,w)^T L}{2} - 1/2 \tag{7.52}$$

Therefore being:

$$w^* = \arg\sup_{w\in[0,2\pi]} \frac{L^T B(X,w) B(X,w)^T L}{2} - 1/2 \tag{7.53}$$

the necessary conditions that any extremal satisfies are:

$$\begin{aligned} \frac{dX}{dV} &= B(X, w^*) B(X, w^*)^T L \\ \frac{dL}{dV} &= -\frac{1}{2} \frac{\partial (L^T B(X,w^*) B(X,w^*)^T L)}{\partial X} \end{aligned} \tag{7.54}$$

The Eq. 7.54 have the property $L^T \frac{dX}{dV} = 1$. Any locally optimal trajectory will be a solution to these equations for an initial co-state. The initial co-states are such that $L_6 = 0$.

If L_0 is an initial co-state that maximizes the pre-Hamiltonian at w^*, then the family of co-states $L = kL_0$ for $k \neq 1$ does not satisfy the transversality conditions $H = 0$. So the variety of the initial co-states is isomorphic to the hyper-sphere S^4 of dimension 4. This property make it easier to study the extremal flow portrait.

7.8.4 Impulsive Solutions

The optimal solutions are constituted by a series of impulses, alternated with coasting arcs. This property is based on the pre-Hamiltonian derivatives. Calling the pre-Hamiltonian:

$$\widetilde{H}(X, L, w) = \frac{L^T B(X,w)B(X,w)^T L}{2} - 1/2$$

The control is a function of the state and the co-state that transforms the pre-Hamiltonian into the Hamiltonian, that is:

$$w^*(X, L) = \arg\sup_{w\in[0,2\pi]} \frac{L^T B(X,w)B(X,w)^T L}{2} - 1/2 \tag{7.55}$$

where we have made evident the functional dependence of the control w^* on the state and co-state. The $\widetilde{H}(X, L, w)$ is an analytic function in all its arguments, in particular it is also periodic in the control w, therefore the *sup* is always attained and we have in addition that the pre-Hamiltonian satisfies, in any point where $w = w^*(X, L)$ and $L_6 = 0$, the following equations:

$$\begin{aligned} &\frac{\partial \widetilde{H}(X,L,w)}{\partial w} = 0 \\ &\frac{\partial^2 \widetilde{H}(X,L,w)}{\partial^2 w} < 0 \\ &\frac{\partial^2 \widetilde{H}(X,L,w)}{\partial w \partial X}\frac{\partial \widetilde{H}(X,L,w)}{\partial L} - \frac{\partial^2 \widetilde{H}(X,L,w)}{\partial w \partial L}\frac{\partial \widetilde{H}(X,L,w)}{\partial X} = 0 \end{aligned} \tag{7.56}$$

The second equation is the Legendre-Clebsch strict inequality, the third is a specific property of the Hamiltonian that can be verified numerically for any $w = w^*(X, L)$. From the first property using the implicit function theorem we derive:

$$\begin{aligned} \frac{\partial w^*}{\partial X} &= -\frac{\partial^2 \widetilde{H}(X,L,w^*)}{\partial w \partial X} \Big/ \frac{\partial^2 \widetilde{H}(X,L,w^*)}{\partial^2 w} \\ \frac{\partial w^*}{\partial L} &= -\frac{\partial^2 \widetilde{H}(X,L,w^*)}{\partial w \partial L} \Big/ \frac{\partial^2 \widetilde{H}(X,L,w^*)}{\partial^2 w} \\ \frac{d}{dV}\frac{\partial \widetilde{H}(X,L,w^*)}{\partial w} &= \frac{\partial^2 \widetilde{H}(X,L,w^*)}{\partial w \partial X}\frac{\partial \widetilde{H}(X,L,w^*)}{\partial L} - \\ &\quad \frac{\partial^2 \widetilde{H}(X,L,w^*)}{\partial w \partial L}\frac{\partial \widetilde{H}(X,L,w^*)}{\partial X} + \frac{\partial^2 \widetilde{H}(X,L,w^*)}{\partial^2 w}\frac{dw}{dV} = 0 \end{aligned} \tag{7.57}$$

So, we derive immediately that $\frac{dw}{dV} = 0$.

If w does not vary along an extremal arc it can only jump from one maxima to another one. This occurs when we have the transition from one firing arc to another one.

7.8.5 The Co-Planar and Co-Apsidal Transfer

The transfer between co-planar orbits with the same apsidal line can be taken as an example of global extremal analysis and at the same time it illustrates the Hohmann two-impulse and bi-elliptic three-impulse transfers in the setting here introduced. This transfer case has been studied by many authors like [10, 13–15] which give a demonstration that an orbital transfer minimal among co-planar co-apsidal orbits is globally optimal in the context of angle free transfers.

Introducing $X_2 = L_2 = X_4 = L_4 = X_5 = L_5 = X_6 = L_6 = 0$, in Eqs. 7.54 and 7.51, we derive that $\mathfrak{l}_1 = \mathfrak{l}_3 = 0$ and $\mathfrak{l}_2$ and $\cos(u)$ can assume only values in the set $-1, 1$. Renaming $\cos(u) = \varepsilon_u \in -1, 1$ and $\mathfrak{l}_2 = \varepsilon_l \in -1, 1$ the two controls, perigee firing occurs when $sign(X_1)\varepsilon_u = 1$ as can be derived from $X_1 = eX_3\cos(g)$.

Introducing the costate $L = [L_1, L_3]^T$, Eqs. 7.51 and 7.52 become

$$\begin{aligned} \varepsilon_l &= B(X,w)^T L = 2X_3^{3/2}\tfrac{\varepsilon_u L_1 - L_3}{\varepsilon_u X_1 + X_3} \\ H &= \sup_{\varepsilon_u \in -1,1} \tfrac{1}{2}(2X_3^{3/2})^2\left(\tfrac{\varepsilon_u L_1 - L_3}{\varepsilon_u X_1 + X_3}\right)^2 - \tfrac{1}{2} \end{aligned} \tag{7.58}$$

By the maximization of the Hamiltonian the switching condition for the control is derived

$$\begin{aligned} \left|\frac{L_1 - L_3}{L_1 + L_3}\right| &> \left|\frac{X_1 + X_3}{X_1 - X_3}\right| \Rightarrow \varepsilon_u = 1 \\ \left|\frac{L_1 - L_3}{L_1 + L_3}\right| &< \left|\frac{X_1 + X_3}{X_1 - X_3}\right| \Rightarrow \varepsilon_u = -1 \end{aligned} \tag{7.59}$$

The Eq. (7.54) of the state and costate become

$$\begin{aligned} \frac{dX_1}{dV} &= \frac{\partial H}{\partial L_1} &= \frac{2X_3^{3/2}\varepsilon_u \varepsilon_l}{\varepsilon_u X_1 + X_3} \\ \frac{dX_3}{dV} &= \frac{\partial H}{\partial L_3} &= \frac{-2X_3^{3/2}\varepsilon_l}{\varepsilon_u X_1 + X_3} \\ \frac{dL_1}{dV} &= -\frac{\partial H}{\partial X_1} &= \frac{\varepsilon_u}{\varepsilon_u X_1 + X_3} \\ \frac{dL_3}{dV} &= -\frac{\partial H}{\partial X_3} &= -\frac{3}{2X_3} + \frac{1}{\varepsilon_u X_1 + X_3} \end{aligned} \tag{7.60}$$

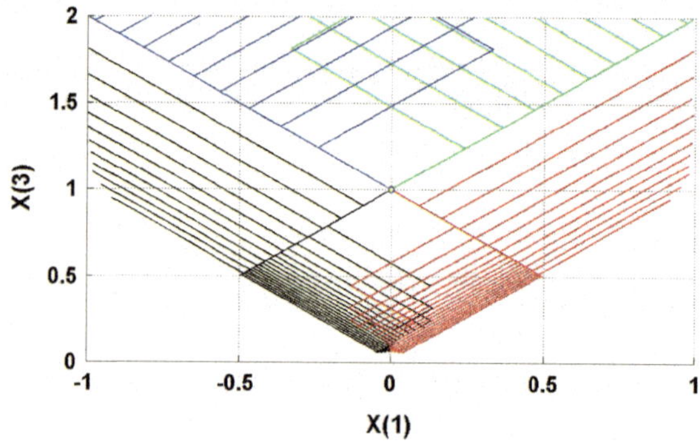

Fig. 7.4 Extremals in planar coapsidal transfer, $e_0 = 0$

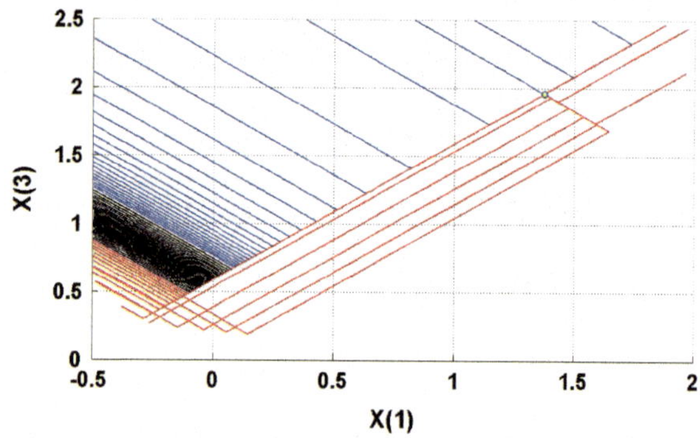

Fig. 7.5 Extremals in planar coapsidal transfer, $e_0 = 0, 7$

These equations are integrable along a constant control curve:

$$
\begin{aligned}
&\varepsilon_u X_{01} + X_{03} = \varepsilon_u X_1 + X_3 \\
&X_1 = X_{01} - \varepsilon_u (X_3 - X_{03}) \\
&L_1 = L_{01} + \varepsilon_u \varepsilon_l (X_3^{-1/2} - X_{03}^{-1/2}) \\
&L_3 = L_{03} - \varepsilon_u (L_1 - L_{01}) - \varepsilon_l (\varepsilon_u X_{01} + X_{03})(X_3^{-3/2} - X_{03}^{-3/2})/2 \\
&X_3^{-1/2} = X_{03}^{-1/2} + \varepsilon_l V / (\varepsilon_u X_{01} + X_{03})
\end{aligned}
\tag{7.61}
$$

All the state initial points are parametrized in the eccentricity e_0, so that $X_{01} = e_0/(1 - e_0^2), X_{03} = 1/(1 - e_0^2)$ and all extremals starting from any of such points of the X_1, X_3 plane can be generated using Eq. (7.60) or their integrated version Eq. (7.61) where the controls ε_u and ε_l are defined by Eqs. (7.59) and (7.58). All the initial co-states are in a locus isomorphic to S^1. Having chosen a point in S^1, the normalization condition of the Hamiltonian $H = 0$ will identify the modulus of the costate for that direction.

Four sets of trajectories are identified by the initial costate component signs, this is highlighted in Fig. 7.4 that shows the extremals departing from $X_{01} = 0,\ X_{03} = 1$ with a different color for each sign combination of $L_{01},\ L_{03}$. The propagation is stopped at $V_f = 0.54$.

Figure 7.4 shows in principle all the possible solutions of the angle free problem with null initial eccentricity. The extremals invade all the $X_3 > |X_1|$ domain apart from a small region where $X_3 \leq 0.064$, below this value, which is the ratio between the semi-major axes of the initial and final orbits, no local minimum exists. Similar conclusions can be derived from Fig. 7.5 where the starting point is $X_{01} = 0.7/(1 - 0.7^2)$, $X_{03} = 1/(1 - 0.7^2)$, in this case the region where no extremal exists is wider.

When the trajectory is such that $X_3 - |X_1| < X_{03} - |X_{01}|$ the first firing is a perigee one followed by an apogee one, the firing sequence is reversed if $X_3 - |X_1| > X_{03} - |X_{01}|$. The cone $X_3 - |X_1| = X_{03} - |X_{01}|$ separates the two regions of extremals and contains the points that can be reached by a single firing.

Some simple properties can be derived by analyzing the extremal portraits like Figs. 7.4 and 7.5:

- Due to the chosen coordinates, the trajectories satisfy $\frac{dX_1}{dX_3} = -\varepsilon_u \in -1, 1$. Therefore, the extremal trajectories in the X_1, X_3 plane are a sequence of lines at ± 45 deg. The apogee firings are the lines where $sign(\Delta X_3/\Delta X_1) sign(X_1) = 1$ and the perigee firings are the lines with $sign(\Delta X_3/\Delta X_1) sign(X_1) = -1$.
- The extremals are contained in the cone $X_3 > |X_1|$ whose boundary is the locus of the parabolic transfers.
- The extremals may contain any number of firing arcs. The same initial and final points can be connected by different extremals with two or more than two firings. It is then impossible to limit the number of firing arcs for an optimal trajectory on the basis of the necessary conditions for minimality.
- The extremals follow a symmetry rule: for any extremal with X_1, X_3, L_1, L_3 exist also the extremal $-X_1, X_3, -L_1, L_3$.
- The final points such that $X_{f3} - |X_{f1}| = X_{03} - |X_{01}|$ are reached by an extremal with a single firing. For the final point reached by a two firing extremal, the optimal sequence consists of a perigee firing followed by an apogee firing when $X_{f3} - |X_{f1}| < X_{03} - |X_{01}|$ and an apogee firing followed by a perigee one when $X_{f3} - |X_{f1}| > X_{03} - |X_{01}|$.

The characterization of the minimal solutions for the open angle problem has been studied also by Hazelrigg (see [13]) with a different methodology that has some analogies with the author's one. The state variables used in [13] are $X = X_3 + |X_1|$ the inverse of the apogee radius and $r_p = 1/(X_3 - |X_1|)$ the perigee radius.

These state variables are invariants during the coasting arcs, therefore the optimal trajectories are described by a sequence of vertical and horizontal lines—the perigee and apogee firings—connecting the terminal points, instead of the sequence of ± 45 deg lines shown in Figs. 7.4 and 7.5.

In the coordinate system introduced in [13] the locus of parabolas is the straight line $X = 0$ and the locus of circular orbits is the hyperbola $Xr_p = 1$ (see [13] Fig. 1). These two locus are mapped in Fig. 7.4 with the cone $X_3 = |X_1|$ and the straight line $X_1 = 0$.

In the Hazelrigg method the cost is written as a line integral connecting the terminal points. Such procedure is applicable only to a subset of trajectories where dX and dr_p keep a constant sign. Using Green's theorem it is then possible to compare

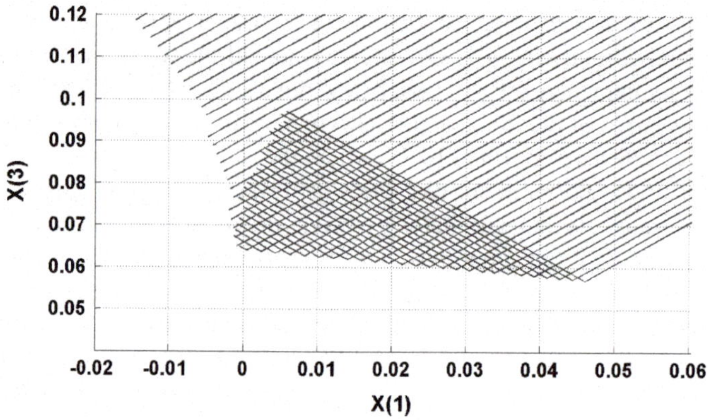

Fig. 7.6 Two and three firing extremals

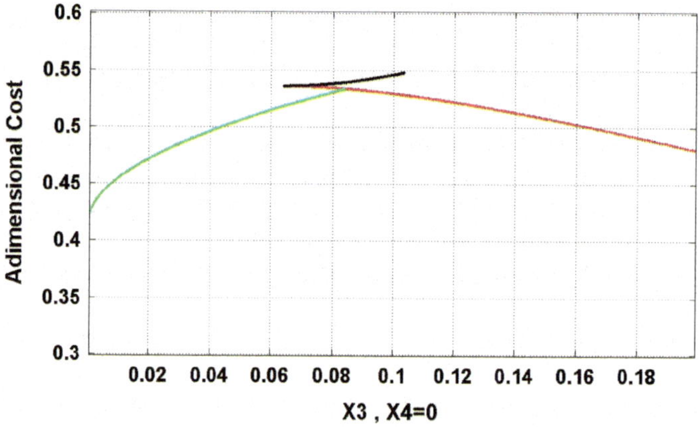

Fig. 7.7 The cost of two (*red*), three (*black*) firing extremals and bi-elliptical parabolic limit (*green*) transfers

the cost of two different trajectories and select the one that has the lower cost. Using some basic trajectories where Green's theorem is applicable Hazelrigg demonstrates by inductive reasoning equivalent rules (see Figs. 4–5 in [13] and the relevant explanations) of those listed previously on the basis of a geometrical inspection of the extremal portrait.

Magnifying in Fig. 7.6 the region of the extremal portrait close to $X_{f1} = 0, X_{f3} = 0.064$ a result that is well established in the literature (see [10]) can be verified. In the set $X_{f1} = 0, X_{f3} = [0.064...0.11]$ three firing bi-elliptical extremals and two firing extremals cross at the same terminal point. The extremal solutions with 2 firings are locally minima and exist for $X_{f3} > 0.064$. The three firing extremals have a higher cost than the two firings, this can be verified in Fig. 7.7 where the cost of the 2 firing extremals is in red and the one of the three firing extremals is in black.

The family of three firing transfers is a sequence of perigee-apogee-perigee firings, targeted to a given circular orbit ($X_3 < 1,\ X_1 = 0$), and parametrized to the height of the apogee of the first firing. By sending the apogee of the first firing to infinity, the family includes in the limit a parabolic transfer ($X_3 = |X_1|$).

It is found that in the set $X_{f3} \in [0.064, 0.084]$ the three firing extremals are locally maxima in the set of the bi-elliptical transfers reaching the same X_{f3}, see Fig. 7.7 where the cost of the three firing solutions reaching the parabolic limit is in green.

This result confirms [16] that have demonstrated the existence of three firing bi-elliptical solutions in $X_3 < 0.084$ that are better than the two firing extremals. Any of such bi-elliptical transfers is not an extremal because it can be improved by raising the apogee of the transfer orbit until the parabolic transfer limit is reached, such limit transfers flow along the cone $X_3 = |X_1|$.

In terms of global minimum the solutions with two firings are the best in $X_{f3} > 0.084$, while the bi-elliptical parabolic limit is the best in $X_{f3} < 0.084$.

Below the point $X_{f3} = 0.064$ and in the surrounding region no extremal exists.

7.8.6 The General Orbital Transfer

The method introduced allows a numerically efficient solution of the *two boundary value problem* (TBVP) for any orbital transfer. The approach introduced uses the following features:

- the integrability of the state and the co-state allows to have algebraic expression of their evolution during the firing (see Appendix E).
- the use of a Particle Swarm optimization algorithm (PSO see [17], Chap. 10) that allows to find the initial co-state that minimizes the distance between the extremal and the target orbital state. Such algorithm is most efficient in being the application $L_0 \rightarrow X_f$ not differentiable or continuous as we have seen in the previous section.
- multiple extremals exist that cover the same region. This can produce the convergence of the swarm to an extremal which is not globally optimal. To avoid that, we use a version of the PSO that is constrained to have the maximum cost below a fixed value that we establish a priori to be the value of the optimal 2 firing solutions. Such two firing optimal solutions can be found numerically by iteratively solving the Lambert problem (see [18]), we will call these solutions L_2 and their cost V_{L_2}, their existence is guaranteed but not their optimality.

The free time optimal orbital transfer TBVP can be posed as follows:

Problem (*Free Time Optimal Orbital Transfer TBVP*) Given $X_i(0), i = 1..6$, find the initial co-states: $L_i(0), i = 1..5, L_6(0) = 0$ such that the Hamiltonian is null and via the state and co-state equations, the final state $X_i(V_f), i = 1..5$ is reached at an unspecified $V_f \leq V_{L_2}$.

In the above proposition V_{L_2} is the cost of the Lambert problem solution, that is the minimal cost among all the 2 firing solutions attaining the final orbit. The

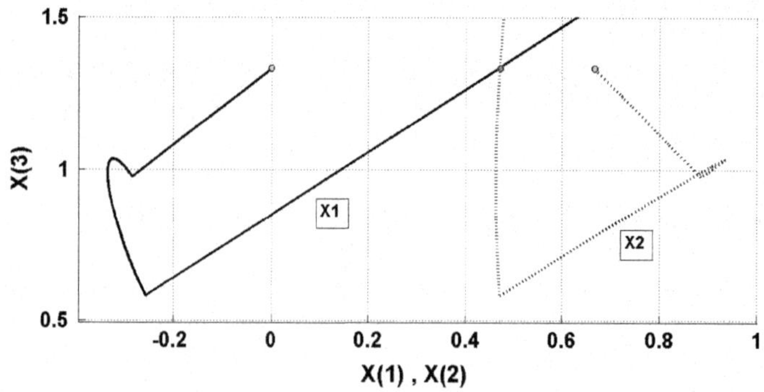

Fig. 7.8 General orbital transfer: in-plane state evolution

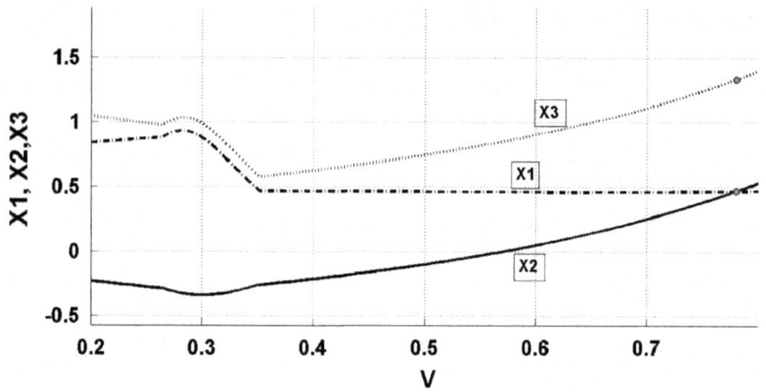

Fig. 7.9 General orbital transfer: in-plane state evolution versus total cost

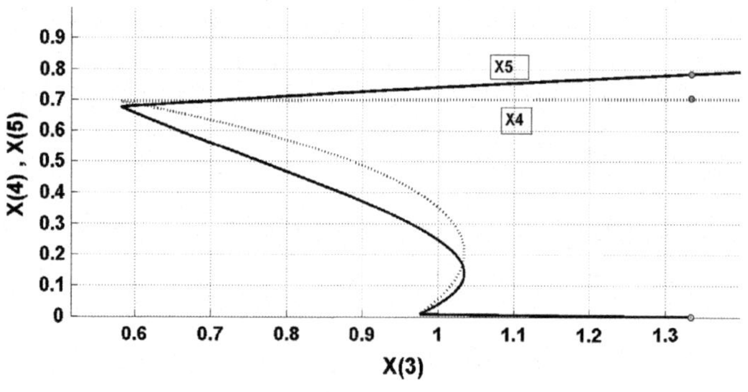

Fig. 7.10 General orbital transfer: out-of-plane state evolution

constraint introduced in the PSO allows to avoid the convergence of the algorithm versus uninteresting extremals providing inefficient cost solutions.

An example of optimal transfer between two elliptic orbits using three impulses is provided in Figs. 7.8, 7.9, 7.10, 7.11 and 7.12. The parameters of the initial and

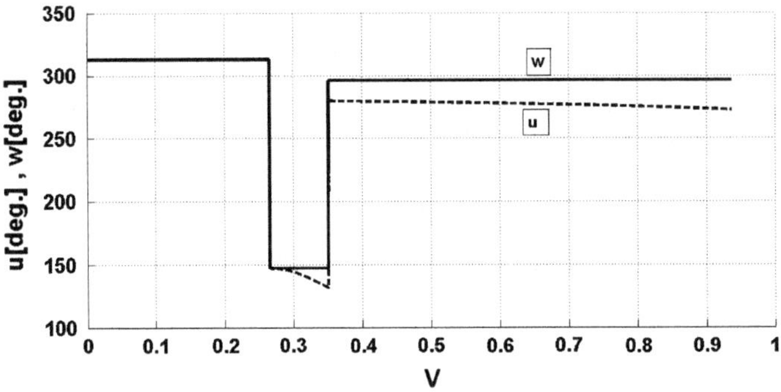

Fig. 7.11 General orbital transfer: firing anomaly u, w versus total cost

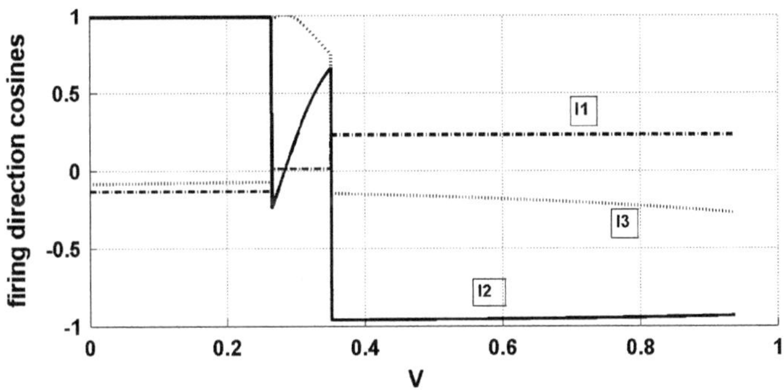

Fig. 7.12 General orbital transfer: firing direction in **ORF** versus total cost

final orbit are:

$$
\begin{aligned}
a_0 &= a_f \\
e_0 &= 0.5, e_f = 0.5 \\
g_0 &= 0, g_f = 45\,\text{deg} \\
I_0 &= 90, I_f = 45\,\text{deg} \\
h_0 &= 0., h_f = 45\,\text{deg}
\end{aligned}
$$

That is, in non-dimensional variables:

$$
\begin{aligned}
X_0 &= (0.6666, 0, 1.3333, 0, 0, 0)^T \\
X_f &= (0.4714, 0.4714, 1.3333, 0.7071, 0.7854, any)^T
\end{aligned}
$$

The initial co-state that makes the state crossing the target is:

$$L_0 = (0.3041, -0.1582, -0.2713, -0.0559, 0.2289, 0)^T; \qquad (7.62)$$

The target is attained after three firings with a total $V_f = 0.77996$ while $V_{L_2} = 0.82490$ in non-dimensional units.

One can wonder what is the maximum number of firing arcs for a generic optimal transfer. This problem has been analyzed for the angle-open case by [14] who shows that any solution with more than three firing arcs can be replaced by another with less firing arcs. In addition Prussing [19] has obtained a result, in the context of linear control but applicable to any sort of linearized transfer problems, that the firing arcs required for an overall minimal solution are equal to the number of final conditions that are imposed. Therefore, by this rule, for a time-open problem we have a maximum of 5 firings in a spatial transfer and 3 firings for a plane transfer.

7.9 Optimal Orbital Transfer: Finite Thrust

7.9.1 Introduction

The finite thrust optimization problem, incorporates the infinite thrust case. In addition today, this problem is receiving more attention in modern satellite design, because of the current trend to use low thrust, *plasmic propulsion systems* (PPS) for the telecommunication satellite transfer to the geosynchronous equatorial orbit (GEO).

In the low thrust case, as we will see, the finite thrust requires a larger V_f, however, because of their much higher specific impulse (see Chap. 8), the low thrust plasmic motors compared to classic monopropellant or bipropellant chemical systems, allow to reduce considerably the consumed propellants. This approach allows to reduce the launchers' size and cost for a given telecommunication satellite payload size. The GEO satellites designed with a PPS may have a useful payload mass up to $\frac{1}{5}$ of the satellite launched mass while those designed with a fully chemical propulsion have a useful payload mass lower than $\frac{1}{10}$ of the launched mass. The use of PPS is going to reduce significantly the cost per transponder of the GEO telecommunication satellites and will probably affect the policy of the launching systems in the near future, in favor of lighter, restartable and lower cost launchers. Many recent papers like [20, 21] deal with such an issue and the European Space Agency has recently implemented a study reported in [22, 23] to use *plasmic propulsion systems* for the in-orbit delivery of the Galileo second generation constellation.

The new approach implies that the satellites must be transferred from a lower, perhaps inclined, orbit to the GEO or MEO operational orbits with *plasmic propulsion*. These transfers may take many months, as shown in [24], that describes the long orbit raising of a GEO satellite realized by the author and the Artemis team to recover from a launcher failure.

The mission analysis problem of said transfers consists in determining the trajectory that minimizes transfer duration or propellant consumption for a given fixed transfer duration.

The selection of the objective function to be used in the design usually depends on exogenous economic factors like the time to market that the service provider needs for his business: currently we expect that such transfer time may vary from three months to one year depending on the market context.

We search for the optimal solution by using the first order necessary conditions of extremality from the Maximum Principle, which results in a two boundary value problem. The difficulty relies in *shooting* the correct boundary conditions with an algorithm, which is not globally convergent in such a highly non-linear context.

The analysis of the extremal equations for low thrust transfers is treated in modern papers like [25, 26] and classic papers like [27–29]. The usual approach which is followed in these papers, is to find approximate solutions implementing a control parametrization, then averaging the dynamical equations and finally using the *Maximum Principle* to optimize the parametrized control. Such approaches allow, in some cases, to achieve analytical solutions which are very useful in the preliminary analysis and also as initial guess of the shooting methods but may not be very accurate, especially for long transfers.

In this chapter we will follow the approach of [9], without treating however the technical aspects linked to the eclipse constraint.

Having identified the necessary conditions for optimality, the averaging operator is applied to the Hamiltonian in a form suitable to apply the averaging theorems that state the ε closeness with the exact optimal solution in a ε^{-1} non-dimensional time scale. The averaging techniques are the only ones that can provide fast and converging computing methods for long orbital transfers.

7.9.2 The Maximum Principle for the Finite Thrust Case

The dynamic system that will be used for the orbital transfer optimization includes Eq. 7.42 and the time as additional variable in order to be in condition to have constraints also on the final time. The variable to be minimized is the non-dimensional characteristic velocity defined by the Eq. 7.43. In order to obtain the perturbation structure of the system in the parameter ε, time will be represented in the state with the variable $T = \varepsilon\tau$, being ε the maximum possible value for the non-dimensional control force (see Eq. 7.40):

$$f = \frac{|\dot{m}|v_e}{m_0 \frac{\mu_e}{a_0^2}} \tag{7.63}$$

$$f \leq \varepsilon \tag{7.64}$$

The state (X, V, T) evolves as:

$$\begin{aligned}\frac{dX}{dw} &= fB(X, w + X_6)\mathfrak{l}e^{V/v_e}\frac{\sqrt{X_3}}{Y_1^2}\\ \frac{dV}{dw} &= fe^{V/v_e}\frac{\sqrt{X_3}}{Y_1^2}\\ \frac{dT}{dw} &= \varepsilon\frac{\sqrt{X_3}}{Y_1^2}\end{aligned} \tag{7.65}$$

where the controls are: f and $\mathfrak{l}$. The non-dimensional control force is limited by the following inequality: $0 \le f \le \varepsilon$.

With this setting we introduce the pre-Hamiltonian $\tilde{H}$ and the co-state $(L \in R^6, L_V, L_T)$, following the standard Maximum Principle method (see Sect. 5.4.1 or among many authors [30]):

$$\tilde{H} = f(L^T B(X, u)\mathfrak{l} + L_V)e^{V/v_e}\frac{\sqrt{X_3}}{Y_1^2} + \varepsilon L_T\frac{\sqrt{X_3}}{Y_1^2} \tag{7.66}$$

By maximizing $\tilde{H}$ in the controls $\mathfrak{l}, f$ we get the Hamiltonian.

$$\begin{aligned}H &= \sup_{f\in[0,\varepsilon],\,\mathfrak{l}\in|\mathfrak{l}|=1}\tilde{H}\\ H &= \varepsilon\kappa(S)S(X, L, L_V, u)e^{V/v_e}\frac{\sqrt{X_3}}{Y_1^2} + \varepsilon L_T\frac{\sqrt{X_3}}{Y_1^2}\end{aligned} \tag{7.67}$$

where the functions introduced S and $\kappa(S)$ have the following definitions:

$$\begin{aligned}S(X, L, L_V, u) &= \sqrt{L^T B(X, u)B(X, u)^T L} + L_V,\\ \kappa(S) &= 1, \forall S > 0\\ \kappa(S) &= 0, \forall S < 0\end{aligned} \tag{7.68}$$

The control actions derived maximizing the pre-Hamiltonian are:

$$\begin{aligned}\mathfrak{l} &= \frac{B(X,u)^T L}{\sqrt{L^T B(X,u)B(X,u)^T L}}\\ f &= \varepsilon\kappa(S)\end{aligned} \tag{7.69}$$

The control action f is not defined when $S(X, L, L_V, u) = 0$, but this happens in a single isolated point, so that, the state and co-state curves are unique, unless we have the non generic case of a singular arc which can be studied using higher order derivatives (see [30]). When the state and co-state curves cross the surface $S(X, L, L_V, u) = 0$ we have the on/off switching of the thruster, therefore we will call S the *switching function*. The necessary conditions that any extremal satisfies are expressed by the following system of differential equations derived from the Hamiltonian:

$$
\begin{aligned}
\frac{dX}{dw} &= \varepsilon\kappa(S)\frac{B(X,u)B(X,u)^T L}{\sqrt{L^T B(X,u)B(X,u)^T L}} e^{V/v_e}\frac{\sqrt{X_3}}{Y_1^2} \\
\frac{dV}{dw} &= \varepsilon\kappa(S)e^{V/v_e}\frac{\sqrt{X_3}}{Y_1^2} \\
\frac{dT}{dw} &= \varepsilon\frac{\sqrt{X_3}}{Y_1^2} \\
\frac{dL}{dw} &= -\varepsilon\kappa(S)\frac{\partial\sqrt{L^T B(X,u)B(X,u)^T L}}{\partial X} e^{V/v_e}\frac{\sqrt{X_3}}{Y_1^2} - \varepsilon(\kappa(S)Se^{V/v_e} + L_T)\frac{\partial\frac{\sqrt{X_3}}{Y_1^2}}{\partial X} \\
\frac{dL_V}{dw} &= -\varepsilon\kappa(S)S\frac{e^{V/v_e}}{v_e}\frac{\sqrt{X_3}}{Y_1^2} \\
\frac{dL_T}{dw} &= 0
\end{aligned}
\tag{7.70}
$$

The above equations have to be integrated with suitable boundary conditions.

7.9.3 Boundary Conditions

The boundary conditions that will be considered are specific for the study of long term orbital transfer between a LEO and GEO/MEO orbits: minimization of the transfer time and minimization of the transfer velocity at a fixed transfer time. However, different classes of problems like rendez-vous are easily implementable with the same basic equations. If we fix the initial and final orbit and consider open the arrival anomaly we have (refer to Sect. 5.4.1):

$$
X_k(0) = X_{0k}, k = 1..6, X_k(w_f) = X_{fk}, k = 1..5 \tag{7.71}
$$

Two transversality conditions are introduced: the first derived from the fact that $X_6(w_f)$ is not specified because we are not interested in the final anomaly, the second derived by the fact that w_f is free:

$$
L_6(w_f) = 0, H(w_f) = 0 \tag{7.72}
$$

The additional transversality conditions depend on the specific optimization problem. We envisage two kinds of problems:

Problem (*Tmin*) The minimum time problem, with additional final conditions:

$$
L_V(w_f) = 0, L_T(w_f) = -1 \tag{7.73}
$$

Problem (*Vmin*) The minimum velocity at fixed final time problem, where the fixed time must be longer than the one resulting in the minimum time problem. In this case the additional final conditions are:

$$
L_V(w_f) = -1, T(w_f) = T_f \tag{7.74}
$$

It is possible to verify that under the minimum time problem conditions, the *switching function* is always positive definite and the optimal solution has a continuous firing. In fact, the derivative of L_V is always negative semi-definite and the final condition $L_V(w_f) = 0$ ensures that $L_V(w_f) \geq 0$ which implies that the *switching function* is always semi positive.

7.9.4 The Averaging Method

Equation 7.70 have their second members multiplied by

$$\varepsilon = \frac{|\dot{m}|v_e}{M_0 \frac{\mu_e}{a_0^2}} \tag{7.75}$$

which in most of low thrust applications can be considered a small parameter in the range of $\approx 1.e-3, 1.e-4$. The second members are periodic functions of the independent variable. Therefore, the system is set in the *perturbation standard form* (see [31], Sect. 1.6) and we can use the averaging technique to implement an approximate solution. Let us introduce the averaging operator:

$$\begin{aligned} \mathscr{E}_w[H(X,L,w)] &= \frac{1}{2\pi}\int_0^{2\pi} H(X,L,w)dw \\ \mathscr{E}_w[H(X,L,w)] &= \bar{H}(X,L) \end{aligned} \tag{7.76}$$

We note that the averaging is performed in the independent variable w. The state and co-state including the variable T, the so-called slow time, are considered fixed in the averaging integral that produces the averaged Hamiltonian:

$$\bar{H}(X,L) = \varepsilon\mathscr{E}_w[\kappa(S)S\frac{\sqrt{X_3}}{Y_1^2}]e^{V/v_e} + \varepsilon L_T\mathscr{E}_w[\frac{\sqrt{X_3}}{Y_1^2}] \tag{7.77}$$

The averaged Hamiltonian is then independent of w and X_6. The averaged equations can be obtained using the $\mathscr{E}_w[.]$ on the second members of the Eq. 7.70 or deriving the Hamiltonian flow from the averaged Hamiltonian Eq. 7.77:

$$\begin{aligned} \frac{dX}{dw} &= \frac{\partial\bar{H}}{\partial L} \\ \frac{dL}{dw} &= -\frac{\partial\bar{H}}{\partial X} \end{aligned} \tag{7.78}$$

We have that $\dfrac{\partial\bar{H}}{\partial X_6} = 0$.

The resulting averaged equations are:

$$
\begin{aligned}
\frac{dX}{dw} &= \varepsilon\mathscr{E}_w[\kappa(S)\frac{B(X,u)B(X,u)^T L}{\sqrt{L^T B(X,u)B(X,u)^T L}}\frac{\sqrt{X_3}}{Y_1^2}]e^{V/v_e}\\
\frac{dV}{dw} &= \varepsilon\mathscr{E}_w[\kappa(S)e^{V/v_e}\frac{\sqrt{X_3}}{Y_1^2}]e^{V/v_e}\\
\frac{dT}{dw} &= \varepsilon\mathscr{E}_w[\frac{\sqrt{X_3}}{Y_1^2}]\\
\frac{dL}{dw} &= -\varepsilon\mathscr{E}_w[\kappa(S)\frac{\partial\sqrt{L^T B(X,u)B(X,u)^T L}}{\partial X}\frac{\sqrt{X_3}}{Y_1^2}]e^{V/v_e}\\
&\quad -\varepsilon\mathscr{E}_w[(\kappa(S)Se^{V/v_e}+L_T)\frac{\partial\frac{\sqrt{X_3}}{Y_1^2}}{\partial X}]\\
\frac{dL_V}{dw} &= -\varepsilon\mathscr{E}_w[\kappa(S)S\frac{\sqrt{X_3}}{Y_1^2}]\frac{e^{V/v_e}}{v_e}\\
\frac{dL_T}{dw} &= 0
\end{aligned}
\tag{7.79}
$$

These equations can be associated with the boundary conditions Eqs. 7.71 and 7.72 where $H(w_f)$ is substituted by $\bar{H}(w_f)$ and Eq. 7.73 or 7.74 depending on the specific problem.

We will call any curve in $D = (X \in \mathbb{R}^8 | \sqrt{X_1^2 + X_2^2} < X_3, X_3 \in \mathbb{R}_+, X_4 \in [-1, 1], X_7 \in \mathbb{R}_+, X_8 \in \mathbb{R}_+) \otimes (L \in \mathbb{R}^8 - 0)$, satisfying the differential equations Eq. 7.79 and one set of boundary conditions of the minimum time problem or of the minimum velocity problem an *averaged extremal*. The *averaged sxtremal* has various properties which simplify its solution and shed light on the solution and the structure of the non-averaged problems.

- the regularity of $\bar{H}(X, L)$ in X, L depends on the regularity of $H(X, L, w)$ in the state and the co-state variables, provided that the $H(X, L, w)$ is limited and integrable in the w variable. The averaged Hamiltonian is derivable infinitely many times in X, L in the domain D;
- due to the final conditions on the averaged Hamiltonian Eq. 7.72, its value is constantly null on any extremal trajectory;
- the averaged Hamiltonian is independent on X_6, therefore L_6 is constant and due to Eq. 7.72, $L_6 = 0$ along any *averaged extremal*;
- from the *averaging theory* (see Ref. [31], *Theorem2.8.1) the averaged extremal solution of Eq.* 7.79 $\bar{X}(w), \bar{L}(w)$, *satisfies the following property.*
  Property: Averaging $\exists\varepsilon_{max} \geq 0, C_1 \geq 0, C_2 \geq 0$, *so that,* $\forall\varepsilon \leq \varepsilon_{max}$ *there exists a solution of the not averaged differential system (Eq.* 7.70*)* $X(w)$, $L(w)$, *that satisfies the same initial conditions at* w_0 *and verifies the following inequality* $|\bar{X}(w) - X(w)| + |\bar{L}(w) - L(w)| \leq C_1\varepsilon, \forall\ (w - w_0) \leq \frac{C_2}{\varepsilon}$.
  So the efficacy of the *averaged extremal* is limited for problems that require a transfer not longer than the $\frac{w}{\varepsilon}$ scale; this means in general many hundreds of satellite orbits;
- the *averaged extremal* satisfies a scaling rule when orbital perturbations are negligible. In fact, in such cases the parameter $\varepsilon = \frac{mv_e}{M_0\frac{\mu_e}{a_0^2}}$ is a multiplicative factor of the averaged Hamiltonian which is independent of w; if we have two problems

having the same boundary conditions but a different ε the solutions are the same apart from a transformation of the independent variable $w_1 = w_2\frac{\varepsilon_2}{\varepsilon_1}$. This means that, increasing the initial mass and keeping constant the thrust, the time to the final orbit will increase proportionally keeping constant the V.

The search of the solution of Eq. 7.79 is numerically faster than the solution of Eq. 7.70, and using the closeness of the two trajectories, the solution of Eq. 7.70 can be easily solved numerically starting from the averaged one. This implies the calculations of the short terms of the solution. To do so, the initial co-state found by the averaged solution, is changed of a ε small quantity in order to satisfy the terminal conditions at a final w_f which is close to the one of the *averaged extremal.*

The terminal condition on the averaged Hamiltonian $\bar{H}(w_f) = 0$ guarantees that the close by extremal has a null Hamiltonian $H(w_f^*) = 0$ at some point of the same orbit w_f^*.

7.9.5 Orbital Perturbation

The effect of orbital perturbations may be extremely important in long transfers. In particular the rotation of the node has a significant effect on LEO to MEO/GEO transfers. When the perturbations produce a node regression we shall leave the node free to rotate in the boundary conditions in order not to lose propellant trying to keep it fixed. Such condition is expressed by setting $L_{f5} = 0$ and keeping X_{f5} free. Any kind of orbital perturbation can be easily considered in the equations, by adding a term to the Hamiltonian. If $A_p(X, w)$ is a perturbation field such that Eq. 7.42 become:

$$\frac{dX}{dw} = B(X, u = w + X_6)(A + A_p(X, w)) \tag{7.80}$$

then the Hamiltonian of the optimal control problem is:

$$H = \varepsilon\kappa(S)S\kappa(E)e^{V/v_e}\frac{\sqrt{X_3}}{Y_1^2} + \varepsilon L_T\frac{\sqrt{X_3}}{Y_1^2} + L^T B(X, u)A_p(X, w)$$

In the averaged version we need to add to the previously defined averaged Hamiltonian (Eq. 7.77) the term $L_T\bar{X}_p$ where $\bar{X}_p(X) = \mathscr{E}_w[B(X, u)A_p(X, w)]$.

We evaluate the perturbation terms for the J_2 case.

The J_2 potential is $F_2 = -J_2\frac{\mu_e}{r}(\frac{R}{r})^2(\frac{3\sin(\beta)^2-1}{2})$, where β is the satellite latitude in **ECI**.

The three accelerations are calculated from the gradient of the potential dividing it by $\frac{\mu_e}{a_0^2}$ to obtain non-dimensional accelerations (we recall that $r = a_0/Y_1$, $\sin(\beta) = \sin(I)\sin(u)$):

$$\begin{aligned}
A_r &= \frac{\partial F_2}{\partial r}\frac{a_0^2}{\mu_e} &&= 1.5(\tfrac{J_2R^2}{a_0^2})Y_1^4(3(1-X_4^2)\sin(u)^2-1)\\
A_t &= \frac{\partial F_2}{r\partial\beta}\frac{\sin(I)\cos(u)}{\cos(\beta)}\frac{a_0^2}{\mu_e} &&= -3(\tfrac{J_2R^2}{a_0^2})Y_1^4(1-X_4^2)\sin(u)\cos(u)\\
A_n &= \frac{\partial F_2}{r\partial\beta}\frac{\cos(I)}{\cos(\beta)}\frac{a_0^2}{\mu_e} &&= -3(\tfrac{J_2R^2}{a_0^2})Y_1^4\sqrt{1-X_4^2})\sin(u))
\end{aligned} \tag{7.81}$$

The following step is to apply the averaging $\bar{X}_p(X) = \mathscr{E}_w[B(X,u)A\frac{\sqrt{X_3}}{Y_1^2}]$. This can be done by quadratures, calling $j_2 = (\frac{J_2R^2}{a_0^2})$ we obtain:

$$\begin{aligned}
\bar{X}_{p1} &= -\tfrac{3}{4}j_2X_3^2X_2(5X_4^2-1)\\
\bar{X}_{p2} &= \tfrac{3}{4}j_2X_3^2X_1(5X_4^2-1)\\
\bar{X}_{p3} &= 0\\
\bar{X}_{p4} &= 0\\
\bar{X}_{p5} &= -\tfrac{3}{2}j_2X_3^2X_4\\
\bar{X}_{p6} &= \tfrac{3}{2}j_2X_3^2X_4^2
\end{aligned} \tag{7.82}$$

The co-state equations can be obtained adding to the r.h.s. of $\frac{dL}{dw}$ in Eq. 7.79 the expression $-\frac{\partial L^T\bar{X}_p}{\partial X}$.

7.9.6 Using New Coordinate Systems

The system of equations selected becomes singular when the orbital inclination approaches zero. In this case it is necessary to transform the orbital state and co-state into a non-singular one. This mapping can be done easily using canonical transformations. Suppose we want to pass from $X \to Z(X)$, with inverse $X = X(Z)$. Following the standard Hamiltonian theory (see Appendix C.1) we can introduce the generating function $F = \lambda^T Z(X)$ where λ is the co-state of the new variables Z. The new co-state can be obtained from $\frac{\partial F}{\partial X} = \frac{\partial \lambda^T Z(X)}{\partial X} = L$, which can be solved using the inverse transformation of the state variables as $\lambda = \frac{\partial L^T X(Z)}{\partial Z}$.

We can use the Z, λ coordinates anytime we have a problem with a low or null inclination to avoid the singularity in the X equations.

7.9.7 Applications

We will show in this chapter some numerical results that have been obtained with a code developed by the author named SOFTT (satellite optimal finite thrust transfer). All the results will be provided in keplerian parameters in order to simplify the understanding.

Mission	a in [km]	a fin [km]	I in [km]	I fin [km]	Time [days]	DV [m/s]	DM [kg]	DV imp [m/s]	DM imp [kg]
LEO to MEO	7378	29601	56	56	230	3 681	252	3 301	1 012
LEO to GEO	7378	42166	0	0	264	4 276	289	3 654	1 067
LEO to IGSO	7378	42166	56	60	265	4 301	290	3 644	1 065
GTO to MEO	24475	29601	6	56	216	3 441	237	2 425	842
GTO to GEO	24475	42166	6	0	142	2 200	156	1 490	596
GTO to IGSO	24475	42166	6	60	213	3 390	234	2 491	857

Fig. 7.13 Table with many different mission results

The table in Fig. 7.13 summarizes the results of many minimum time, orbital transfers from LEO and GTO to MEO, GEO and IGSO (inclined geosynchronous orbits).

For each case, eclipse and J_2 effects are not taken into account. The satellite properties are $m_0 = 1500$ kg, $|\dot{m}|v_e = 254$ mN, $gv_e = 2076$ s, the perturbation parameter for the Missions to GEO is $\varepsilon = 7.5532\ 10^{-4}$ and for the missions to MEO $\varepsilon = 3.722\ 10^{-4}$.

In the same table we have reported the DV_{imp} necessary to implement an impulsive minimum DV strategy. As expected they are always less than the DV requested by the Tmin solutions. We have also evaluated the propellant Dm_{imp}necessary for the impulsive strategy using $I_{sp} = gv_e = 300$ s, typical of a good bipropellant chemical propulsion system. We can see that for all the transfers analyzed the propellant consumption is in favor of the low thrust plasmic approach. In all the missions starting from LEO, the mass consumption of a chemical bipropellant propulsion system is always higher than 65 % of the initial mass, so that these kind of transfers have never been implemented with bipropellant, because they are useless for practical applications. On the contrary, using a low thrust approach, the propellant consumption is always ~15 % of the initial mass, which is an acceptable propellant ratio for the applications. Obviously, the duration of the transfer must be carefully considered in the mission feasibility analysis. The duration can be shortened by using either more motors or more powerful ones (see Chap. 8).

The following analysis is a GTO to MEO transfer. We calculate the minimum DV solutions for many different transfer times. We want to show that varying the time of the minimum DV solutions, we tend on the one hand towards the minimum time solution and on the other hand towards the infinite thrust solution.

For this case the J_2 and eclipse effect are not taken into account. The satellite properties are the following: $m_0 = 1500$ kg, $|\dot{m}|v_e = 254$ mN, $gv_e = 2076$ s, the perturbation parameter is $\varepsilon = 3.7224\ 10^{-4}$.

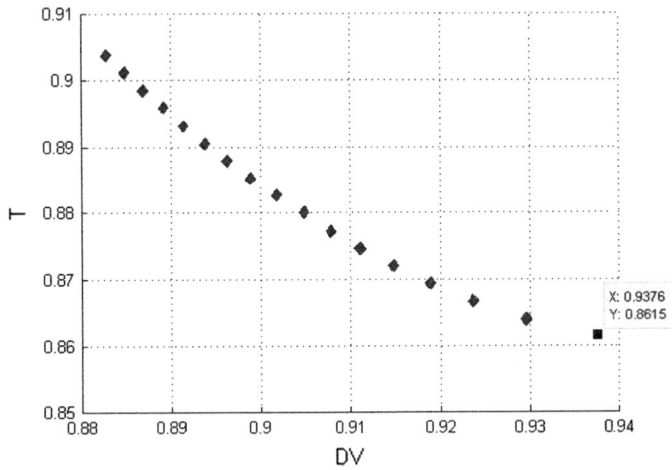

Fig. 7.14 Time versus DV for Vmin analysis, *black square* point is the Tmin solution

GTO reference: $a = 24475\,\text{km},\ e = 0.7292, I = 6\,\text{deg}$
MEO reference: $a = 29601\,\text{km},\ e = 0.0,\ I = 56\ \text{deg}$

For this mission the minimum transfer time is 216 km and $DV = 3.441$ km/s.

Figure 7.14 shows the different required $DV/\sqrt{\mu_e/a_{MEO}}$ in function of transfer time, the black square point represents the Tmin solution; for longer values of transfer time, the DV required is lower than the DV of the Tmin solution.

As we give more time to perform the optimization, the Vmin solution tends to the impulsive, infinite thrust, solution. This kind of analysis should be applied when, having designed a platform for a given time requirement, we want to minimize the usage of propellant using all the transfer time available margins to improve the satellite's operational life.

From the scaling rule illustrated in Sect. 7.9.4, it can be inferred that increasing ε - by, for example, increasing the thrust—the curve in Fig. 7.14 will be proportionally shifted towards lower final times, while the maximum DV, which is related to the Tmin solution, and the minimum DV, related to the infinite thrust solution, will remain the same.

A last analysis deals with a GTO to MEO orbital transfer with the variation of the final right ascension. This example will show that we can have for a given problem more than one extremal solution. For this example the satellite and the PPS characteristics are: $m_0 = 1800\,\text{kg},\ |\dot{m}|v_e = 0.240\,\text{mN},\ gv_e = 1500\,\text{s}$, the perturbation parameter is $\varepsilon = 2.9310\ 10^{-4}$. The general boundary conditions are the following:

- The final orbit conditions are fixed;
- The minimum time solution is considered so the *switching function* is always $S > 0$ and then the PPS works all the time. $L_{f,7} = 0,\ L_{f,8} = -1$, while the other costates are free.

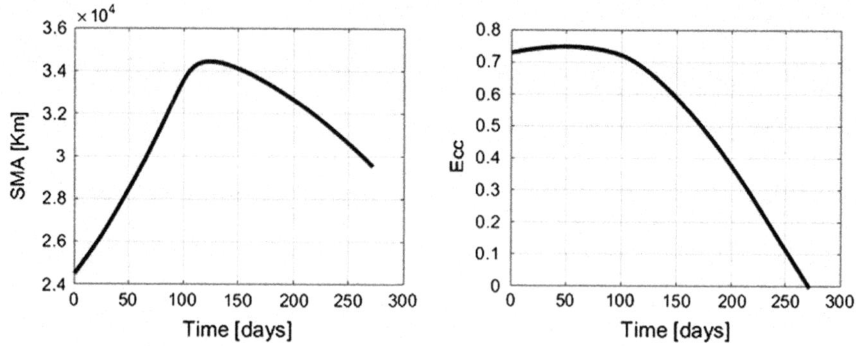

Fig. 7.15 Evolution of the semi-major axis and eccentricity

In this optimization the J_2 perturbation effects and the eclipse effects are not considered so the PPS does not switch off during the eclipse phase.

In the first simulation $D\Omega = 0$,

$$
\begin{aligned}
&\text{GTO reference: } a = 24475\,\text{km},\ e = 0.7292,\ I = 6\,\text{deg},\ \omega = 178\,\text{deg},\ \Omega = 0\,,\\
&\text{MEO reference: } a = 29601\,\text{km},\ e = 0.0,\ I = 56\,\text{deg},\ \Omega = 0\,.
\end{aligned}
$$

The GTO to MEO orbital transfer results using SOFTT are:

$$
\begin{aligned}
&\text{transfer time} = 271\,\text{km},\\
&\text{transfer } DV = \ 3.508\,\text{km/s},\\
&\text{number of orbits} = \ 433,\\
&\text{transfer } Dm = \ 382\,\text{kg}.
\end{aligned}
$$

The evolution of the Keplerian parameters is reported in Figs. 7.15 and 7.16.

These figures show how most of the element variations occur during the first period of the transfer (150 days) when the eccentricity is higher, in fact the inclination passes

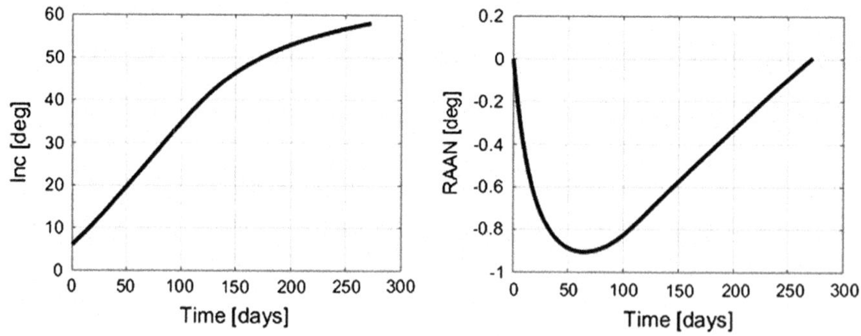

Fig. 7.16 Evolution of the inclination and right ascension

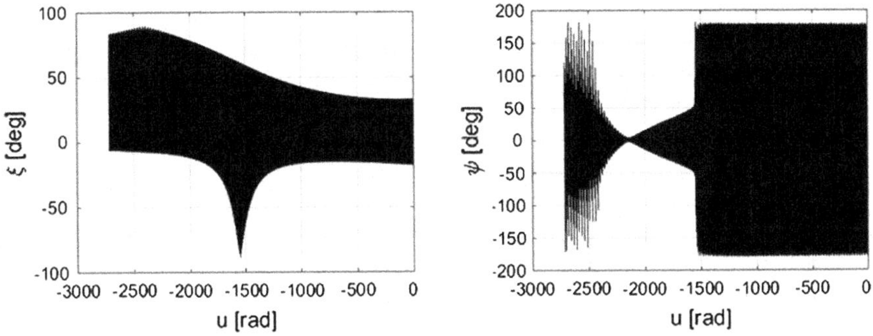

Fig. 7.17 Evolution of the maneuver angles ξ and ψ

from 6 deg to 45 deg and the semi-major axis arrives at the maximum value, about 35000 km.

The evolution of the firing vector, represented with the angles ξ and ψ is reported in Fig. 7.17: ξ is the angle between the firing vector and the orbital plane, while the ψ angle is between the firing vector projection on the orbital plane and the velocity.

Using the same orbital transfer parameters a study of the evolution of the DV as a function of the change of the final right ascension has been done. In particular starting from $D\Omega = 0$ deg we increased the final right ascension step by step.

The evolution of the DV as function of the change of the final right ascension is reported in Fig. 7.18. The circled solutions will be analyzed later in more detail.

Figure 7.18 shows that there are two families of extremal solutions and that we cannot expect to have a single extremal for a given problem. In particular starting from the solution for $D\Omega = 0$ deg and increasing the $D\Omega$, the DV increases until $D\Omega \approx 100$ deg, starting from the solution for $D\Omega = 180$ deg and decreasing the $D\Omega$,

Fig. 7.18 Evolution of the DV cost as function of the change of the $D\Omega$

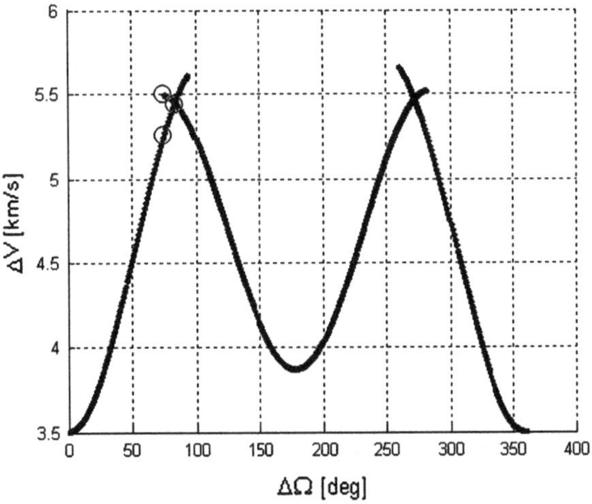

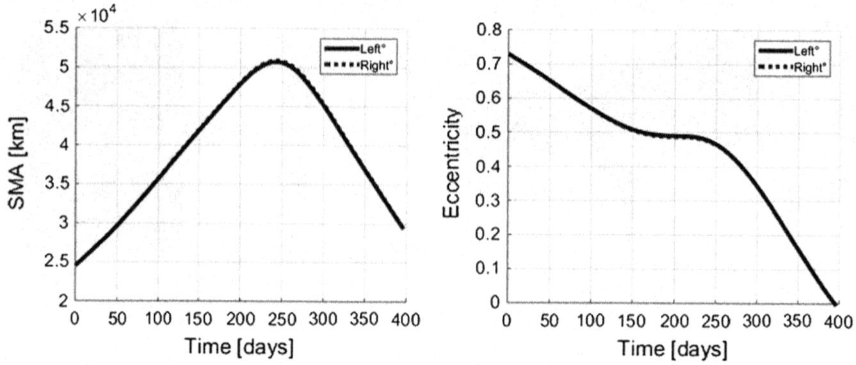

Fig. 7.19 Evolution of the semi-major axis and eccentricity

the DV increases until $D\Omega \approx 70\,\text{deg}$ crossing the previous family for $D\Omega = 84\,\text{deg}$ The same evolution occurs nearby $D\Omega \approx 270\,\text{deg}$ where the families of extremals cross each other.

We will see that the two families implement two different firing strategies which are both locally extremal but in terms of global optimization in some conditions one is better than the other.

In the next figures we illustrate the main difference in terms of evolution of Keplerian parameters and maneuver angles, between the left family (coming from $D\Omega = 0$) and the right one (coming from $D\Omega = 180\,\text{deg}$) at the cross point $D\Omega = 84\,\text{deg}$.

Note that, in all the following figures, the dotted line represents the right family while the continuous line shows the behavior of the left family.

From Fig. 7.19 we see that the evolution of the semi-major axis and the eccentricity are very close in the two families, instead from Fig. 7.20, the inclination and the right ascension dynamics look different.

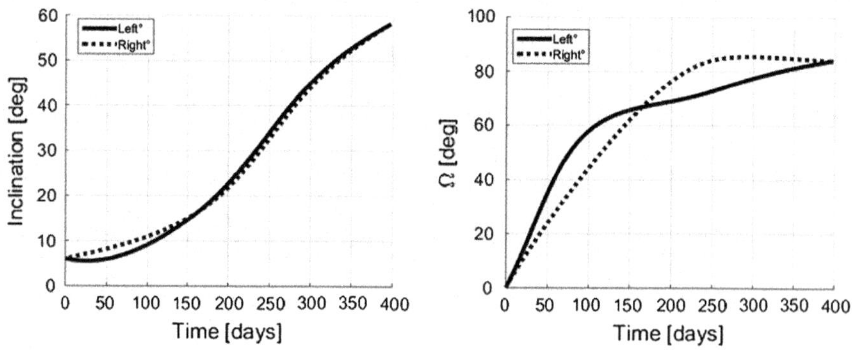

Fig. 7.20 Evolution of the right ascension

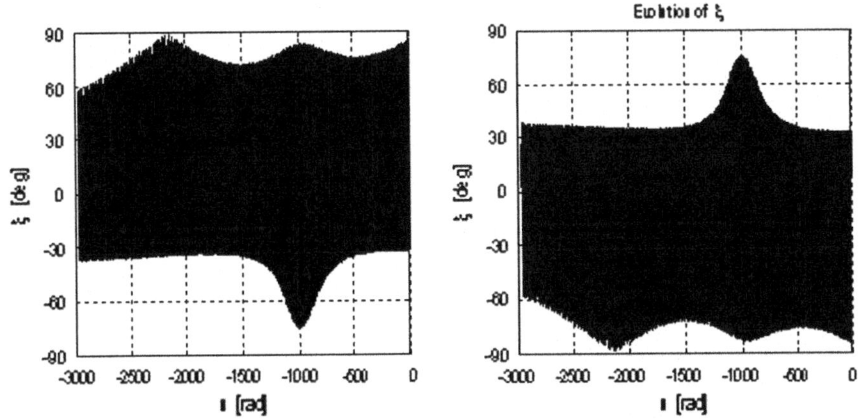

Fig. 7.21 Evolution of the maneuver angle ξ, of the *left* family (*left* figure) and of the *right* family (*right* figure)

For what concerns the inclination, the left solution starts decreasing while in the right solution the inclination increases immediately; for what concerns the right ascension, the left solution starts with a strong slope for 150 days, after that the slope slows down while still increasing. In the right solution, the RAAN increases with a strong slope for 250 days, overcomes the target value and then decreases until the end of the orbit transfer.

The evolution of the maneuver angles ξ and ψ are reported in the following images.

Looking at Fig. 7.21 the evolution of the ξ angle is quite different between the two families, this angle peaks in any orbit in the opposite direction between the two families. We analyze this behavior for the orbit around $u = -2200$ rad in Fig. 7.23: it is clear that the firing vector out-of-plane strategy is antisymmetric with respect to

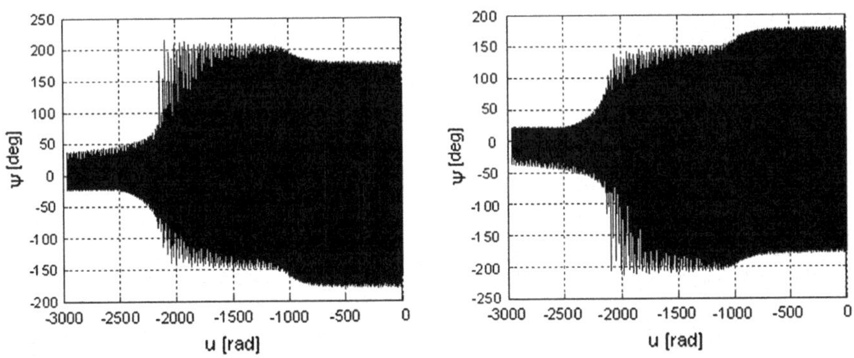

Fig. 7.22 Evolution of the maneuver angle ψ, of the *left* family (*left* figure) and of the *right* family (*right* figure)

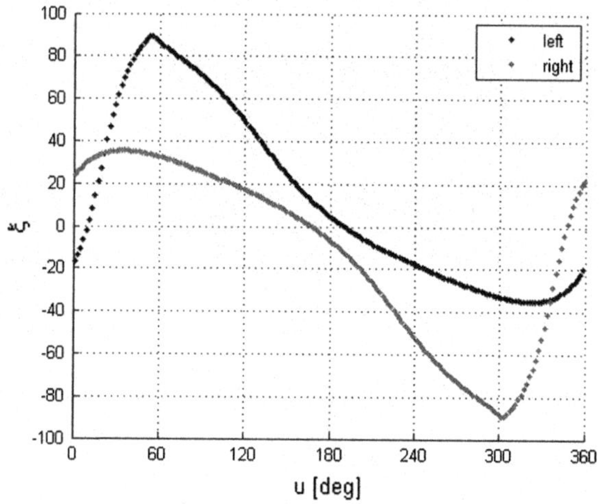

Fig. 7.23 Evolution of the maneuver angle ξ in one orbit around $u = -2200$ rad

$u = 180$ deg. The evolution of the ψ angle is similar in the two families and it is not reported (Fig. 7.22).

Let us study the two families when they are outside the crossing points (the solutions analyzed are those which are circled in Fig. 7.18 in $D\Omega = 78$ deg). The two solutions are both extremals for the Tmin problem, but the left solution is better than the right one for this $D\Omega$. In fact for the left family solution we have:

transfer time = 387.3 km,
transfer DV = 5.316 km/s,
number of orbits = 482,
transfer Dm = 545.8 kg,

for the right family solution we have:

transfer time = 397.7 km,
transfer DV = 5.488 km/s,
number of orbits = 460,
transfer Dm = 560.4 kg.

It is remarkable that the right solution implements a lower number of orbits but more time to reach the target. A comparison between the keplerian elements of these two different strategies is provided in Figs. 7.24 and 7.25.

The evolution of the firing angles ξ and ψ is reported in the following images:

Looking at Fig. 7.26 the evolution of the ξ angle recalls the one found at the crossing point. This is illustrated in Fig. 7.28 for $u = -2200$ rad: the firing vector out-of-plane strategy is almost antisymmetric with respect to $u = 180$ deg. The evolution of the ψ angle in this orbit is similar in the two families and it is not reported (Fig. 7.27).

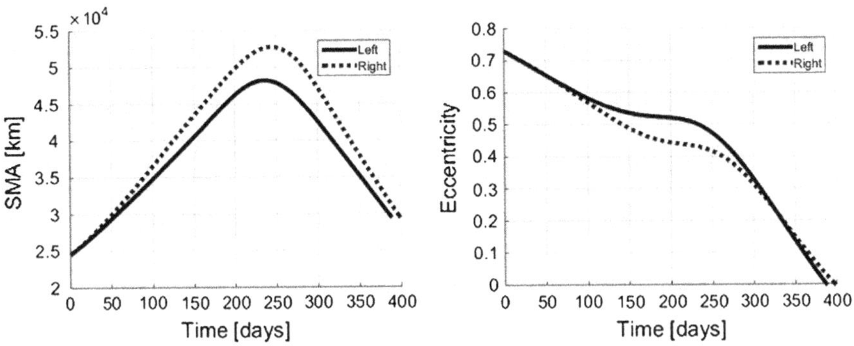

Fig. 7.24 Evolution of the semi-major axis and eccentricity

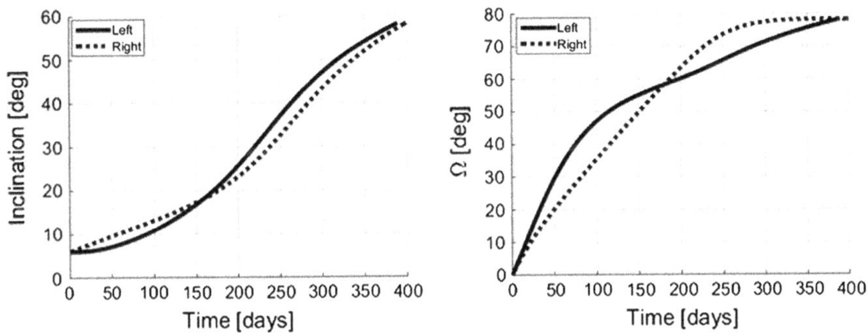

Fig. 7.25 Evolution of the inclination and right ascension

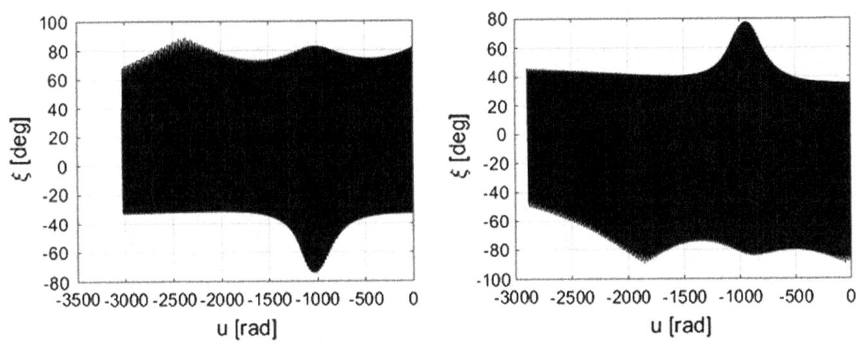

Fig. 7.26 Evolution of the maneuver angle ξ, of the *left* family (*left* figure) and of the *right* family (*right* figure)

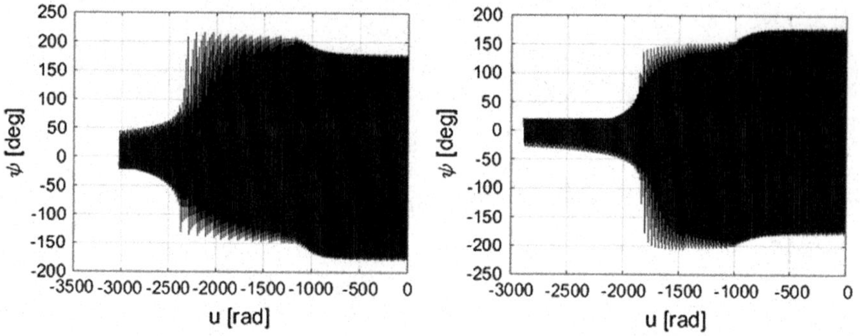

Fig. 7.27 Evolution of the maneuver angle ψ, of the *left* family (*left* figure) and of the *right* family (*right* figure)

Fig. 7.28 Evolution of the maneuver angle ξ in one orbit around $u = -2200$ rad

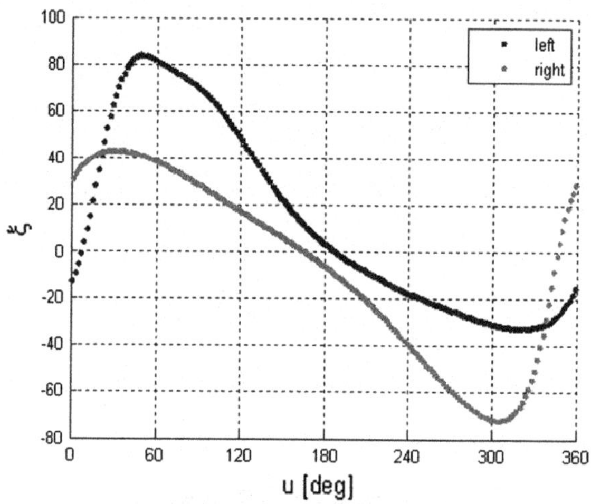

The evolution of the Keplerian parameters for the two solution families are quite different:

- in the right family the semi-major axis peaks to a much higher value with respect to the left family;
- both the eccentricity evolutions decrease quickly but the inflection points occur at different anomalies;
- the inclination of the left solution increase with a slower rate at the beginning, to recover with respect to the right solution just before the mid point;
- the right ascension of the left solution starts with a stronger slope for 100 days, after that, the slope slows down but the RAAN continues to increase, while in the right solution, the RAAN increases for 250 days, when it arrives near the target value, then remains quite constant until the end of the orbit transfer.

References

1. E.M. Soop, *Handbook of Geostationary Orbits* (Kluwer Academic Publishers, 1994)
2. D. Boccaletti, G. Pucacco, *Theory of Orbits*, vol. 2, Astronomy and Astrophysics Library (Springer, Berlin, 1996)
3. O. Montenbruck, E. Gill, *Satellite Orbits Method and Applications* (Models, Springer, 2005)
4. A.E. Roy, *Orbital Motion* (Adam Hilger Ltd, 1982)
5. A. Celletti, C. Gales, On the dynamics of space debris: 1:1 and 2:1 resonances. J. Nonlinear Sci. **24**, 1231–1262 (2014)
6. J.P. Carrou, *Spaceflight Dynamics* (Cepadues Editions, Toulouse, 1995)
7. R.M. Georgevic, Mathematical model of the solar radiation force and torques acting on the components of a spacecraft. NASA Tech. Memorandum 33–494 (1971)
8. L. Mazzini, Time open orbital transfer in a transformed Hamiltonian setting. J. Guidance Control Dyn. **36**(5) (2013)
9. L. Mazzini, Finite thrust orbital transfers. Acta Astronaut. AA-D-13-00699 (2014)
10. D.F. Lawden, *Optimal Trajectories for Space Navigation* (Butterworths, 1963)
11. D.R. Glandorf, Lagrange multipliers and the state transition matrix for coasting arcs, AIAA J. 7, **2**, 363–365 (1969)
12. P.M. Lion, M. Handelsman, Primer vector on fixed time impulsive trajectories, optimal two and three impulse fixed time rendezvous in the vicinity of a circular orbit. AIAA J. **6**(1) (1968)
13. G.A. Hazelrigg, Globally optimal impulsive transfers via Green's theorem. J. Guidance Control **7**(4), 462–470 (1984)
14. L. Ting, Optimun orbital transfer by impulses. ARS J. 1013–1018 (1960)
15. L. Ting, Optimun orbital transfer by several impulses. Astronaut. Acta **6**(5), 256–265 (1960)
16. R.F. Hoelker, R. Siber, *The Bielliptical Transfer Between Circular Coplanar Orbits*. Army Ballistic Missile Agency, Redstone Arsenal, Ala. Report No. DA-TM-2-59 (1959)
17. B.A. Conway, *Spacecraft Trajectory Optimization* (Cambridge University Press)
18. R.H. Battin, R.M. Vaughan, An elegant Lambert algorithm. J. Guidance Control Dyn. **7**, 662–670 (1984)
19. J.E. Prussing, Optimal two and three impulse fixed time rendezvous in the vicinity of a circular orbit. J. Spacecraft Rockets **40**(6) (2003)
20. A. Dutta, P. Libraro, N.J. Kasdin, E. Choueiri, Design of next generation all-electric telecommunication satellites, in *AIAA International Communication Satellite Systems Conference*. Florence, Italy (2013)
21. O.B. Duchemin et al., Ariane 5-ME and electric propulsion: GEO insertion options, in *47th AIAA/ASME/SAE/ASEE Joint Propulsion Conference and Exhibit* 31 July–03 Aug., San Diego, California, AIAA 2011-6084 (2011)
22. R. Biesbroek et al., Advanced galileo injection into LEO/GTO using electrical propulsion, AGILE 1, ESA Study Report, CDF-120-C, June (2013)
23. R. Biesbroek et al., Advanced galileo injection into LEO/GTO using electrical propulsion, AGILE 2, ESA Study Report, CDF-133-C, June (2013)
24. L. Mazzini, V. Tamilia, The Artemis satellite unexpected trip to GEO orbit. Space Technol. **24**(1), 43–56 (2004)
25. C.A. Kluever, Low thrust trajectory optimization using orbital averaging and control parametrization, in *Spacecraft Trajectory Optimization*, ed. by B.A. Conway (Cambridge University Press), pp. 112–138
26. J.A. Kechichian, Analytic representations of optimal low thrust transfer in circular orbits, in *Spacecraft Trajectory Optimization*, ed. by B.A. Conway (Cambridge University Press), Chap. 6, pp. 139–177
27. T.N. Edelbaum, Propulsion requirements for controllable satellites. ARS J. **31**, 1079–1089 (1961)
28. J.A. Kechichian, The reformulation of Edelbaum's low thrust transfer problem using optimal control theory. AIAA 92–4576

29. J.E. Pollard, Simplified approach for assessment of low-thrust elliptical orbit transfers. IEPC-97–160, 979–986
30. B. Chachuat, *Non linear and dynamic optimization, from theory to practice*, La-Teaching, 2007-001, Ecole Polytechnique Federale de Lausanne
31. J.A. Sanders, F. Verhulst, J. Murdock, *Averaging methods in Nonlinear Dynamical Systems*, 2nd edn., vol. 59 (Springer - Applied Mathematical Sciences) (2000)
32. R.R. Bate, D.D. Mueller, J.E. White, *Fundamentals of Astrodynamics* (Dover Publications Inc., New York, 1971)
33. J.P. Marec, *Optimal Space Trajectories* (Elsevier, Amsterdam, 1979)
34. K. Aksnes, A second-order artificial satellite theory based on an intermediate orbit. NASA Tech. Rep. 32–1507 (1970)
35. I.G. Izsak, A note on perturbation theory. Astron. J. **68**, 559 (1963)
36. K.G. Eckel, Optimal impulsive transfer with time constraint. Acta Astronaut. **9**(3), 139–146 (1992)
37. M. Pontani, Simple methods to determine globally optimal orbital transfers. J. Guidance Control Dyn
38. J.E. Prussing, Optimal impulsive linear systems: sufficient conditions and maximum number of impulses. J. Astronaut. Sci. **43**(2), 195–206 (1995)
39. D.J. Jezewski, H.L. Rozendaal, An efficient method for calculating optimal free space N-impulse trajectories. AIAA J. 6(11), 2160–2165 (1968)
40. M. Pontani, B.A. Conway, Particle swarm optimization applied to impulsive orbital transfers. Acta Astronaut. **74**, 141–155 (2012)
41. O. Abdelkhalik, D. Mortari, N-impulse orbit transfer using genetic algorithms. J. Spacecraft Rockets **44**(2) (2007)
42. F.H. Clarke, Maximum principles without differentiability. Bull. Am. Math. Soc. **81**(1) (1975)
43. F.H. Clarke, M.R. de Pinho, The non smooth maximum principle. J. Control Cybern. **38** (4A) (2009)
44. A.E. Bryson, Y.C. Ho, *Applied Optimal Control* (Hemisphere Publishing Co., Washington, 1975)
45. L. Sackett, H.L. Malchow, T.N. Edelbaum, Solar electric geocentric transfer with attitude constraints: analysis. NASA CR-134927, 112–138 (1975)
46. S. Geffroy, R. Epenoy, Optimal low thrust transfers with constraints—generalization of averaging techniques. Acta Astronaut. **41**(3), 133–149 (1997)
47. Z.B. Tarzi, Optimum low thrust elliptic orbit transfer using numerical averaging. Ph.D. thesis, University of California, Los Angeles (2012)
48. G. Colasurdo, L. Casalino, Optimal law-thrust maneuvers in presence of earth shadow,in *AIAA/AAS Astrodynamics Specialist Conference and Exhibit*, 16–19 Aug 2004, Providence, Rhode Island (2004)
49. C. Ferrier, R. Epenoy, Optimal control for engines with electro-ionic propulsion under constraint of eclipse. Acta Astronaut. **48**(4), 181–192 (2001)
50. T.N. Edelbaum, Optimum power limited Orbit Transfer in strong gravity fields, in *AIAA/ION, Astrodynamics Guidance and Control Conference*, 24–26 Aug 1964, AIAA paper 64–666 (1964)
51. J.F. Rosenblueth, Equality-inequality mixed constraints in optimal control. Int. J. Math. Anal. **3**(28), 1369–1387 (2009)
52. M.d.R.de Pinho, A maximum principle for optimal control problems with mixed constraints. IMA J. Math. Control Inf. 18-189-205 (2001)
53. A. Buica, J. Llibre, Averaging methods for finding periodic orbits via Brouwer degree. Bull. Sci. Math. **128**, 7–22 (2004)
54. D. Brouwer, Solution of the problem of artificial satellite theory without drag. Astron. J. **64**(1274), 378–396 (1959)

Chapter 8
Spacecraft Propulsion

Abstract The mass ejection systems called also propulsion systems are used both for attitude and orbit control. The propulsion systems currently available or in development operate using a wide spectrum of propulsion methods. In this chapter we review the main propulsion systems with a specific attention to chemical and plasmic propulsion. The theory of these propulsion systems is explored in order to characterize the main performance parameters at system level.

8.1 Introduction

The purpose of the spacecraft propulsion system is twofold: first, it serves to transfer the spacecraft from one orbit to another (i.e. provides a *DV*); second, it provides the thrust that is needed to create an external torque for the attitude control.

The space propulsion systems currently available or in development operate using a wide spectrum of propulsion methods. It is important for the AOC designer, to qualify them in terms of a simple set of performance parameters.

Thrusters propel the spacecrafts by accelerating mass and ejecting it from the vehicle. The ejected mass may be in the form of either burning products or energetic charged particles.

Whatever the propulsion mechanism is, the propellant mass needed to increase the spacecraft's velocity of an amount DV can be computed by the *Rocket Equation*. Be $m = m_d + m_p$ the total mass of the spacecraft, m_d is the dry satellite mass (the mass of the satellite without propellants), and m_p is the propellant mass, which decreases in time, i.e. $\frac{dm_p}{dt} < 0$. The total momentum of the system is the sum of that confined in the satellite plus that expelled with the thruster. Thus, being the vector v_e the exhaust velocity of the propellant—that is the relative velocity between the propellant molecules being ejected and the satellite—projected in **ECI**, Eq. 7.1 considering the action of thruster becomes:

$$m\frac{dv}{dt} + v_e|\frac{dm_p}{dt}| = -\mu_e m\frac{x}{|x|^3} + F_{env} \tag{8.1}$$

where F_{env} represents all the forces excluding the Keplerian and the thrusters' ones. This equation represents the conservation of the Momentum Eq. 3.42 when

© Springer International Publishing Switzerland 2016
L. Mazzini, *Flexible Spacecraft Dynamics, Control and Guidance*,
Springer Aerospace Technology, DOI 10.1007/978-3-319-25540-8_8

we include also the contribution of the thruster's ejected gas; it can also be written:

$$m\frac{dv}{dt} = v_e\frac{dm}{dt} - \mu_e m\frac{x}{|x|^3} + F_{env} \tag{8.2}$$

The action of a thruster can be modeled as a vector force F_{th} in **ECI** of strength and direction:

$$F_{th} = \frac{d}{dt}(m_p v_e) = \dot{m}_p v_e = \dot{m} v_e, \tag{8.3}$$

Equation 8.3 is called the *Rocket Equation.*

This equation can be used to calculate all the components of the generalized forces generated by the thruster to be used in the equations of the flexible body Eq. 3.29.

$$f = M^T F_{th} \tag{8.4}$$

$$c = (r_t + \langle\phi(r_t)\varepsilon\rangle) \times (M^T F_{th}) \tag{8.5}$$

$$Q^{\varepsilon} = F_{th}^T M\phi(r_t) \tag{8.6}$$

The deformation functions should be calculated here in the point r_t, the position in **BRF** of the throat of the thruster nozzle where the thruster force is generally applied. Normally, during the propelled phases the thrusters produce an important excitation of the flexible modes. It should be noted that the mass properties of Eq. 3.29 in this case are not stationary, and the CoM drifts during the firing.

It is customary to characterize the mission profile requirements to the propulsion system in terms of the total DV needed, which is either that provided to the vehicle to change orbit or that used to counterbalance the external gravitational effects (orbit keeping). Such DV value is—for a large part—independent of the mass of the satellite and depends only on the mission profile. From Eq. 8.2:

$$\frac{v_e}{m}\frac{dm}{dt} = \frac{dv}{dt} + \mu_e\frac{x}{|x|^3} - F_{env}/m \tag{8.7}$$

Under the hypothesis that the force F_{env} is proportional to the satellite mass (i.e. all forces of gravitational type), the right side of Eq. 8.7 is independent of the satellite mass. Defining the DV_t as the time integral of the acceleration produced by the thrusters until mission time t:

$$DV_t = \int_{t_0}^{t}\frac{|F_{th}|}{m}dt = \int_{t_0}^{t}\left(-\frac{{v_e}^T}{|v_e|}\right)\left(\frac{dv}{dt} + \mu_e\frac{x}{|x|^3} - F_{env}/m\right)dt \tag{8.8}$$

we see that DV_t is independent on the satellite mass for a given mission profile (satellite orbital history). Knowing the total $DV = DV_\infty$ for a given mission we can easily recover the necessary propellant mass at launch $m_{p0} = m_p(t_0)$.

If we call $m_0 = m_d + m_{p0}$ the launch mass and $m_t = m_d + m_p(t)$ the satellite mass at a given mission time:

$$DV_t = -|v_e| \int_{m_0}^{m_t} \frac{dm}{m}, \tag{8.9}$$

which is easily integrated in:

$$DV_t = -|v_e| ln(\frac{m_t}{m_0}) = |v_e| ln(\frac{m_d + m_{p0}}{m_d + m_p(t)}). \tag{8.10}$$

It follows that:

$$m_t = m_0 e^{-\frac{DV_t}{|v_e|}} = m_0 e^{-\frac{DV_t}{gI_{sp}}}, \tag{8.11}$$

where we have defined the specific impulse:

$$I_{sp} = \frac{T}{\dot{m}_p g} = \frac{|v_e|}{g} \tag{8.12}$$

being $T = |F_{th}|$ and $g = 9.8067\,\mathrm{m/s^2}$ the gravitational acceleration.

In terms of consumed propellant Eq. 8.11 becomes:

$$m_{p0} - m_p(t) = m_0 - m_t = m_0(1 - e^{-\frac{DV_t}{gI_{sp}}}), \tag{8.13}$$

When the tanks are empty and we have accomplished the complete mission task $DV = DV_\infty$, the remaining satellite mass is the dry mass m_d and from Eq. 8.13, with $m_p(t) = 0$ we get:

$$m_{p0} = m_0(1 - e^{-\frac{DV_\infty}{gI_{sp}}}) = m_d(e^{\frac{DV_\infty}{gI_{sp}}} - 1). \tag{8.14}$$

We remark that for a given mission with a specified DV and delivered mass m_d, the initial spacecraft m_0 and m_{p0} can be reduced by increasing I_{sp}, which has important implications for the launch vehicle size and cost.

It is clear from Eq. 8.14 that m_{p0} increases exponentially with the required DV_∞. Thrusters that provide a large I_{sp} compared to the mission DV will have a m_{p0} that is a small fraction of the initial spacecraft mass m_0. In other words, once fixed m_0 and DV_∞, the larger the ratio I_{sp}/DV_∞, the larger the available mass for useful payload. As an example, consider a $m_d = 500$ kg of payload with a required $DV_\infty = 5$ km/s. A spacecraft propelled by an engine with typical $I_{sp} = 306$ s (as in the case of chemical propulsion), would require 2146 kg of propellant. In contrast, a thruster with $I_{sp} = 3060$ s (as in the case of electric propulsion), would use only 91 kg of propellant.

Such a substantial difference can explain the interest in high efficiency propulsion systems like the electric propulsion systems. As we will see, the quantities defined above will help to characterize a thruster and quantify its efficiency.

To date, several kinds of propulsion systems do exist which are classified according to the different propellants they use:

Table 8.1 A first comparison between the main propulsion systems

Thruster	Propellant	I_{sp} (s)	Input power (kW)	Efficiency (%)
Cold gas	Various	50–75		
Monopropellant	N_2H_4/H_2O_2	150–225		
Bipropellant	Various	300–450		
Resistojet	N_2H_4	300	0.5–1	65–90
Arcjet	N_2H_4	500–600	0.9–2.2	25–45
Ion thruster	Xenon	2500–3600	0.4–4.3	40–80
Hall thruster	Xenon	1500–2000	1.4–4.5	36–60
Pulsed plasma thruster	Teflon	850–1200	$\lesssim 0.2$	7–13

- chemical propulsion systems, which burn a chemical propellant (either solid or liquid) and exhaust the combustion products;
- electric propulsion systems, which use a gaseous propellant that is ionized and accelerated;
- cold gas systems, where a highly pressurized gas is left free to flow in an accelerating nozzle and then expelled.

In Table 8.1 we show a first comparison between the typical performances of the main propulsion systems. In particular, we provide I_{sp}, the input power needed, and the efficiency (only in the case of the electric thrusters, see Sect. 8.4).

This chapter is organized as follows.

In Sect. 8.2 we introduce the topic of the selection of an appropriate thruster architecture in function of the AOCS requirements. In Sect. 8.3 we overview the most important chemical propulsion systems, pointing out the differences between monopropellant and bipropellant ones in terms of performance. Section 8.4 is devoted to the electric propulsion systems. Cold gas devices will not be treated in this book, as despite their cheapness and safety they are mainly used as auxiliary systems. The interested reader may consult an extensive literature, like [1].

8.2 Propulsion Architecture

According to the requirements of the attitude and orbit control it is possible to select different reaction control thrusters (RCTs) architectures. Many solutions exist: thrusters may be fixed or gimballed on 2 d.o.f. mechanisms (a solution proposed for plasmic thrusters), can be used only for the orbital maneuvers or also for the attitude control.

A possible architectural solution, using chemical thrusters, capable of performing the orbital correction maneuvers along **BRF** $\pm X$ and $-Y$ axis and the attitude control is shown in Fig. 8.1.

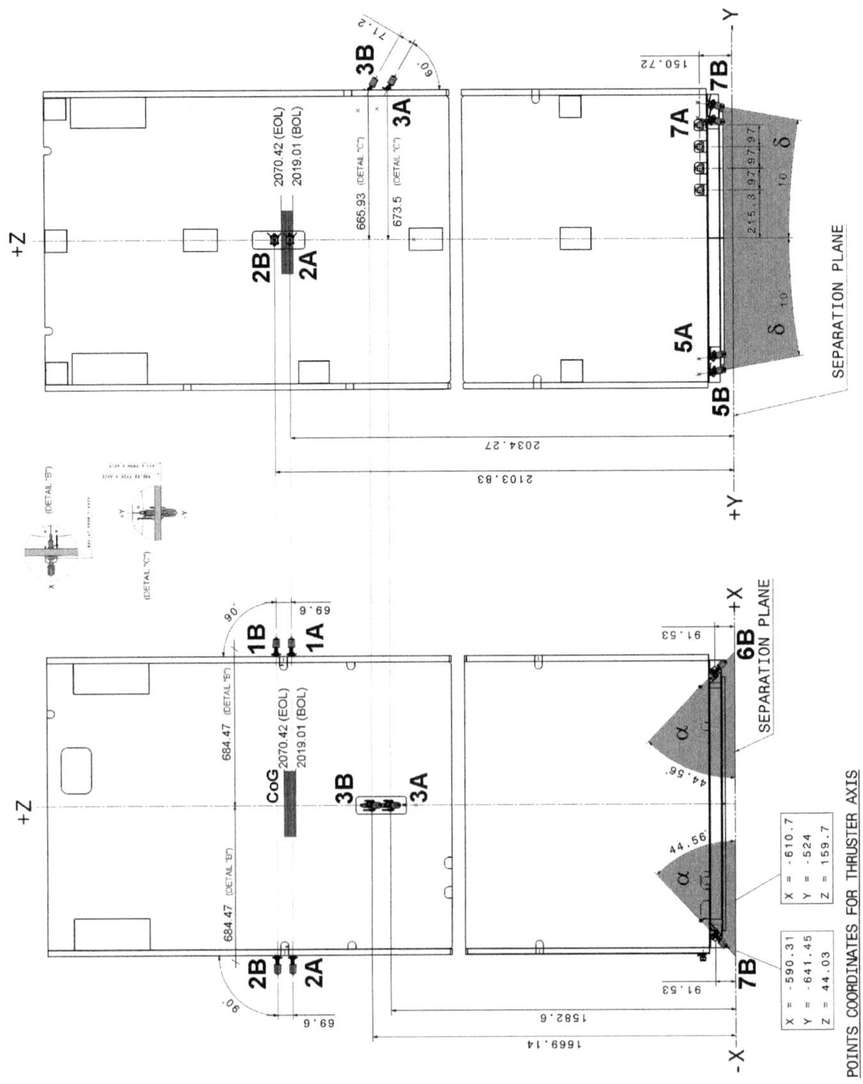

Fig. 8.1 RCT accommodation

This propulsion system uses 6 (3 main + 3 redundant) RCTs for orbital control. The thrusters of the main branch are labeled with A, those of the redundant are labeled with B. The three couples are placed on the $+X$ (1A/B), $-X$ (2A/B) and $-Y$ (3A/B) S/C side to perform both in-plane and out-of-plane orbit control. This configuration allows to perform the orbital maintenance maneuvers keeping the nominal attitude.

Table 8.2 RCT torque capability

T = 1 N RCT	Polarity	Torque value wrt CoM
4A + 5A	+X	$c_x = +2T\sin(\alpha)[y_{RCT,M}\cos(\delta) - (z_{RCT} - z_G)\sin(\delta)]$
6A + 7A	−X	$c_x = -2T\sin(\alpha)[y_{RCT,M}\cos(\delta) - (z_{RCT} - z_G)\sin(\delta)]$
4A + 6A	+Y	$c_y = -2T[x_{RCT}\cos(\delta)\sin(\alpha) + (z_{RCT} - z_G)\cos(\alpha)]$
5A + 7A	−Y	$c_y = +2T[x_{RCT}\cos(\delta)\sin(\alpha) + (z_{RCT} - z_G)\cos(\alpha)]$
4A + 7A	+Z	$c_z = +2T[x_{RCT}\sin(\delta)\sin(\alpha) + y_{RCT,M})\cos(\alpha)]$
5A + 6A	−Z	$c_z = +2T[x_{RCT}\sin(\delta)\sin(\alpha) + y_{RCT,M})\cos(\alpha)]$
4B + 5B	+X	$c_x = +2T\sin(\alpha)[y_{RCT,R}\cos(\delta) - (z_{RCT} - z_G)\sin(\delta)]$
6B + 7B	−X	$c_x = -2T\sin(\alpha)[y_{RCT,R}\cos(\delta) - (z_{RCT} - z_G)\sin(\delta)]$
4B + 6B	+Y	$c_y = -2T[x_{RCT}\cos(\delta)\sin(\alpha) + (z_{RCT} - z_G)\cos(\alpha)]$
5B + 7B	−Y	$c_y = +2T[x_{RCT}\cos(\delta)\sin(\alpha) + (z_{RCT} - z_G)\cos(\alpha)]$
4B + 7B	+Z	$c_z = +2T[x_{RCT}\sin(\delta)\sin(\alpha) + y_{RCT,R})\cos(\alpha)]$
5B + 6B	−Z	$c_z = +2T[x_{RCT}\sin(\delta)\sin(\alpha) + y_{RCT,R})\cos(\alpha)]$

The RCTs mounted on the $\pm X$ S/C side are oriented in the velocity direction and increase (or decrease) the satellite's velocity to compensate for the drag that tends to reduce the satellite's altitude or to perform phasing maneuvers.

The two RCTs on $-Y$ side of the spacecraft, dedicated to out-of-plane maneuvers, are oriented so as to point the CoM when firing. The squint angle minimizes the disturbing torques but produces a cross-coupling parasite force along the −Z **BRF** axis has an effect on the orbit which must be considered in the mission analysis mass budget.

The RCT architecture is able also to provide the attitude control actuation torque with the use of eight (4 main + 4 redundant) additional RCTs in −Z satellite side. Figure 8.1 reports the orientation of main (4A-5A-6A-7A) and redundant (4B-5B-6B-7B) branches with respect to the **BRF**.

This configuration allows the thrusters to provide the torque around all the three body axes as shown in Table 8.2. The location (i.e. x_{RCT}, y_{RCT}, y_{RCT}) and the angles (i.e. α, δ) are typically selected in order to maximize the produced torques with respect to the CoM (positioned in $x_G = 0$, $y_G = 0$, z_G) around the three axes.

The attitude RCTs when firing for attitude control produce also parasite forces which have an effect on the orbit, these effects which have to be minimized when designing the RCT architecture, must be considered in the mission analysis mass budget.

The RCT architecture which has been illustrated is one of the possible solutions for LEO missions, that require orbit corrections and attitude control. To accomplish these tasks, monopropellant chemical propulsion systems are usually employed.

In geostationary (GEO) missions, a high-thrust engine, so-called apogee boost engine (ABM), is also required in order to provide the necessary DV to reach the final orbit. Due to their higher I_{sp}, bipropellant devices are usually employed both for ABM and RCTs. A specific important issue of these missions is that high thrust ABM can produce unbalance torques due to thrust direction misalignment that must be carefully considered in sizing the attitude RCTs architecture.

8.3 Chemical Propulsion

8.3.1 General Principles

In describing chemical propellers, we will strictly refer to liquid propellants, as they are mostly used for satellite thrusters. Chemical propellers convert the heat generated by chemical reactions into kinetic energy of the exhaust gas, producing thrust. The basic configuration consists of:

- a tank with a high pressure inert gas to pressurize the fluidic system (in conjunction with a pressure regulator)
- a tank with a liquid fuel
- a tank with an oxidizer (only in bipropellant systems)
- a combustion chamber, into which the fuel (and possibly an oxidant) is injected;
- a nozzle which converts the high-pressure gas into a high velocity flow.

An example of a typical liquid-propellant engine is shown in Fig. 8.2.

The combustion which provides thrust is in principle any chemical exothermic reaction. From the chamber, the energy produced is converted into velocity as the gas expands and cools by passing through the nozzle, where the stream is accelerated by the reaction of the walls.

A typical requirement for engines is that they be restartable. To accomplish this task, electric spark igniters can be used, but they release a relatively low energy. Another possibility is the use of the so-called *hypergolic propellants*. They consist in mixtures which spontaneously ignite when their constituents are kept in contact. Typically they are composed by a fuel and an oxidizer. The ignition of the thruster can be controlled by regulating the flow of the hypergolic propellants into the combustion chamber through valves. Hypergolic propellants have the advantage of being easily stored as liquids at room temperatures, but have the disadvantages of being toxic and corrosive, and thus difficult to handle. The most used hypergolic propellants are

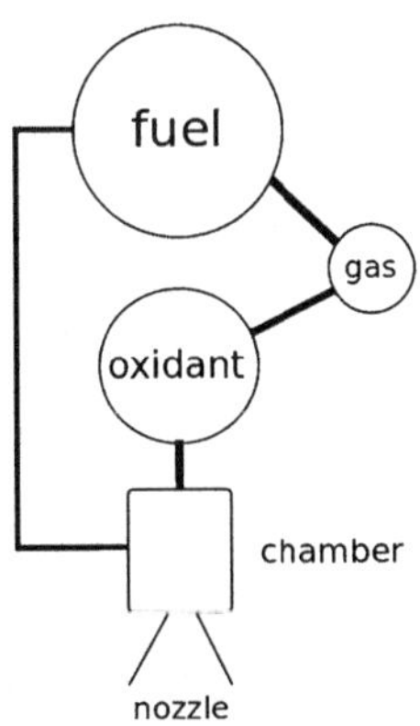

Fig. 8.2 Scheme of a chemical propulsion system with liquid propellant

the mixtures of dinitrogen tetroxide,[1] and hydrazine (N_2H_4) and/or unsymmetrical dimethylhidrazine (UDMH). As an example, the mixture N_2O_4 and Aerozine 50 is often used, where the latter is a compound of 50 % UDMH and 50 % N_2H_4. Regarding the performance of such propellants, they provide a relatively low specific impulse, which is typically about one half that achievable with liquid hydrogen or liquid oxygen. A design issue to take into account when considering chemical propulsion systems is the temperature. In the combustion chamber, temperatures as high as ~3000 K are reached, well above the melting point of most metals (up to ~2000K). Thus, both the combustion chamber and the nozzle must be cooled. This can be accomplished in several ways. The first one is the so-called *film cooling*. Part of the propellant is directed along the inner surface of the chamber and inside the nozzle. Its evaporation results in a cool layer of gas between the walls and the hot gas. The injector is placed on the wall of the chamber, then it has to inject parallel to the wall. The efficiency of the cooling decreases with distance from the injector, thus the nozzle is not cooled as well. As a consequence, a refractory material is used, which allows the nozzle to withstand the high temperature. The whole process just now described is termed *dump cooling*. Alternatively, the gas can be fed back into the combustion chamber and burned to contribute to the main thrust. The latter proceeding turns out to be much more convenient. In fact, after cooling the temperature of the gas is much lower than that in the combustion chamber, so that originates a much smaller thrust. When injected again into the chamber much more energy can be released. This process is called *regenerative cooling*.

In order to give an idea of the importance of the choice of propellants, in the following we define two important quantities. We define the *thrust coefficient*, T^*, which depends on the properties of the nozzle, as the ratio between the thrust and the pressure force (also known as *notional force*), which is defined as the product of the combustion chamber pressure p_c and the nozzle's throat area Σ, namely

$$T^* = \frac{T}{p_c \Sigma}. \tag{8.15}$$

We also define the *characteristic velocity*, v^*, which is a measure of the efficiency of conversion of thermal energy into high-velocity exhaust gas. It depends on the properties of the propellant and the combustion, and corresponds to the ratio between the notional force and the mass flow rate $\dot{m}$. An expression for v^* can be found in [2], and reads

$$v^* = \frac{p_c \Sigma}{\dot{m}} = \left[\gamma \left(\frac{2}{\gamma + 1} \right)^{(\gamma+1)/(\gamma-1)} \frac{\bar{M}}{RT_c} \right]^{-1/2}, \tag{8.16}$$

where the γ index is the ratio between the specific heat of the exhaust gas at constant pressure and constant volume, respectively; $\bar{M}$ is the molecular weight of the exhaust gas; p_c and T_c are, respectively, the pressure (taken at the entrance of the

[1] We note that dinitrogen tetroxide, N_2O_4, is also indicated as nitrogen tetroxide, *NTO*.

exhaust nozzle) and the temperature of the gas in the combustion chamber. It can be demonstrated [2] that the exhaust velocity and the thrust are related to v^* and T^* by the equations

$$v_e = T^* v^* \tag{8.17}$$

$$T = \dot{m} T^* v^*. \tag{8.18}$$

As we can see from Eqs. 8.17 and 8.18, the choice of different propellants significantly affects the performance of the chemical propulsion. In fact, the performance of a given thruster (in terms of the quantities v_e and T) depends on the molecular weight, the combustion temperature, and the ratio of the specific heats of the exhaust gas, which are related to the chemical energy released by the reaction. From Eq. 8.16, thrust is proportional to the combustion temperature and inversely depends on the molecular weight. Thus high exhaust velocities are associated with high combustion temperatures and low molecular weights. In the following section we focus on the performance of monopropellant and bipropellant propulsion systems.

8.3.2 *Liquid Propellants*

8.3.2.1 Monopropellant Thrusters

Monopropellant thrusters often consist of just a pressurized tank and a valve. The main task issue in designing a monopropellant thruster is to manage the onset of the controlled reaction in a thrust chamber, preventing it from taking place elsewhere in the system. This is mostly addressed to by using catalysts combined with propellants characterized by reaction activation energies too large to be reached by means of thermally-induced ignition processes at operational temperatures.

Hydrazine and its derivatives are propellants very commonly used for propulsion. The density of Hydrazine is 795 kg/m^3 at ambient conditions. Hydrazine (N_2H_4) dissociates exothermically due to a catalyst (typically iridium or platinum) and produces a hot mixture of hydrogen (66 %), nitrogen (33 %), and a little percentage of ammonia. As an example, on iridium catalysts hydrazine decomposes in two steps:

$$3N_2H_4 \rightarrow 4NH_3 + N_2, \tag{8.19}$$

with an exothermic release of ~335 kJ/mol, and

$$4NH_3 \rightarrow 2N_2 + N_2 + 6H_2, \tag{8.20}$$

with an endothermic reaction involving ~184 kJ/mol. Of course, the endothermic decomposition of ammonia should be minimized in order to increase the efficiency. This can be achieved with suitable choices of design geometry and chamber pressure. The most important advantage in using hydrazine is that there is no need for another propellant, and thus propulsion systems are simple. The combustion temperature is quite low, $T_c \simeq 880$ K, and the mean molecular weight is also low, being $\bar{M} = 11$, thus the exhaust velocity available is $v_e \simeq 1.7$ km/s. The simplicity of systems involving a single propellant, together with the absence of ignition and restarting problems, and the relatively easy handling properties, contribute to the safety and reliability of such systems, which were successfully used for a large number of space missions throughout history.

The typical performance of hydrazine monopropellant thrusters is a maximum specific impulse $I_{sp} \sim 250$ s. The mean specific impulse for monopropellant thrusters is $\langle I_{sp} \rangle \simeq 230$ s. The maximum thrust achieved is $T \simeq 3800$ N for high mass space vehicles.

We remark that many other monopropellants have been investigated, but only a few have found continued applications.

8.3.2.2 Bipropellant Thrusters

Bipropellant propulsion systems for satellite injections and orbit corrections have been used since the late 1960s, when the French-German satellite *Symphonie* operated with a bipropellant 400 N thruster for the apogee injection, and seven 10 N thrusters for orbit correction maneuvers. It was soon recognized that bipropellant propulsion systems offered a higher specific impulse ($I_{sp} \approx 300$ s) respect to the traditional solid/monopropellant systems and allowed to increase the payload capability.

An example of a typical bipropellant thruster is given in Fig. 8.2. Two tanks are present: one filled with fuel, the other with an oxidizer. The combination in a mixture gives rise to the burning which converts the chemical energy into thermal energy. The combustion products are then accelerated by a nozzle to the thruster's exit, providing the needed exhaust velocity.

As well as monopropellant, hydrazine can also be used as the main component of a bipropellant system together with liquid oxygen, or with nitrogen-based oxidizers (e.g. nitrogen tetroxide, with density 1440 kg/m^3 at ambient conditions). In some cases, monomethyl-hydrazine (CH_3NHNH_2, with density 880 kg/m^3 at ambient conditions) is even safer and easier to manage than hydrazine. The bipropellant system made of CH_3NHNH_2 and nitrogen tetroxide has been successfully used for space applications like the Space Shuttle and the lunar rover in Apollo missions. Another compound, UDMH ($(CH_3)_2NNH_2$), is typically used with nitrogen tetroxide and other oxidizers. Due to its liquid phase between 216 and 336 K, it is very easy to handle (more than hydrazine). Together with monomethyl-hydrazine (MMH) it can increase the thruster's performance.

8.3.2.3 Green Propulsion

Monopropellant propulsion systems are relatively simple, reliable and cheaper than bipropellant systems. In particular, hydrazine (N_2H_4) has been extensively used in the last decades for attitude control and maneuvers with a relatively low cost in terms of DV. Despite hydrazine having an excellent safety track record, spanning six decades and thousands of spacecrafts, its toxicity and flammability led to the search for *green* monopropellant systems, which have also lower management costs, due to a simplification of propellant handling operations. In the next years the space legislations are likely to forbid or limit the use of hydrazine. To date, some of the most important known green propellants are:

Hydrogen Peroxide. Hydrogen peroxide with <20 % of water dilution can reach a I_{sp} about 40 % less than that of hydrazine, but is 40 % denser and low-cost. However, its vapors has irritant effects on the eyes and respiratory system, and may be dangerous for health.
Nitrous Oxide. It is very low-cost, but has a 40 % lower performance (in terms of ρI_{sp}) respect to hydrazine. Other disadvantages lie in the high and steep dependence of performance on temperature vapor pressure, which requires high tank mass and additional system complexity. Moreover, Nitrous Oxide presents an asphyxiation hazard and a well known anesthetic effects. Strict compatibility requirements are also needed.
Nitrous Oxide + Hydrocarbon. This blend was tested at first during World War II, then abandoned, and finally developed again from 2003. It can achieve $I_{sp} \simeq 310\,\text{s}$ and $\rho I_{sp} \simeq 20\,\%$ greater than hydrazine. It offers performance better than nitrous oxide with similar low-costs and a low-toxicity. However, mass impacts and asphyxiation hazards are increased. Explosive decomposition is also more likely to occur.
Ammonium Dinitramide. Developed first in Russia in the 1970s and re-tested in the 2000s, ammonium dinitramide allowed to achieve I_{sp} similar to hydrazine, with a density ~30 % higher. It represents a high health hazard even at brief exposures. Exposure to air must be avoided to prevent evaporation of ammonia, which stabilizes the blend. To date, 5 and 22 N thrusters have been developed. This kind of technology seems to have more expensive production costs respect to propellants whose constituents are of industrial use.
Hydroxylammonium Nitrate. It can achieve a $I_{sp} \simeq 250\,\text{s}$ and a 50 % higher ρI_{sp} than hydrazine. It is characterized by a low-toxicity and high stability.

8.4 Electric Propulsion

Electric propulsion systems have been intensively studied since the seventies, but only in the last years there has been a global trend to their use. This is ultimately due to the higher I_{sp} they can provide with respect to chemical propulsion.

For this reason new solutions involving electric thrusters have been employed in several missions as a valid alternative to chemical thrusters, at first for station-keeping maneuvers in GEO satellites, then for orbital steering and even for exploration missions.

We now define some general performance parameters of these thrusters that will be used in the following sections. In electric thrusters the thrust is due to the kinetic energy of ions. For singly ionized species the specific impulse defined in Eq. 8.12, being $\dot{m}_i$ the ionic mass rate, becomes:

$$I_{sp} = \frac{T}{\dot{m}_p g} = \frac{\dot{m}_i v_i}{\dot{m}_p g}. \tag{8.21}$$

We can have an indication of how much propellant is ionized by defining the ratio of ions flow to that of propellant.

$$\eta_i = \frac{\dot{m}_i}{\dot{m}_p} = \frac{I_b m_i}{e\dot{m}_p}. \tag{8.22}$$

We now relax the hypothesis of singly ionized species (with associated beam current $I_b = I^+$). Suppose we have also twice ionized species (with associated current I^{++}), so that $I_b = I^+ + I^{++}$. In this case, Eq. 8.22 becomes, for both single and twice ionized species

$$\eta_{i,m} = \frac{I^+ m_i}{e\dot{m}_p} + \frac{I^{++} m_i}{2e\dot{m}_p} = \frac{I^+ m_i}{e\dot{m}_p}\left(1 + \frac{I^{++}}{2I^+}\right). \tag{8.23}$$

The correction factor for twice ionized atoms is then

$$\alpha_\eta = \frac{\eta_{i,m}}{\eta_i} = \frac{I^+\left(1 + \frac{I^{++}}{2I^+}\right)}{I^+ + I^{++}}, \tag{8.24}$$

and of course

$$\eta_{i,m} = \alpha_\eta \eta_i, \quad \dot{m}_i = \alpha_\eta \eta_i \dot{m}_p \tag{8.25}$$

We now define the efficiency of an electric propulsion device. The *electrical efficiency* is the beam power P_b to the total input power, P_{tot}:

$$\eta_e = \frac{P_b}{P_{tot}}, \tag{8.26}$$

where the total power accounts for any kind of electrical cost, like for example production and maintenance of plasma and acceleration of ions.

The *discharge loss* is the ratio between the power needed to produce ions, P_d, and the produced ion current, I_b:

$$\eta_d = \frac{P_d}{I_b}. \tag{8.27}$$

This parameter characterize the motor in terms of plasma generation efficiency.

Generally speaking, we can define the *jet power* as the kinetic thrust power of the beam, namely

$$P_j = \frac{dE_K}{dt} = \frac{d}{dt}\left(\frac{1}{2}m_p v_e^2\right) = \frac{\dot{m}_p v_e^2}{2} = \frac{T^2}{2\dot{m}_p}, \tag{8.28}$$

where E_K is the kinetic energy, and we have used Eq. 8.3. This relationship shows that the increase of the thrust without an increase of the propellant flow rate, corresponds to an increase of the jet power. The *total efficiency* of an electrically powered thruster is the jet power to the total electric power:

$$\eta_{tot} = \frac{P_j}{P_{tot}} = \frac{T^2}{2\dot{m}_p P_{tot}}, \tag{8.29}$$

where we have used Eq. 8.28.

By using Eqs. 8.21 and 8.29, we may compute the thrust per unit power as

$$\frac{T}{P_{tot}} = \frac{2\dot{m}_p \eta_{tot}}{T} = \frac{2\eta_{tot}}{g I_{sp}}, \tag{8.30}$$

so that for a given total input power and given thruster efficiency, by increasing I_{sp} a lower T is obtained.

The fraction of the input power which is lost, i.e. does not contribute to the thrust, must be radiated into space. The *dissipated power* can be defined as

$$P_{diss} = P_{tot}(1 - \eta_e). \tag{8.31}$$

Actually, several concepts of electric thrusters exist, which use different methods to produce the thrust. They can be divided into (1) electro-thermal, (2) electrostatic, on which we will focus, and (3) electromagnetic thrusters.

8.4.1 Electro-Thermal Propulsion

Electro-thermal thrusters, like all the electric thrusters, are characterized by a very high effective exhaust velocity v_e (and thus a very high specific impulse I_{sp}) along with a low thrust, respect to chemical propulsion systems. Unlike other propulsion systems, electrothermal thrusters are free of constraints in the choice of the propellant, which represents a great advantage. The most used propellant is hydrogen, which has the lowest $\bar{M}$, but its performance is not so high, as it dissociates at high temperatures in monoatomic hydrogen, subtracting energy from the exhaust stream. Helium is better, as it is already monoatomic and does not dissociate. Other possible candidates are water and hydrazine, which however have higher $\bar{M}$ and can dissociate. Electro-thermal thrusters are of two types:

(1) Resistojet thrusters. The propellant is heated by passing through a resistively heated chamber or element, before entering a downstream nozzle. I_{sp} is typically limited to $\lesssim$500 s.

(2) Arcjet thrusters. The propellant is heated passing through a high current arc in line with the nozzle feed system. The neutral gas is initially insensitive to the electric field. As the electric potential is raised to $\Delta V \simeq$ 200–300 V, ionization occurs, and the gas becomes a conductor. The neutral gas is heated by the collisions with ions and electrons, and the rate of ionization increases. The gap in the anode allows the propellant to flow. The produced ions hit and erode the electrodes, which are made of tungsten due to the high melting point of this material. This limits the life of the electrodes and the current that can pass through the device. The I_{sp} is limited to $\lesssim$700 s.

8.4.2 Electrostatic Propulsion

The most important electric propellers are the electrostatic ones, in which strong electric fields move the ions from the combustion chamber to the thruster's exit.

In electrostatic engines the thrust is determined by the flow rate of ions, whose velocity greatly exceeds that of the neutral species

$$T = \dot{m}_p v_e \simeq \dot{m}_i v_i, \tag{8.32}$$

being m_i and v_i the ion mass and velocity, respectively. For the conservation of energy, the kinetic energy of accelerated particles should be equal to the work done by the electrostatic field in accelerating them, e.g.

$$\frac{1}{2} m_i v_i^2 = q \Delta V, \tag{8.33}$$

where ΔV is the voltage applied to produce the electrostatic field. It follows

$$v_i = \sqrt{\frac{2q\Delta V}{m_i}}. \tag{8.34}$$

The ion beam current can be defined as $I_b = q_{tot}/t$, where q_{tot} is the total charge $q_{tot} = N_+ q$ (N_+ is the number of ions flowing in time t). It follows that the number of ions flowing in a unitary time is $Q = N_+/t = q_{tot}/(q\,t) = I_b/q$ and the flow rate is:

$$\dot{m}_i = \frac{I_b m_i}{q}. \tag{8.35}$$

By substituting Eqs. 8.34 and 8.35 in 8.32, we find

$$T = \frac{I_b m_i}{q}\sqrt{\frac{2q\Delta V}{m_i}}. \tag{8.36}$$

By posing $q = e$ we find the important result for unidirectional, singly ionized and monoenergetic beam:

$$T = I_b\sqrt{\frac{2\Delta V m_i}{e}}, \tag{8.37}$$

which states that the thrust is proportional to the ion current and the square root of the applied voltage. Moreover, the heavier the species, the higher the thrust.

We now relax the hypothesis of unidirectional beam. In the more general case, part of the produced thrust vector is deviated of an angle θ respect to the axis of the thruster body. Thus, the effective thrust corresponds to the nominal corrected by a factor $\cos\theta$. For more complicated geometries, the correction must be integrated over the grid profiles.

We also relax the hypothesis of singly ionized species. If we suppose to have also twice ionized species, Eq. 8.37 becomes

$$T_m = I^+\sqrt{\frac{2\Delta V m_i}{e}} + I^{++}\sqrt{\frac{2\Delta V m_i}{2e}} = I^+\sqrt{\frac{2\Delta V m_i}{e}}\left(1 + \frac{I^{++}}{\sqrt{2}I^+}\right). \tag{8.38}$$

The correction factor for twice ionized atoms can be defined as

$$\alpha = \frac{T_m}{T} = \frac{I^+\left(1 + \frac{I^{++}}{\sqrt{2}I^+}\right)}{I^+ + I^{++}}, \tag{8.39}$$

and Eq. 8.37 is generalized in

$$T = \alpha I_b\sqrt{\frac{2\Delta V m_i}{e}}\cos\theta. \tag{8.40}$$

Taking into account Eqs. 8.21, 8.35, 8.40 and the definitions 8.22 and 8.25, we obtain

$$I_{sp} = \frac{\alpha\alpha_\eta e\eta_i I_b\sqrt{\frac{2\Delta V m_i}{e}}\cos\theta}{I_b m_i g} = \frac{\alpha\alpha_\eta\eta_i}{g}\sqrt{\frac{2\Delta V e}{m_i}}\cos\theta. \tag{8.41}$$

The total efficiency of electrostatic thruster

$$\eta_{tot} = \frac{T^2}{2\dot{m}_p P_{tot}} = \frac{T I_{sp}}{2gP_{tot}} \tag{8.42}$$

may be evaluated from Eqs. 8.24, 8.25, 8.29, 8.35 and 8.40:

$$\eta_{tot} = \frac{\alpha^2 \frac{2m_i \Delta V}{e} \cos^2\theta \alpha_\eta \eta_i I_b{}^2}{2P_{tot} I_b m_i / e} = \frac{\alpha^2 \alpha_\eta \Delta V \eta_i I_b}{P_{tot}} \cos^2\theta. \tag{8.43}$$

From Eq. 8.26 with $P_b = I_b \Delta V$, Eq. 8.43 becomes

$$\eta_{tot} = \alpha^2 \alpha_\eta \eta_e \eta_i \cos^2\theta. \tag{8.44}$$

For typical ion thrusters $\eta_{tot} \sim 70\,\%$, thus the device converts $\sim$70 % of the total input electric power in the product of thrust and I_{sp} as results from Eq. 8.42. From the this equation we derive also that for a given power and efficiency we can obtain more thrust for less I_{sp} or viceversa.

8.4.2.1 Gridded Ion Thrusters (GITs)

In GITs an array of grids is used to electro-statically (at voltages $\Delta V \sim 10\,\text{kV}$) extract and accelerate to high velocity the ions generated from the propellant. GITs have a very high total efficiency ($\sim$60–80 %) and very high specific impulses ($\sim$2000–$10,000\,\text{s}$).

GITs can be divided into three components: (1) a plasma generator, (2) an array of ion accelerator grids, and (3) a neutralizer cathode.

Starting from the plasma generator, three different technologies are available: (i) Direct Current (DC), (ii) Radio Frequency (RF), and (iii) Micro-Wave (MW) discharge.

We focus on DC *plasma generators* following [3], as we have seen the generator efficiency is characterized by the *discharge loss* $\eta_d = P_d/I_b$.

In a DC generator (represented in panel (a) of Fig. 8.3), a resistive cathode heater raises the temperature of a thermo-ionic emitter at a level sufficient to produce electrons, while neutral gas is injected into the anode chamber. Electrons extracted by the hollow cathode are injected into the chamber to ionize the propellant gas.

As the plasma discharge is ignited, the heater is switched off in order not to neutralize the ions produced in the chamber. A neutralizer provides electrons in order to avoid the charging of the surfaces at the exit of the chamber. The ions flow to the accelerator grids in a Bohm current (see [3])

$$I_i = \frac{1}{2} n_i e v_a A, \tag{8.45}$$

where n_i is the ion density, $v_a = \sqrt{kT_e/m_i}$ is the Bohm velocity of ions, and A is the total ion loss area (the grid area). The ion beam current is $I_b = I_i T_g$, where T_g is a factor giving the effective grid transparency, and where we neglected the current lost to the grids.

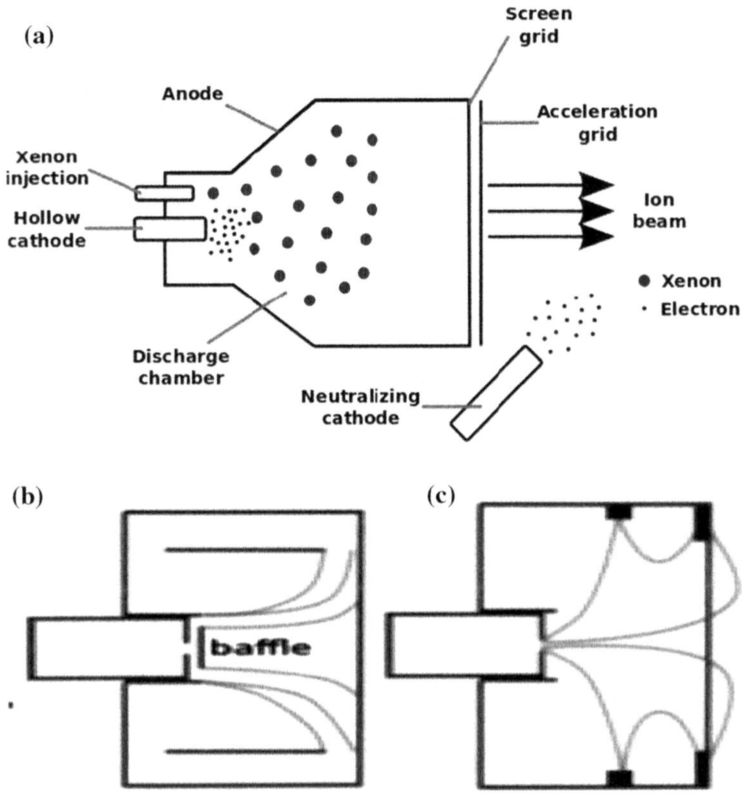

Fig. 8.3 *Panel* **a** Simplified representation of a GIT; *Panel* **b** Geometry of the Kaufman's discharge chamber; *Panel* **c** Geometry of the Sovey's discharge chamber. *Grey lines* depict the magnetic field lines used for the plasma confinement

The rate of ionization of neutrals by the energetic electrons can be written as $I_p = n_0 n_e e \langle \sigma_{en} v_e \rangle V$, where n_0 is the neutral gas density, n_e is the electron density, σ_{en} is the ionization cross section, V is the volume filled with plasma, and v_e is the velocity of electrons. The term in brackets gives the ionization rate averaged over a Maxwellian distribution of electron velocities.

When computing the total discharge power supplied we have to consider the power going into production of electrons, heating of neutrals, ionization of neutrals, heating of plasma, and the power carried to the walls and the grids.

In formula, the total discharge power may be written as

$$P_d = I_p U^+ + I^* U^* + I_i \varepsilon_i + \frac{n_e V}{\tau} \varepsilon_e, \tag{8.46}$$

where U^+ and U^* are the ionization and the excitation potential of the propellant gas, respectively, τ is a typical electron confinement time, ε_i and ε_e are the ion and electron energy carried to the walls, respectively.

The quantity I^* represents the excited neutral production rate, which is given by the expression

$$I^* = \sum_j n_0 n_e e \langle \sigma_* v_e \rangle_j V, \tag{8.47}$$

being σ_* the excitation cross section rate coefficient averaged over the electron ensemble and summed over all the possible excited states j. It follows that Eq. 8.46 can be written as

$$P_d = n_0 n_e e \langle \sigma_{en} v_e \rangle V \left[U^+ + \frac{\langle \sigma_* v_e \rangle_j}{\langle \sigma_{en} v_e \rangle} U^* \right] + I_i \varepsilon_i + \frac{n_e V}{\tau} \varepsilon_e. \tag{8.48}$$

By assuming that electrons and ions flow at the same rate, we have $I_i = n_i eV/\tau$. Following [3], the energy that ions and electrons transfer to the walls may be written as:

$$\varepsilon_e = 2\frac{kT_e}{e} + \phi \tag{8.49}$$

$$\varepsilon_i = \frac{kT_e}{2e} + \phi, \tag{8.50}$$

being ϕ the difference between the potential of the plasma and that at the walls. An expression for ϕ given in the same reference reads

$$\phi = \frac{kT_e}{e} ln \left(\frac{A_a}{A} \sqrt{\frac{2m_i}{\pi m_e}} \right), \tag{8.51}$$

where A_a is the electron loss area, m_e the electron mass.

A characteristic in common for all plasma generators is the discharge loss, defined as (see Eq. 8.27)

$$\eta_d = \frac{P_d}{I_b}. \tag{8.52}$$

η_d is a measure of the efficiency of the plasma generator, and it can be demonstrated that by taking into account all the contributions in power for idealized thrusters we have (see [3]):

$$\eta_d = \frac{2n_0 \langle \sigma_{en} v_e \rangle V}{v_a A T_g} \left[U^+ + \frac{\langle \sigma_* v_e \rangle}{\langle \sigma_{en} v_e \rangle} U^* \right] + \frac{1}{eT_g} \left[2.5kT_e + 2kT_e ln \left(\frac{A_a}{A} \sqrt{\frac{2m_i}{\pi m_e}} \right) \right], \tag{8.53}$$

with the following notation (we repeat it here for the sake of clarity): v_e is the electron velocity, T_e is the temperature of electrons, m_i is the ion mass, m_e is the electron mass, U^+ is the ionization potential of the propellant gas, U^* is the excitation potential of the gas, σ_* is the excitation cross section, and A_a is the reduced anode area. In the result of Eq. 8.53, the first term is due to the production of ions and excited neutrals after the interaction with electrons, the second contributions is ascribed to the heating of electrons that are lost to the anode walls.

As we can see, the grid transparency T_g has an important role in the discharge loss.

The ratio between the rates of excitation to ionization in the second term of Eq. 8.53 indicates that inefficiency increases when the rate of excitation is higher than that of ionization. In Xenon atoms this happens because for electron temperatures $T_e \lesssim 8\,\text{eV}$ the lowest excitation potential ($\simeq$10 V) is below the ionization potential ($\simeq$12 V). As a consequence, a significant part of the input energy goes into the heating of neutrals, and part is lost through radiation to the walls.

In real systems, when considering all the energy losses, at least 7.5 more energy than the nominal is needed to extract and move an ion to the main thruster's beam.

In many cases the efficiency of ion thrusters can be improved by providing a magnetic confinement to the ionization chamber. In modeling the efficiency, this is equivalent to introducing a reduced anode area, $A_a < A$.

The most important magnetic configurations are those introduced by Kaufman [4] and shown in panel (b) of Fig. 8.3, and the permanent magnetic ring-cusp thruster introduced by Sovey [5] and shown in panel (c) of the same figure. In the former, the electrons from the cathode follow their path to the anode and undergo collisions. The introduction of a strong divergent magnetic field reduces the effective loss area and makes the electron and ion beam profile more uniform. The baffle in front of the cathode opening inhibits on-axis electrons, limiting the increase of effective loss area. In the latter, alternating polarity permanent magnets produce a magnetic field with a gradient in strength, which causes the bounce of electrons close to magnets (see Appendix F.3), until they lose energy by colliding with neutrals or discharge on the anode surfaces. In both cases the overall result is the reduction of η_d.

Let us analyze the *ion acceleration*. The ions generated in the discharge chamber are accelerated from the grids and form the thrust beam. Acceleration is provided by applying a voltage between the grids separated by a distance d, thus creating and electric field of strength $E = V/d$. From Poisson's Equation (F.1), the electric field between the grids is

$$E(x) = \frac{q}{\varepsilon_0} \int_0^x n_i(x')dx' + E_{SC}, \tag{8.54}$$

where ε_0 is the dielectric constant in vacuum, q is the ion charge and n_i is the ion number density, and E_{SC} is the electric field on the screen grid. The surface charge density of the screen grid is $\sigma = \varepsilon_0 E_{SC}$. As inside a perfect conductor the electric field is zero (Coulomb theorem), the force per unit area acting on the screen grid is equal to σ times the average field outside and on the conductor

$$F_{SC} = \frac{\sigma E_{SC}}{2} = \frac{\varepsilon_0 E_{SC}^2}{2}. \tag{8.55}$$

With the same reasoning, but taking into account that the charge on the acceleration grid is opposite in sign, the force per unit area acting on the acceleration grid is

$$F_{acc} = \frac{-\sigma E_{acc}}{2} = \frac{-\varepsilon_0 E_{acc}^2}{2}, \tag{8.56}$$

where E_{acc} and F_{acc} are the electric field and the force on the acceleration grid, respectively. The net thrust can be written as the sum of the force per unit area acting on both the grids:

$$T = F_{SC} + F_{acc} = \frac{1}{2}\varepsilon_0 \left(E_{SC}^2 - E_{acc}^2 \right). \tag{8.57}$$

The link between the force acting on the ions and the thrust can be found by computing the electrostatic force per unit area to which the distribution of ions between the grids is subjected:

$$F_i = q \int_0^d n_i(x) E(x) dx. \tag{8.58}$$

By differentiating Eq. 8.54 and substituting n_i in Eq. 8.58 we find

$$F_i = \varepsilon_0 \int_{E_{SC}}^{E_{acc}} E(x) dE(x) = \varepsilon_0 \frac{E_{acc}^2 - E_{SC}^2}{2} = -T. \tag{8.59}$$

This demonstrates how thrust is transferred from the ions to the thruster body by the electrostatic force.

The modern era of GITs began in 1994, with the engine produced by Mitsubishi onboard the Japanese test satellite ETS-6. This thruster provided a maximum nominal thrust of $T \simeq 20$ mN and $I_{sp} \simeq 2400$ s. The first successful commercial application was the *Xenon ion propulsion system* (XIPS) onboard the Hughes PAS-5 satellite, launched in 1997. XIPS mounted two identical ion engines (for redundancy) devoted to station keeping, which produced $T = 18$ mN and $I_{sp} = 2500$ s, with a total efficiency of $\eta_{tot} = 50\,\%$. NASA's facility NSTAR was launched in 1998 and to date is the most tested device ever. It was operated at several input powers (from $P_{tot} \simeq 577$ to $P_{tot} \simeq 2567$ W). Thrust ranges from 20.7 to 92.7 mN; I_{sp} from 1979 to 3127 s; the total efficiency ranges from 42 to 61.8 %. The second generation of XIPS, which had a larger effective area, was launched in 1999 and for the first time was also devoted to the orbit raising. The high-power mode ($P_{tot} \simeq 4.5$ kW) is able to achieve a thrust $T \simeq 165$ mN and a $I_{sp} \simeq 3500$ s. In low-power mode ($P_{tot} \simeq 2.3$ kW), for station-keeping maneuvers, the achieved thrust is $T \simeq 79$ mN, with $I_{sp} \simeq 3400$ s. In 2001 the ESA mission Artemis was launched, which had a GIT propulsion systems with two EITA (electron-bombardment ion thruster assembly) and two RITA (radio frequency ion thruster assembly) which were able to raise the orbit from 31,000 km to the geostationary height after a failure of the launcher's upper stage. This singular and interesting salvage was resolved with the contribution of the author who has described the story and the technical problems encountered in his paper [6].

The EITA maximum thrust achieved was $T \simeq 18\,\text{mN}$ and a $I_{sp} \simeq 3200\,\text{s}$, with an efficiency of $\eta_{tot} \simeq 55\,\%$. In RIT-10 achieved a thrust $T \simeq 15\,\text{mN}$ and $I_{sp} \simeq 3400\,\text{s}$, with an efficiency $\eta_{tot} \simeq 51\,\%$. The Japanese agency JAXA launched ECR ion thrusters on-board the mission Hayabusa in 2003, which was designed for a sample return mission on a near-Earth asteroid, with a $T \simeq 8.1\,\text{mN}$ and $I_{sp} \simeq 3090\,\text{s}$, and a total efficiency $\eta_{tot} \simeq 36\,\%$. In 2006 a new Mitsubishi test satellite with ion thrusters was launched, which achieves a nominal thrust from 20.9 to 23.2 mN and a specific impulse from 2402 to 2665 s, with a total efficiency $\eta_{tot} \simeq 46\,\%$ to $\eta_{tot} \simeq 50\,\%$.

8.4.2.2 Radio Frequency Ion Thrusters (RFITs)

One of the technologies available for plasma generators is based on radio-frequency discharge, for which a RF coil with N turns supplied with power is wrapped around a chamber with a gas injected. By assuming cylindrical coordinates, the current generates a variable and axial magnetic field with strength

$$B_z = \frac{NI}{\mu_0} e^{i\omega t}, \tag{8.60}$$

where I is the current in the coil, μ_0 the permittivity in the vacuum, ω is the cyclic frequency of the RF, and t is time. From Faraday's law Eq. F.3, an electric field is induced in poloidal $\hat{\boldsymbol{\theta}}$ direction and such to produce a magnetic field that is opposite to the former. The electric field strength is

$$E_\theta = -\frac{i\omega r}{2} B_{z0} e^{i\omega t}, \tag{8.61}$$

where r is the distance from the axis and B_{z0} is the peak axial magnetic field generated by the RF coil. The electrons in the discharge chamber are accelerated by the electric field, and thus can collide with the neutral gas injected into the chamber and are heated. It can be demonstrated that the minimum pressure at temperature T needed to heat the electrons is given by

$$p_{min} = \frac{-ln(1-P)}{\sigma_{en} L} kT, \tag{8.62}$$

where P is the probability that in covering a path of length L an electron collides with a neutral, and we used the equation of state $p = nkT$.

The chamber is connected to a system of two accelerator grids. As the plasma is generated and flows through the first grid, a strong voltage is applied between the grids and accelerates ions, which form the beam. The collision between the neutrals and secondary plasma electrons can support the process and allow to reduce the operating pressure p_{min}, thus increasing the efficiency of the thruster. The electrons to be heated can be provided from (i) the free electrons naturally present inside the chamber (but there are just a few); (ii) the induced electric field inside the chamber; (iii) a small cathode or a neutralizer cathode.

8.4.2.3 Microwave Ion Thrusters (MWITs)

Another way to generate plasma is to use electromagnetic fields at microwave frequencies.[2] By linearizing Maxwell's equations Eqs. F.3, F.4, and after some manipulation, the dispersion relation

$$(\omega^2 - c^2k^2)\tilde{E} = -\frac{i\omega}{\varepsilon_0}\tilde{J} \tag{8.63}$$

is obtained, where the fluctuating vectors $\tilde{E} = Ee^{i(kx-\omega t)}$ and $\tilde{J} = Je^{i(kx-\omega t)}$ (modeled as plane waves propagating in the $\hat{x}$ direction) are introduced.

At microwave frequencies the time scales are fast enough to neglect the inertial response of ions respect to that of electrons. So that only electrons contribute to the perturbed current density, and for the plasma in the discharge chamber we have $\tilde{J} = -n_e e\tilde{v_e}$ and $\tilde{v_e} = v_e e^{i(kx-\omega t)}$.

The linearized first-order equation of motion for electrons in the case $B_0 = 0$ or $B_0||\tilde{E}$ can be written as $-i\omega m_e\tilde{v_e} = -e\tilde{E}$, from which it can be isolated $\tilde{v_e}$ and computed $\tilde{J}$. Equation 8.63 becomes

$$(\omega^2 - c^2k^2)\tilde{E} = \frac{i\omega}{\varepsilon_0}\frac{e^2n_e}{i\omega m_e}\tilde{E} = \frac{e^2n_e}{m_e\varepsilon_0}\tilde{E}. \tag{8.64}$$

The quantity $e^2n_e/m_e\varepsilon_0$ corresponds to the square of the *plasma oscillation electron frequency* ω_e (see F.7), which can be approximated with the plasma frequency ω_p. It follows that Eq. 8.64 can be written $\omega^2 = c^2k^2 + \omega_p^2$. The dispersion relation can be rewritten $\lambda = c/\sqrt{\nu_p^2 - \nu^2}$. When $\nu_p \gg \nu$, λ is short, and energy can be efficiently transferred from the electromagnetic wave to the plasma. When $\nu \rightarrow \nu_p$ the wavelength approaches infinity, and the oscillation does not interact with plasma. Finally, when $\nu_p \ll \nu$, the solution of the dispersion relation is imaginary, and evanescent waves set,[3] which are back-reflected. We note that at the cutoff frequency corresponds a value of the plasma density, at which a wave can inject power to heat the plasma.

[2]The microwave range is usually referred to as that between 0.3 and 300 GHz. Typically, those used in these thrusters are *UHF* (ultra high frequency, in the range between 0.3 and 3 GHz (decimeter waves), and *SHF* (super high frequency), in the range between 3 and 30 GHz.

[3]The frequency beyond which the wave becomes evanescent is called the *cutoff frequency*.

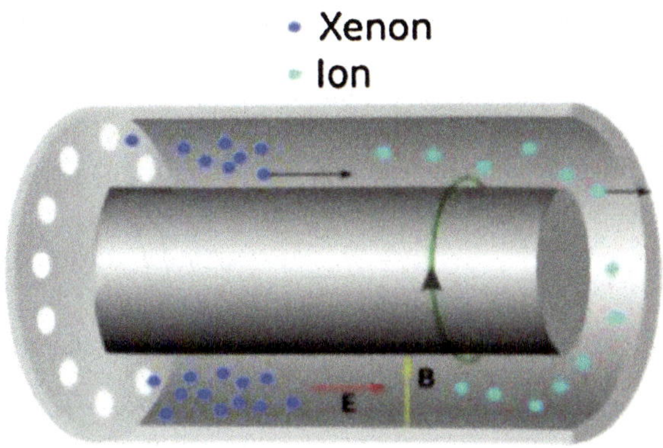

Fig. 8.4 Representation of a HET. In *blue* we represent the Xenon atoms, in *cyan* the ions produced by the collisions with electrons, in *red* the direction of the electric field, in *yellow* the direction of the magnetic field, in *green* the direction of the drift velocity

In the discharge chamber, a microwave can be injected with a frequency corresponding to the Larmor frequency (or gyrofrequency; see Appendix F.3).

The higher the magnetic field strength and the plasma density, the higher ω_p, so that power can be injected into plasma from the electromagnetic wave generated by the gyromotion of charges around the magnetic field. In order to have an effective transfer of energy from the microwave electromagnetic waves to electrons, collisions must occur.For pressures higher than the ignition pressure, the number of collisions is sufficient to allow the electrons to thermalize and sustain the discharge. Due to collisions between electrons and neutrals, the energy is transferred into heating, thus allowing to lower the operational pressure required to generate plasma.

8.4.2.4 Hall Effect Thrusters (HETs)

HETs are electrostatic thrusters which use the Hall effect to generate plasma (see Appendix F.3). The drift induced by the coupling of perpendicular electric and magnetic fields is responsible for the production of ions, which are accelerated and give high exhaust velocities.

The overall geometry of HETs is shown in Fig. 8.4. A radial magnetic field is induced by coils and directed from S to N. The external cathode and the anode inside the discharge region produce an electric field which is perpendicular to the magnetic field. Neutral propellent atoms (typically Xenon) are fed into the discharge channel through the anode, electrons are injected through the cathode. Electrons are subjected to both electrostatic and magnetic fields. The coupling of electric and magnetic fields produces a drift force $F_d = E \times B$ which acts on the electrons attempting to reach the anode (see Appendix F.3), and reduces their mobility in direction parallel to E.[4] Electrons spiral around the thruster axis due to F_d, constituting the *Hall current*. The interaction (through collisions) of electrons and neutral gas produces ions and

[4]The Larmor radius of ions is much larger than that of electrons, thus as a first approximation the magnetic force on ions can be neglected.

additional electrons. Ions are accelerated by the electric field from the anode to the exit. Part of the electron beam is then used to neutralize the ion beam, in order to avoid the charging of the thruster' s surfaces.

Suppose the plasma in the acceleration region to be quasi-neutral ($n_i \simeq n_e$), and the electric and magnetic fields to be uniform. The electrostatic force F_i producing the ion acceleration is

$$F_i = 2\pi q \iint r dr dz n_i E, \tag{8.65}$$

where we have assumed axial symmetry. In presence of uniform magnetic and electric fields electrons drift with velocity

$$v_e = \frac{E \times B}{B^2}. \tag{8.66}$$

The total force on electrons is the Lorentz force made of both the contributions

$$F_e = 2\pi e \iint r dr dz n_e E - 2\pi e \iint r dr dz n_e v_e \times B. \tag{8.67}$$

This quantity vanishes, as under the hypothesis of uniform and unidirectional electric and magnetic fields the equality $v_e \times B = -E$ holds. Moreover, for the quasi-neutrality, the first term on r.h.s. corresponds to the electrostatic force acting on ions (Eq. 8.65). By defining the *Hall current density* $J_H = -en_e v_e$, Eq. 8.67 can be rewritten as $F_i = J_H \times B$. Thus the thrust produced is $T = -F_i = B \times J_H$.

Hall thrusters have been used for space applications since 1971, with the launch of the Russian mission SPT (stationary plasma thrusters), devoted to station-keeping maneuvers in geostationary orbits. In particular, SPT-100 produced a thrust $T \simeq$ 82 mN, with a $I_{sp} \simeq 1600$ s and a total efficiency $\eta_{tot} \simeq 50\,\%$. The first flight on a USA spacecraft was only in 1998, with the D-55 TAL (thruster with anode layer) Hall thruster, manufactured in Russia. The first ESA study of Hall thruster technology for commercial use was possible with the SMART-1 mission (Small Mission for Advanced Research in Technology). It was launched in 2003, and provided with a PPS-1350-G engine manufactured in France. The thruster operated in a range of powers from $P_{tot} \simeq 462$ to 1190 W, producing a maximum thrust $T \simeq 70$ mN and $I_{sp} \simeq 1600$ s. The first Hall thruster for commercial use on a USA satellite was, in 2004, the SPT-100 thruster produced in Russia. Since that time, launches and tests of several HETs have followed, contributing to further develop this technology, which turned out to be very promising. It is foreseen that many commercial and scientific missions in the next years will use Hall thrusters, due to their performance and relatively simple construction and operation.

The performance of HETs is mainly determined by (i) the geometry, (ii) the material of the discharge region, (iii) the magnetic field produced by the coils. Currently used HETs have I_{sp} ranging from ~700 to 3000 s, with maximum thrusts from ~20 to 280 mN, and efficiency from 26 to 58 %. The performance of HETs also depends on the operational input power and the discharge voltage. Despite the fact that they can achieve smaller I_{sp} than ion thrusters, HETs require a lower power to operate. In fact, the higher the voltage (typically up to $\Delta V \sim 400\,\mathrm{V}$) the higher the I_{sp} achieved. On the contrary, the lower ΔV (typically down to $\Delta V \sim 150\,\mathrm{V}$) the lower I_{sp}, but the higher the specific thrust (up to ~80 mN/kW). Currently SNECMA, Aerojet, Fakel, SITAEL and other companies are producing or developing Hall effect motors of various Power size up to 5KW for commercial use.

8.4.2.5 GIT and HET Plumes Constraints

In electrostatic thrusters, the beam of exhausted ions (termed *plume*) deeply interacts with the spacecraft material, giving rise to multiple effects which must be accounted for in order to avoid/limit structural damage. In particular, the energetic ions of the plume can interact with neutral gas (for example that coming from the neutralizer system) producing fast neutral atoms in the beam direction and slow ions. Such slow ions are subjected to the local electric fields associated with the charge distributions at the acceleration region and the neutralizer plasma. The electric field can accelerate the impacted ions backward, or at least deflect them to potentially hit the spacecraft components in the vicinity.

Plumes may cause the deposition of impurities on the components of the thruster. This may be due to the sputtering of the grids in ion thrusters, the erosion of the ceramic channel in Hall thrusters, or also to the sputtering of other electrodes. The deposition of material on the surfaces can modify their properties such as emissivity and transparency.

More in detail, thrusters' plumes may affect the spacecraft's operations in several ways:

- The impact of plumes with the surfaces of the device may transfer momentum (which is the difference between the momentum of the plume and that of the particles leaving the surface) to the spacecraft, thereby exerting a torque. The momentum of the plume is, in principle, the sum of the contributions from ions and neutrals, but the latter can be neglected. In the case of elastic scattering, the transferred momentum p_t is simply twice the incident momentum p_i, as the beam ions conserve their kinetic energy and reverse their motion in the interaction. A fraction of ions can remain longer near the target surface, thermalizing and leaving the surface with a thermal velocity $v_{i,T} = \sqrt{kT/m}$. In such a process, called *surface accommodation*, the transferred momentum is $p_t = (2 - N_{acc})p_i$, where N_{acc} is the fraction of surface-accommodated ions, and ranges from 0 to 1.
- Ions of the plume may cause erosion, and its products may deposit and contaminate the surfaces, changing their physical characteristics. The primary energetic ions of the beam within a cone angle of the thrust direction ~20° wide are the main culprits for this *sputtering* process. Such a cone angle is usually narrower for GIT than for HET. The effect can be limited by acting on the orientation of the thrusters. GITs are also subjected to the sputtering of the grids, which shortens the thruster life and contaminates the spacecraft's surfaces. HETs plumes extend at angles

about twice as wide as those of GITs. On the early life, sputtering is mainly due to the erosion of the ceramic channel walls, and has a little effect on the properties of the surfaces. On the contrary, late life erosion of the magnetic poles is more problematic, and even extremely thin layers of metal impurities can modify the physical properties of the surfaces.

- The charged particles of the plume can connect electrically the thruster to the spacecraft surfaces. It is then necessary to take into account all the possible sources of currents in order to prevent current loops and floating potentials.

8.4.2.6 A Comparison Between GITs and HETs

In the upper panel of Fig. 8.5 we show the maximum thrust achieved as a function of the specific impulse for GITs and HETs. From the plot emerges the presence of separated populations. GITs and HETs have similar thrusts. Only in a few cases the maximum thrust of HETs is up to an order of magnitude higher than that of GITs. On the contrary, GITs tend to have higher specific impulses (up to ~4200 s) than HET thrusters (up to ~3000 s).

In the lower panel of the same figure, the specific thrust as a function of the specific impulse is shown. As we can see, HETs and GITs are complementary in satisfying the requirements of the proposed mission. In fact, if a high specific thrust is required, HETs better fulfill such a requirement than GITs. On the other hand, GITs represent the best choice when the requirement for a high specific impulse must be privileged.

Regarding the efficiency, GITs cover a wide range showing the best (up to ~70 %) and the worst (~35 %) efficiencies. HETs cover a narrower range: their minimum efficiency (~45 %) is well above the GIT minimum efficiency, and their maximum is close to the GIT maximum.

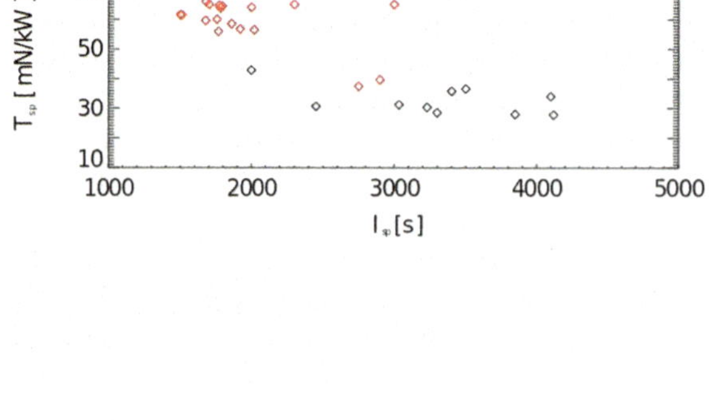

Fig. 8.5 Maximum thrust [*mN*] (*upper panel*) and specific thrust [*mN/kW*] (*lower panel*) as a function of the specific impulse for GITs (*black*) and HETs (*red*)

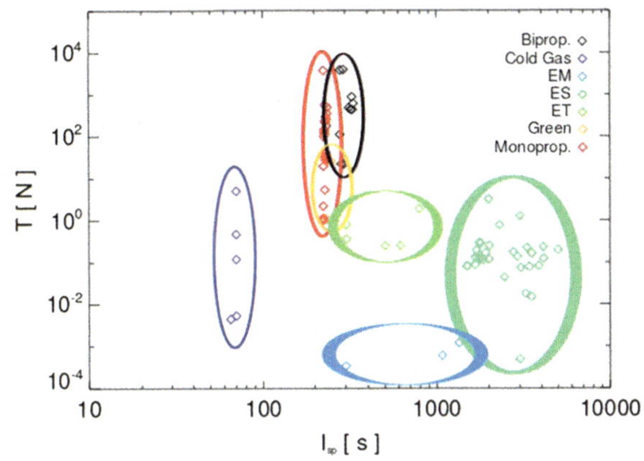

Fig. 8.6 Maximum thrust as a function of the specific impulse for current satellite thrusters. In alphabetical order: bipropellant (*black*), cold gas (*violet*), electromagnetic (EM, *blue*), electrostatic (ES, *cyan*), electrothermal (ET, *green*), green (*gold*), monopropellant (*red*)

8.5 Final Remarks

In Fig. 8.6 we compare the performance of the current satellite thrusters: monopropellant (in red), bipropellant (in black), green propellant (in gold), electrothermal (in green), electrostatic (in cyan), electromagnetic (in blue), cold gas propellant (in violet). As we can see, the range of thrusts achieved is quite uniformly covered from the fraction of mN to a thousand N. Most importantly, the different kinds of thrusters are spread in the specific impulse. The electric thrusters (electrothermal, electrostatic and electromagnetic) cover a relatively wide range of I_{sp} from $\sim$300 to 5000 s. In particular, electrostatic thrusters (mainly GITs and HETs) achieve the highest specific impulses. This allows to reduce the needed propellant mass (and thus the launch costs) or increase the payload maintaining the same mass, which is an enormous advantage respect to traditional monopropellant and bipropellant device systems.

References

1. H. Nguyen, J. Kohler, L. Stenmark, The merits of cold gas micropropulsion in state of the art space missions, IAC-02-S.2.07
2. M.J.L. Turner, in *Rocket and Spacecraft Propulsion: Principles, Practice and New Developments*, 3rd edn., Springer-Praxis Series in Astronomy and Space Science, ed. by M.J.L. Turner (2009)
3. D.M. Goebel, I. Katz, *Fundamentals of Electric Propulsion: Ion and Hall Thrusters*, vol. 1 (Wiley, Hoboken, 2008)
4. H.R. Kaufman, *An Ion Rocket with an Electron-Bombardment Ion* (NASA Technical Note, D-585, 1961)
5. J.S. Sovey, Improved ion containment using a ring-cusp ion thruster. J. Spacecraft **21**, 488–495 (1984)
6. L. Mazzini, V. Tamilia, The Artemis satellite unexpected trip to GEO Orbit. Space Technol. **24**(1), 43–56 (2004)

Chapter 9
Sensors and Actuators Technologies

Abstract This chapter is devoted to illustrate the more commonly used sensors and actuators in satellite AOCS. They will be described focusing their performances and technologies. We will describe the different types of sensors that can be used for attitude and position determination, namely Sun sensors, rate sensors, magnetometers and Star sensors and GNSS receivers. The actuators that will be illustrated are the ones most often used in attitude control: reaction wheels, momentum wheels, control moment gyros and magneto-torquers. Actuation systems which utilize mass ejection for attitude and orbit control have already been described in Chap. 8.

9.1 Introduction

When we analyze different satellite configurations available in the market, it is possible to recognize that attitude and orbit control systems are composed by the following main elements: sensors which measure physical signals used to reconstruct the state of the system; actuators which produce torques and forces necessary to impose the desired state evolution; control algorithms based on different logic and control structures that define the systems' dynamics. According to different mission needs, different AOC systems can be designed using specific sensors and actuators selecting them from a large class of units on the basis of specific trade-off analyses necessary to achieve the best compromise between system performances and cost.

A set of reference unit vectors in **ORF** of known directions (typically Earth, Sun, stars and magnetic field) and a set of measured unit vectors in **BRF** available from Earth sensors, Sun sensors, star tracker sensors and magnetometer sensors can be used as input to attitude determination algorithms to estimate the transformation matrix $M_{E \hookleftarrow B}$. If we include in the previous list of sensors also gyros, we can have a quite complete set of attitude sensors commonly used in the satellite attitude determination algorithms. In Sect. 4.3.5 a coarse and a fine attitude determination algorithms using these sensors have been shown to estimate the orientation of the satellite in a **BRF** with respect to **ORF**.

In Chap. 4 a reference AOC architecture and a set of control algorithms (i.e. AOC Modes) has been shown describing also the main objectives and expected

© Springer International Publishing Switzerland 2016
L. Mazzini, *Flexible Spacecraft Dynamics, Control and Guidance*,
Springer Aerospace Technology, DOI 10.1007/978-3-319-25540-8_9

performances of each mode. Each AOC mode can use different actuators properly sized to implement the desired control torque necessary to achieve the final satellite attitude with specified pointing and stabilization performances.

The most commonly used actuators in attitude control are reaction wheels, momentum wheels, control moment gyros, magneto-torquers and the actuation systems which utilize mass ejection (RCTs) which have been described in Chap. 8.

In this chapter all these instruments for actuation and measurement will be described in terms of the available performances and technologies. We will not treat important subjects like: components hardenization and hardenization by design, equipment qualification for the laucher and the space environment, design validation methods, all issues that have a very import role in the cost and reliability of space equipment and therefore are a basic topics in AOCS design.

We remark that a limited number of suppliers of space sensors and actuators are mentioned in this chapter, they are the suppliers best known to us during our working experience. The reader is advised that many other suppliers exist offering very competitive products possibly more suitable to its design or study needs.

9.2 Attitude Sensors

This section shows different types of sensors used in attitude determination algorithms. The different types of sensors (gyro, sun sensor, earth sensor, star tracker and magnetometer) discussed in this section are commercially available. For all of them different suppliers exist capable to offer products based on the same technology and a wide offer in terms of accuracy, *field of view* (FOV), mass, power consumption and cost. The basic mechanisms behind all these sensors will also be mentioned.

9.2.1 Gyroscopes

Gyroscopes (gyros) are inertial sensors measuring the angular rate or the incremental angular rotation about its measurement axis. Two basic types of gyro exist: the rate gyro and the rate integrated gyro.

The classic mechanical gyro technology is based on a rapidly spinning mass that senses changes in the inertial orientation of the gyro spin axis. Any satellite rotation about the gyro's input axis realizes a torque on the gyro output axis that causes an angular rotation or another physical reaction which is measured in output.

The rate gyro sensors (RG) measure the spacecraft rotation rate. They are relatively inexpensive and they can be used for the attitude stabilization (e.g. in a PID controller) or the rate damping modes (see Chap. 4). All these sensors are affected by different error sources like angular drift errors that can be estimated and then calibrated using optical attitude sensor measurements such as those provided by Earth-Sun sensors or star trackers. The rate integrating gyros (RIG) measure the angular displacement.

They are more expensive than rate gyro sensors and are used in attitude determination algorithms.

A rate integrated gyro output can be differentiated to obtain the average rotation rate and also in this case the sensor can be used also as a rate gyro.

In addition to the mechanical gyro two other technologies exist to derive the angular information: the optical gyro and the vibrating gyro.

The optical gyros utilize a different physical principle to measure the satellite angular rotation: the Sagnac effect. This technology allows to achieve an accuracy one or two orders of magnitude better than the mechanical gyro technology.

In the space gyro commercial market we have many suppliers (e.g. Honeywell, Selex ES, Sagem, Airbus Defence and Space, Northrop Grumman and many others) producing both classes (e.g. RG and RIG) of products using different technological principles. Two specific RIG sensor products with different technologies will be illustrated as an example.

In the RG class, the market is led by the MEMS (micro electro mechanical systems) gyros using a miniaturized vibrating gyro technology which can nowadays provide performances in the range of 1–10°/h drift error at very low cost and reduced sensor weight and dimensions.

9.2.1.1 Rate Integrating Sensor (Fiber Optic Gyro)

This paragraph shows a typical commercial RIG based on fiber optic technology. The RIG is configured with a single 4-channel redundant package, where the fourth channel is able to provide backup measurements for the three nominal ones. This way a redundancy scheme is available depending on the HW unit selected for the nominal behavior (i.e. 3 out of 4 redundancy scheme). The gyro package sensor data is acquired by the on-board computer directly via Mil 1553 bus. This bus is one of the most used command and control standard I/F available for space application.

During the satellite's nominal mission phase, the gyro measurement is usually used together with one or two star tracker measurements to feed the on board attitude estimation filter (see Sect. 4.3.5.2). If two star tracker sensors (STR) are mounted on the spacecraft and one of the two is in blind condition, full attitude performance can still be achieved by filtering together one STR and the gyro measurements. In case only one star tracker is used in conjunction with the gyro, if the star tracker is in blind condition the estimation filter can propagate the attitude using the gyro data only.

Usually, the gyro is also used during the initial mission phase to provide the rate measurement for rate damping, and during the Sun search phase, to provide the rate information to drive the S/C angular rotation.

Fiber optic gyro (FOG) technology (i.e. continuously operating for a long period of time, typically 15 years) for space applications can provide very good performances in all the main requirements:

- a fine resolution (e.g. better than 0.01 arcsec);
- a linear scale factor (e.g. better than 10 ppm);

- a bias stability better than 0.01° h;
- a low angular random walk (e.g. better than 0.002° $\sqrt{(h)}$);
- high measurement bandwidth;
- no high frequency noise measurement;
- solid-state reliability and robustness.

Considering the relatively high cost of the high perfromance gyro solutions, lower cost solutions with down-scaled performances are also available.

Among the possible suppliers of FOGs are Astrium Defence and Space(Astrix) and Northrop Grumann with a product embedded in its IMU (the LN-200S, an Inertial Measurement Unit containg also 3-axes accellerometers)

Airbus Defence and Space has designed a FOG solution called ASTRIX® 120 that features two main elements, as illustrated in Fig. 9.1.

- a Sagnac interferometer assembly (SIA), mainly constituted by a fiber optic coil ended by an integrated optical circuit called COI, which detects rotation rate around the fiber optic coil axis;
- an electronic module, (named FEM for FOG Electronic Module) featuring processing electronics to extract the inertial information from the optical signal coming out of the interferometer, interface electronics to be able to communicate with spacecraft's and ground test facility's systems and power distribution.

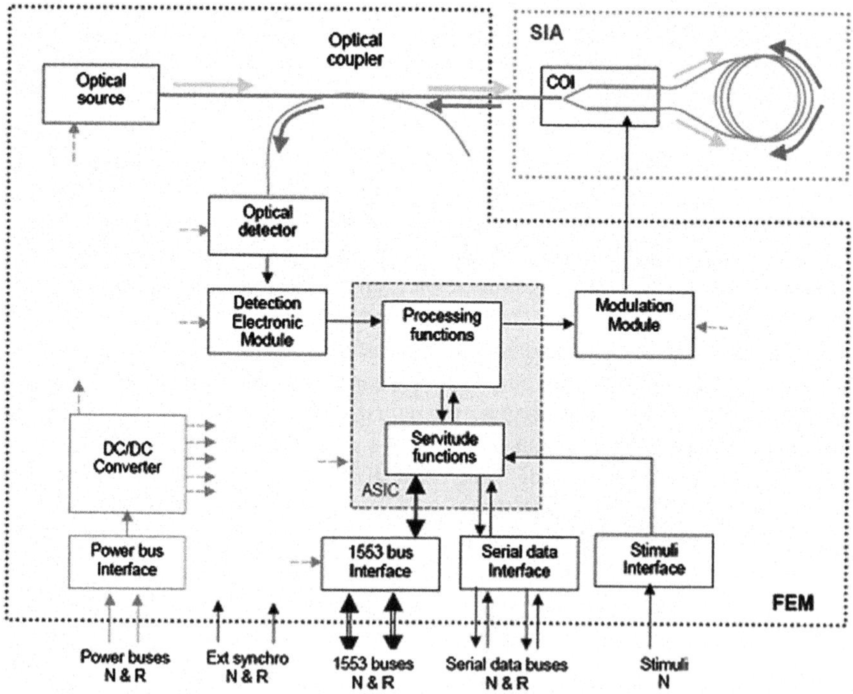

Fig. 9.1 FOG channel architecture, courtesy of airbus defense and space

Fig. 9.2 ASTRIX® 120 assembly view: courtesy of airbus defence and space

Figure 9.2 shows the ASTRIX® 120 with its 4 independent gyro channels.

The fiber optic gyroscopes work according to a physical phenomenon discovered by Sagnac in 1913:

Principle: Sagnac *an optical interferometer enclosing a surface located on a rotating support will detect a phase difference of the optical signal, which is proportional to the angular rate and to the surface enclosed by the optical path.*

Therefore the FOG gyro channel requires the following optical architecture:

- an optical path enclosing a maximal surface: this is realized through a long optical fiber, which is wound many times round, in order to make a reasonably sized interferometer;
- a source, to deliver the light to the interferometer, with high spatial coherence and low temporal coherence. The high spatial coherence ensures a good coupling efficiency of the light into the fibers;
- a multifunction COI (integrated optical circuit), which closes the interferometer, to share the light between the two extremities of the fiber optic coil, to filter undesirable polarization, and to allow high bandwidth phase modulation of the light sent into each extremity of the fiber coil;
- a coupler to extract the optical signal coming back from the ring interferometer, which is the signal carrying the Sagnac inertial information, and to direct it towards an optical detector.

The typical signal processing architecture is the following:

- a detection module to convert the light signal carrying the inertial information into an electrical signal, ready to be processed;
- a closed-loop signal processing to increase the dynamic range and remove the effect of fluctuations of the optical power and of the gain of the detection chain,

allowing an easy auto-calibration of the system and excellent scale factor stability and linearity;
- a biasing modulation allowing the measurement of very low rotation rates as well as the sign of those rotations.

9.2.1.2 Rate Integrating Sensor (Laser Gyro)

The typical specifications of a 3-channel rate integrating laser gyro (RLG) are:

- compensated angle and velocity outputs;
- Mil-Std-1553B interface;
- 15 year life;
- 0.02–0.05° h bias stability over 1 year;
- 5–30 ppm scale factor stability over 1 year;
- 0.005–0.01° $\sqrt{(}$h) angular random walk;
- radiation tolerance: > 15 Yr GEO, 12.5 Yr LEO.

Figure 9.3 provides various details on the functional principle of the RLG. In each sensing unit, two laser beams are projected along a path (i.e. triangular), each in the opposite direction to the other (see Fig. 9.3). The beam is emitted by a laser device, and the path is made triangular by deflecting the beams off special mirrors located at the vertices of the triangle. There is a single laser beam reception point which coincides with one of the mirrors, whose transparency allows a photoelectric cell located behind the transparent mirror to take measurements with an optical interferometer.

The RLG is based, as the FOG, on Sagnac's principle. When the device spins about an axis perpendicular to the plane of the triangular closed path, the light rotating in the direction opposite to the spin experiences a path length slightly shorter than the path length experienced by the light beam moving in the direction of the rotation. The

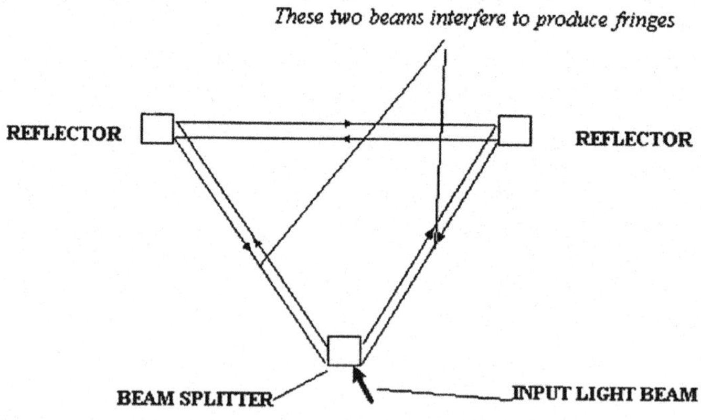

Fig. 9.3 RLG functional principle

first beam encounters the mirror moving toward it, while the second beam encounters the mirror moving away. If we consider the speed of light and the speed at which the satellite spins about its axis, we find that the difference between the two paths is in the order of 1 part in 10^9.

This is the reason why a monochromatic laser is used: when the two beams combine they produce an interference figure which is made up of light and dark bands on the plane of each mirror, and these bands indicate the points where the two paths differ by half a wavelength.

When the device rotates of a well defined quantum of rotation corresponding to an half wavelength, the light and the dark bands on the surface of the mirror shift by one band width. This shift is detected using a photo-diode.

Space qualified RLG units are commercialized by Honeywell, L-3, Kearfott and others.

9.2.2 Sun Sensor

The Sun sensors are the most commonly used attitude sensors in safe modes (see in Chap. 4 the SAM-EM mode). When the mission is not very demanding in terms of accuracy the sun sensors can be selected also for the NPM attitude determination. The Sun sensor measures the incident sunlight direction with respect to the sensor optical reference plane. The design is generally simple and reliable and its algorithms are robust, simple and inexpensive. There are two main types of Sun sensors: the analog and the digital sensors.

An analog sun sensor typically has an analog output signal that is a function of the Sun incidence angle. Analog sensors are based on photocells that have a current output proportional to the cosine of the angle between the Sun's direction and the perpendicular to the detectors.

The digital Sun sensors are based on different types of detectors. The most common detectors are CCD or APS. The sensor is typically equipped with a measurement unit designed around digital devices like ASIC or FPGA, able to calculate the Sun's incidence angle relative to the perpendicular of the sensor mounting surface when the Sun is in the sensor's field of view (FOV) and to provide this information to the AOCS. Moreover, the sensor provides other information like Sun presence, detector temperature, secondary voltage and other housekeeping information that can be used by the AOCS to determine the attitude of the spacecraft and to detect sensor failures.

The digital Sun sensor is more expensive than the analog Sun sensor but ensures a more accurate performance (i.e. a few hundredths of degree instead of a few tenths of degree).

In the commercial market many suppliers provide very reliable and low cost products (e.g. Selex ES, Jena Optronik, Adcole, TNO, Moog Bradford, Goodrich and many others) producing both classes (e.g. analog and digital) using different designs and different technological principles. Here a specific digital sensor and an analog sensor will be illustrated as an example.

9.2.2.1 Analog Sun Sensor

Commercial Analog Sun sensor units may be both internally or externally redundant. The sensor is designed to provide the Sun unit vector when the Sun is detected in the field of view of the unit.

The typical Sun sensor field of view (FOV) is $\pm 64^\circ \times \pm 64^\circ$ and the measuring accuracy is typically better than 0.3–0.5° The sensor has a rectangular FOV:
around the x-direction (in α): with $-64^\circ \leq \beta \leq 64^\circ$
around the y-direction (in β): with $-64^\circ \leq \alpha \leq 64^\circ$

In the most used design the sunlight illuminates a photo diode arrangement passing through a fissure located above and parallel to the plane of the photodiodes. Depending on the incident angle α of the sunlight and the distance h between the fissure and the photo diode array a well-defined portion of the array is illuminated. The illuminated area of the photo diodes and thus the generated photo currents are a function of the incident light angle α. A dark current reference diode can be used to monitor temperature changes of the photo diodes. Temperature changes of the dark current of the measurement diodes can be compensated to minimize offset errors.

The dark current reference diode can be used to get information about the health status of the Sun sensor. A sudden increase of the dark current, monitored in orbit operation, can be used to determine malfunctions or damage to the sensor.

To improve the sensor's accuracy each unit is calibrated through an on-ground measurement campaign in order to provide a set of spline calibration coefficients. Moreover a proper algorithm has to be implemented, in the sensor data processing, to filter the albedo effects.

9.2.3 Digital Sun Sensor

The digital Sun sensor technology will be shown by illustrating a commercial product APSS produced by Selex-ES. This digital Sun sensor is designed to be used in LEO, MEO or GEO orbits and it is based on an active pixel sensor technology (APS).

The active pixel Sun sensor (APSS) is a two axis solar sensor for accurate attitude measurement. The sensor measures the Sun's direction in quite a large FOV ($\pm 64^\circ \times \pm 64^\circ$). The electronics can also be commanded to use smaller areas of the FOV in order to make the sensor robust to possible stray light effects. The sensor principle is based on the barycenter detection of the incident energy of the Sun, in the visible range onto the detector array APS.

In the APSS the Sun light passes through a pin hole, creating a Sun spot on a bi-dimensional CMOS APS; the electronics evaluate the position of the Sun spot by means of a centroid algorithm and provides the AOCS with the Sun position co-ordinate in the detector reference frame.

The core of the sensor is the CMOS APS detector, with integrated capabilities of direct pixel addressing, analogic to digital conversion, and anti-blooming circuit. An ASIC collects all the intelligent functions of the sensor. Moreover, the APSS is provided with a latch up protection circuit, a power section (that includes a DC/DC converter or a secondary voltage supply) and the RS422 serial data interface used to exchange data with the spacecraft.

The APSS is a medium/high accuracy Sun sensor, wide field of view, small size, low mass and autonomous operation. In Fig. 9.4 the functional reference frame of the APSS on the mechanical housing is shown.

The APSS Electronics consist of two parts: data interface electronics and power supply electronics. The Sun sensor electronics block diagram is shown in Fig. 9.5.

Figure 9.6 shows the physical principle on which the APSS measurement is based. The optical system is made of an incoming pin hole, a detector window positioned below and the focal plane assembly with a 1024 × 1024 APS detector. The pin hole is positioned at a proper distance from the sensitive area in order to allow a ±64° FOV. In this system, the Sun's direction (α_{in}, β_{in}), corresponds to a bi-dimensional position (X, Y) in the APS functional reference frame (**FRF**). The input attitude variables α_{in}, β_{in} are related to the Sun's position in the sensor **FRF**.

The incident light ray falls on the APS detector proceeding from the hole and the detector window to the active area of the APS. Defining θ the Sun's position in the FOV, the displacement d is given by the following expression:

$$d(\theta) = f\ \tan(\theta) + h_\omega \left[\frac{\sin(\theta)}{\sqrt{{n_\omega}^2 - \sin^2(\theta)}} - \tan(\theta) \right] \tag{9.1}$$

From the previous formula, it is possible to write the X and Y position in the detector area as function of α_{in} and β_{in}.

The Sun sensor measures the Sun's position (α_{in}, β_{in}) on the APS detector (X,Y). In order to evaluate the corresponding Sun's position (α, β) linked to the output data

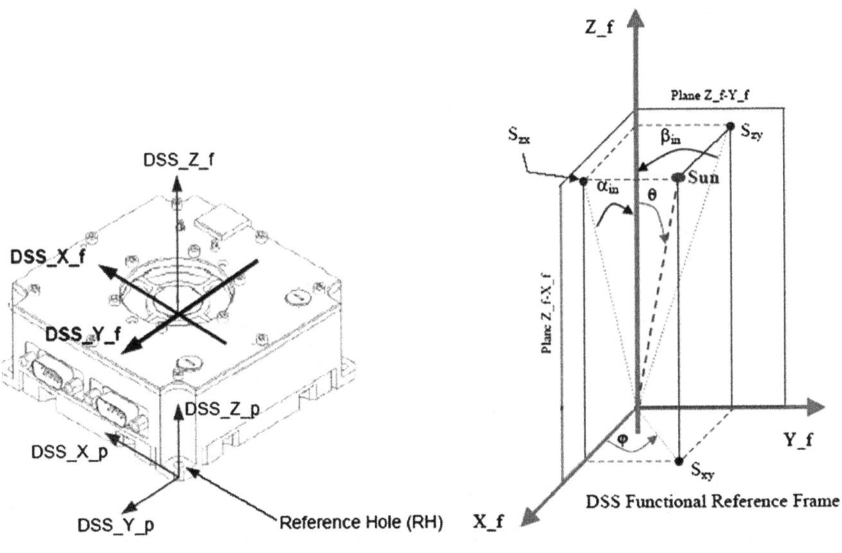

Fig. 9.4 APSS functional reference frame: courtesy of SELEX-ES spa

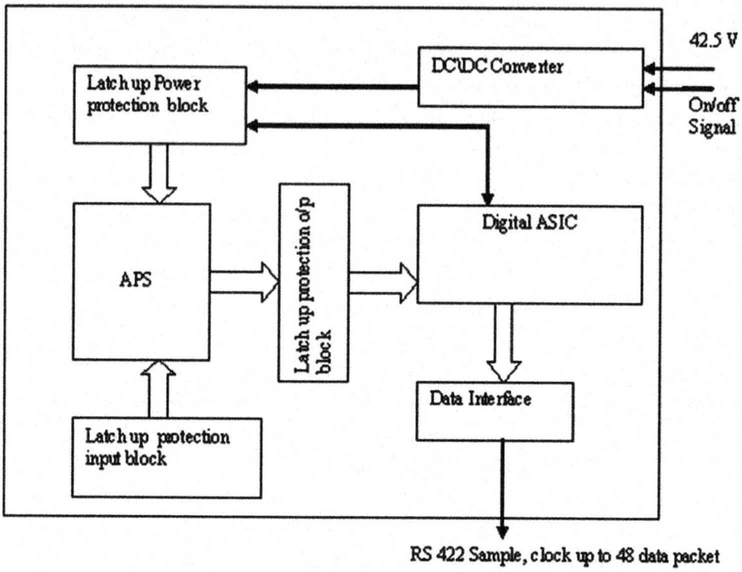

Fig. 9.5 Digital sun analog sensor assembly (layout diagram): courtesy of SELEX-ES spa

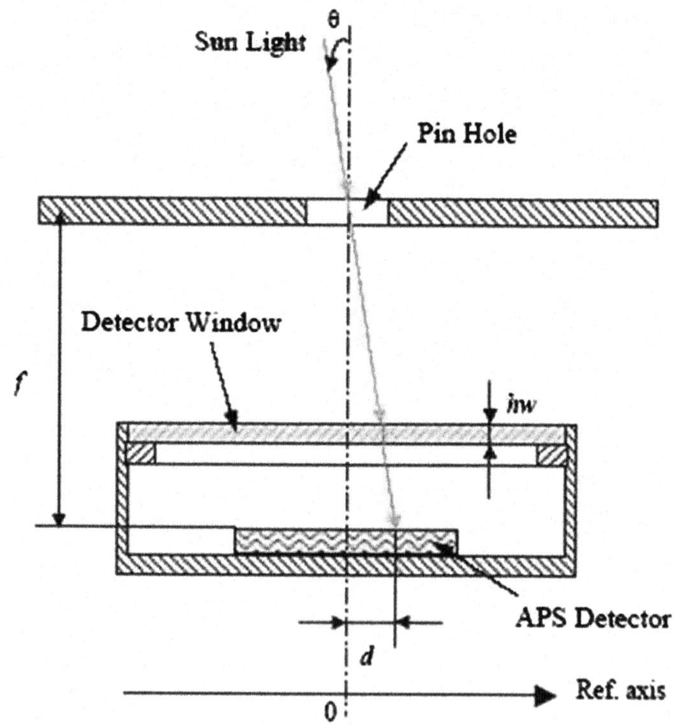

Fig. 9.6 APSS main functional principle (shown for one axis): courtesy of SELEX-ES spa

d	is the displacement of the image on the APS detector from the center	[mm]
f	is the focal length	[mm]
θ	is the angular direction of the incident light ray	[o]
h_ω	is the detector window thickness	[mm]
n_ω	is the refraction index of the detector window (BK7)	[–]

X,Y it is sufficient to apply an inverse formula $X, Y \rightarrow \alpha, \beta$ that takes into account the calibration coefficient necessary to obtain a good level of accuracy (e.g. better than 0.05°).

9.2.4 Earth Sensor

The Earth sensor measures the direction of the Earth center in the satellite reference system (**BRF**). It is often called an horizon sensor because the Earth center is derived from the profile of the Earth. This sensor is capable to give a two axis measurements. Thus, to have a three axis measurement it is necessary to use in conjunction with the Earth sensor another sensor like a Sun sensor or magnetometer (for LEO mission only). In this case using attitude determination algorithms like that shown in Chap. 4 it is possible to estimate the kinematics of satellite.

For many years, the Earth sensors were the primary sensors used in all LEO, MEO and GEO missions. In the last 10 years, the usage of the Earth sensor is progressively decreasing in favor of the more accurate star tracker sensor.

In today market few suppliers exist which commercialize this kind of product (e.g. Selex ES, Sodern, Goodrich). These suppliers apply different designs and different technological principles adaptable to different orbital ranges. Here a specific sensor product suitable for the GEO orbit will be illustrated as an example. This sensor, that has been the principal actor in the story of Earth orbiting satellites, is slowly being replaced by star tracker measurement devices which are more flexible and show similar reliability and cost figures. This is the reason why the number of products on the market is relatively small.

9.2.4.1 Infra Red Earth Sensor for GEO Missions

The infra red Earth sensor (IRES) is used to measure two angles defining the direction of the Earth disk center produced by Selex-ES. The Earth is identified in the infra red electromagnetic band around a wavelength of about 15 μm, corresponding to the CO_2 absorption band of the atmosphere. The IRES is used both for transfer orbit and for geostationary orbit mission phases. The sensor provides pitch and roll attitude measurement information using measurements of the infra-red horizon, being the yaw angle the rotation around the Earth center direction. It can operate in two different modes:

- narrow scan mode, nominally used in geostationary orbit;

- wide scan mode, nominally used during the Earth acquisition mode.

The IRES is mounted typically so that its axes are parallel to the **BRF** axes of the spacecraft, with its bore-sight pointing along the **BRF** +Z axis, nominally looking at the Earth.

The sensor is designed to be used at geostationary altitude, but it provides attitude measurement data for all orbit altitudes from 18,000 to 45,000 km. Once the satellite is on station it will be in a geostationary orbit that is a near-circular equatorial orbit at an altitude of approximately 36,000 km above the Earth's surface; during transfer orbit it will be at different altitudes, therefore the IRES has to provide attitude data over the range of altitudes from 18,000 to 45,000 km.

The IRES measured roll angle is the true roll angle only at geostationary altitude (36,000 km), at different altitudes a gain correction factor must be used to obtain the true roll angle.

Figure 9.7 shows the mounting geometry of an IRES. The figure illustrates that the sensor has an optical axis that is approximately perpendicular to its mounting interface. The sensor provides measurement of attitude errors relative to the Earth nadir about two perpendicular axes (pitch and roll) that form a triad with the unit's optical axis.

Figure 9.8 shows a functional block diagram of the Earth sensor assembly. The block diagram shows that each IRES consists of the following main components:

1. an infrared telescope with four bolometers and amplifiers
2. a scanning electronics and angle read-out
3. a command/data interface unit
4. a power unit
5. a test unit

The IRES uses a cylindrical shaped telescope with a single lens to focus the input radiation onto four bolometers, with internal baffling to avoid spurious reflections.

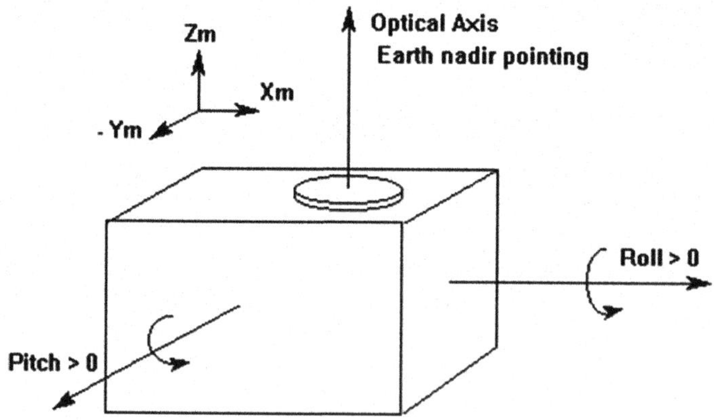

Fig. 9.7 IRES diagram: courtesy of SELEX-ES spa

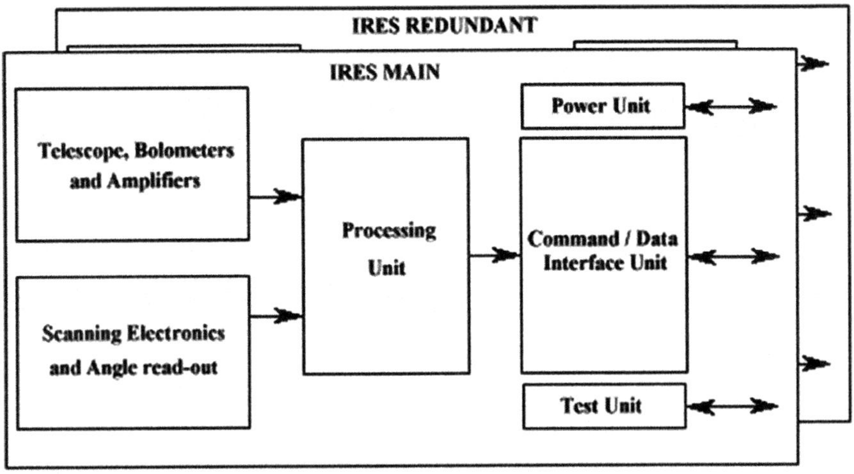

Fig. 9.8 Earth sensor assembly functional block diagram: courtesy of SELEX-ES spa

A scanning mechanism located just behind the telescope realizes an oscillation of the infrared Earth image on the focal plane, each bolometer scans periodically the cold Space and a portion of the Earth.

The four bolometers are mounted on the sides of a pyramidal shaped support, whose axis is the nominal sensor boresight. When the sensor is operational this bore-sight is pointed close to the direction of the Earth center.

When the spacecraft is at geostationary altitude each bolometer detects a space-Earth and Earth-space transition at latitude ±45° East or West.

The thermal radiation from deep space is negligible in comparison with that of the Earth. The two discontinuities in the infrared radiation signal crossing from space-Earth and Earth-space for each bolometer are detected and each generates an impulse.

These pulses are then compared to reference pulses generated by the electronics at the center of the scan path of each bolometer, so that when the Earth is perfectly projected to the center of the focal plane—i.e. null roll and pitch pointing—the reference pulses coincides with the Earth-space crossing generated pulses.

If there is a pointing error a time difference between these pulses occurs, which is converted to an angle by the processing electronics.

The sensor can operate in either narrow scan mode or wide scan mode, in which the composite field of view of the bolometers scans through either ±5.5° or ±11.0° at a frequency of 10 Hz.

9.2.5 Star Tracker

Star sensors represent today the most accurate attitude sensors for demanding satellite missions. The basic principle of a star sensor is based on the accurate measurement of the star directions in **BRF** that is compared with the known accurate star reference direction in **ECI** available in the on board catalog. Using both measured and known star directions an attitude estimation algorithm like QUEST (see [1, 2]) can be used to obtain a three axis measurement of the satellite attitude in **ECI**.

Star sensors are divided in two major categories: scanners and fixed head star trackers.

Scanners, popular 20–30 years ago, can be used when the satellites are spinning. In this case, the sensor uses the spin of the spacecraft for the searching and scanning function when stars pass through multiple slits in the scanner's field of view.

Today the more used star sensors are the fixed-head star trackers (single or multiple heads) that, thanks to the more powerful processors available in today space application, are able to electronically implement a search and track function over a limited field of view. These sensors are able to acquire, select and recognize stars from a *lost-in-space* attitude, with relatively high tumbling rates (few degs/sec.) in few seconds with an high value of probability of success. After the attitude acquisition, the star trackers are today able to maintain autonomously the tracking of stars (up to 15 stars or more for each head) and on this basis are able to provide a very accurate attitude information.

The commercial market of these sensors is rich with many possible suppliers all providing excellent products (e.g. Selex ES, Jena Optronik, Sodern, Ball, Blue Canyon Technology, Goodrich and many others). Even if the basic detection technology is the same among the different products (e.g. CCD or the new APS detectors), different architectural solutions can be implemented (single head or multiple head, centralized or distributed processing). The sensor cost is highly dependent on the level of the requested accuracy which implies different levels of sensor calibrations. A specific product example will be illustrated in the next section.

A high performance attitude determination system requires to use an assembly composed of up to three autonomous star tracker sensors (STR) or a multi-head star tracker equipped with up to three heads and a redundant central electronic unit, to be used during the nominal mission phase, in order to obtain fine attitude measurements to support mission needs.

Three star sensor units or one multi-head configuration provide the needed redundancy. The baseline is to have two star trackers operating while the third one is available for cold redundancy. The nominal usage of two star trackers improves the overall pointing accuracy, since the single-sensor accuracy around its boresight is much worse than the accuracy in the two directions orthogonal to it. The three star trackers' orientation must be chosen in order to have an angular separation between the sensors' boresights, so as to optimize the overall measurement performance, taking into account the sensor's field of view constraints.

As already mentioned, the single STR measurement performance is not isotropic since the sensor provides a more accurate measurement around the axes orthogonal to the boresight than around boresight rotation; in particular, the noise level of the angle of rotation around the boresight is one order of magnitude higher than that around the two orthogonal directions and also the low frequency error component around the sensor's boresight, mainly due to the variation of the focal length with the temperature, is also significantly higher.

The STR 2 over 3 redundancy is a functional redundancy that allows to have an almost isotropic performance in case of one STR failure.

Each optical sensor has a useful field of view (FOV) where it provides the requested output as long as there is no light source (i.e. stray-light) or satellite surface in a certain surrounding angular cone which is often called unobstructed FOV.

Star Trackers are particularly sensitive and they have therefore unobstructed FOV for Moon, Sun and Earth light direct impingement. The satellite surfaces may also reflect part of the light of external sources producing stray-light which can deteriorate the quality of the measurement. For these reasons it is customary to have an unobstructed FOV also for satellite surfaces surrounding the sensor. When the unobstructed FOV conditions are violated we speak of sensor blinding. Blinding depends on two factors of the sensor design: the detector technology (i.e. a robust detector technology like APS that is not affected by blooming effect like CCD technology) and the baffle. In order to protect the optical system from stray-light and direct sources (Sun, Earth, Moon…) the star sensors are equipped with baffles, which are conic shaped optical insulators fixed around the optic. The transmittance achievable by the baffle is defined in order to meet the star sensor requirements in terms of attitude acquisition (i.e. probability of successful acquisition in a given time) and the capability of the sensor to maintain the tracking once the attitude has been determined and the sources are external to a desired unobstructed FOV.

The baffle is typically composed of a certain number of cavities whose function is to capture and reflect back the light not coming from the useful FOV (in mechanical model of the sensor of Fig. 9.9 are visible the baffle cavities and the internal disposition of the optic including the focal plane). The star sensor's envelope, mass and accommodation constraint (unobstructed FOV with the satellite surfaces) are largely dependent on the baffle size, which is affected by the exclusion angles requirements: Sun, Earth limb and Moon exclusion angles whose typical values are 40°, 30° and 20°, respectively, for a commercial medium accuracy sensor (see Fig. 9.10).

The star trackers configuration is often the result of an optimization analysis that is aimed at providing, along the orbit and in nominal attitude, the absence of blinding conditions due to Earth, Sun and Moon.

This analysis is used to finalize the mounting position of each star tracker so that no blinding condition is possible. When the optimization process does not provide a viable solution (e.g. due to large satellite appendages or large slew maneuvers), the AOC design has to consider this condition. In fact different AOC design solutions can be applied using a data fusion approach with other sensors like gyroscopes (to manage the blinding intervals propagating the angular position using the gyro measured rate) or adding additional star tracker heads.

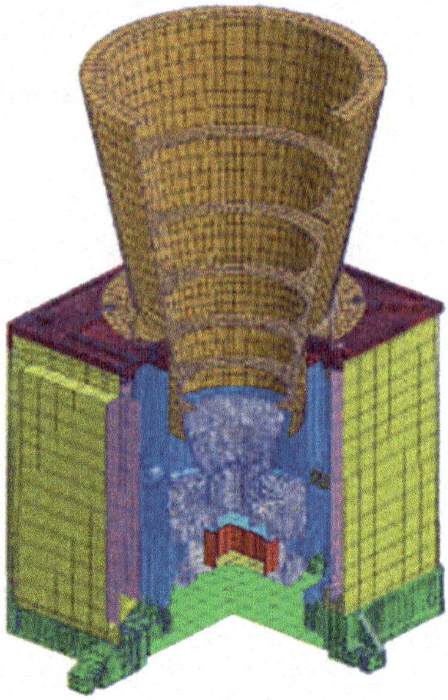

Fig. 9.9 A-STR mechanical model: courtesy of SELEX-ES spa

Characteristics	Values
Exclusion angle (deg)	
• Sun:	40
• Earth :	30
• Moon:	20
System attenuation:	$> 3x10^{-8}$
Dimensions (mm)	180 (L) x 160 (Ø)
Mass (g)	300

Fig. 9.10 A-STR baffle properties: courtesy of SELEX-ES spa

Hereafter we show a commercial autonomous star tracker (A-STR) designed by Selex-ES.

The A-STR is able to provide a spacecraft attitude measurement from lost-in-space conditions, with relatively high S/C rates, to fine tracking performances at low rates, without the intervention of external processing units. It performs the measurement of stars' positions in a two-dimensional frame by means of a dedicated detector and provides a 3-axis attitude determination quaternion w.r.t. the inertial J2000 reference frame. The detectors' signals are amplified and pre-processed to allow the identification of a number of stars of suitable magnitude. Star catalogs and associated pattern recognition algorithms will be implemented in the star sensor to allow the attitude determination with no a priori knowledge of the attitude (lost in space function).

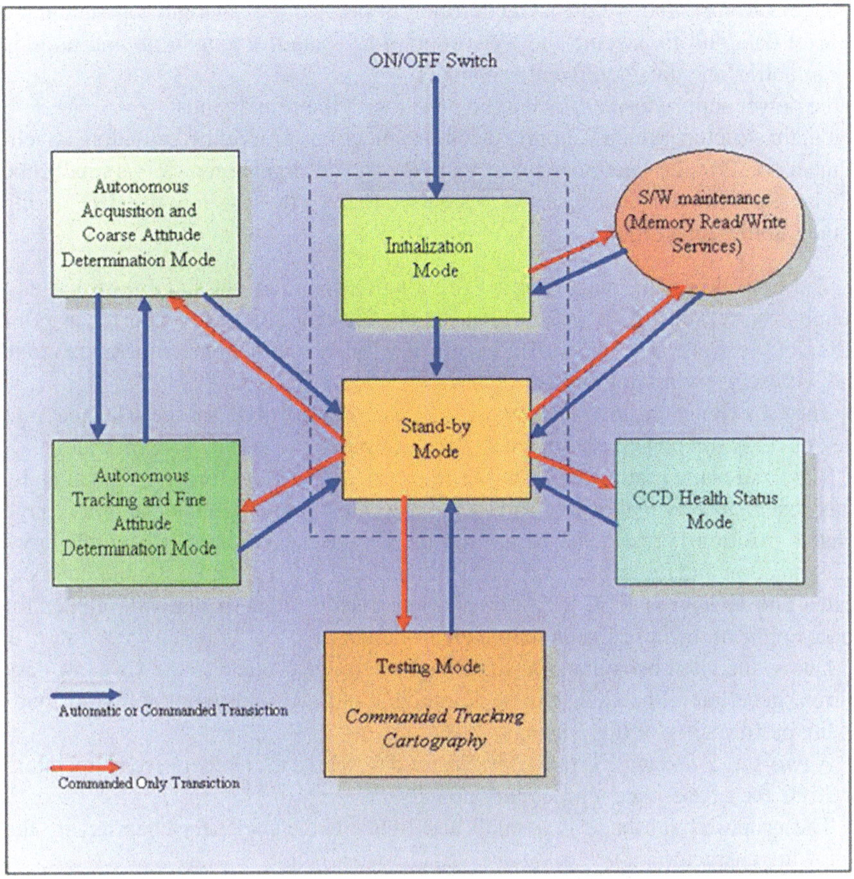

Fig. 9.11 A-STR operative modes diagram: courtesy of SELEX-ES spa

In addition, the sensor provides auxiliary modes for ground or in-flight testing and for SW maintenance. The several operative modes and the relevant transitions are depicted in Fig. 9.11.

The sensor consists of a unique and compact assembly that includes the following modules:

- a sunshade providing appropriate protection against light from the Sun, Earth or other bright objects outside the FOV or reflected by the S/C itself;
- an electro-optical module including the optical system that can observe a zone of the sky on the CCD detector. The CCD temperature is kept to its operative value by a thermo-electric-cooler only used when the temperature exceeds a specified value (+20 °);

- the electronics to drive the CCD detector, to process the analogue signal and the pixel data and to execute the SW program for pattern recognition and attitude estimation and data/command exchange;
- the power supply for interfacing the power with the primary bus;
- a main structure which supports the above mentioned modules, providing attachment points to the satellite mounting surface and an alignment cube—a small cube with super-flat mirror surfaces which is used to align the sensor to the **BRF** during the satellite integration.

The optical system, manufactured in radiation resistant glass and mounted in a titanium barrel, ensures a good optical quality over a large FOV and for stars of different spectral classes, as well as excellent stability over a large temperature range and with severe vibration levels.

The CCD detector, the thermoelectric cooler and the plate for the attachment to the optical barrel constitute the focal plane assembly.

The focal plane is attached directly to the optical system to form a single module. This design approach allows to achieve an extremely high alignment stability as the relative position between the barrel and the CCD sensitive area is not influenced by elongation and deformation of parts interposed between the barrel and the focal plane. The detector is a 512×512 pixel frame transfer CCD specially developed for space applications in severe radiation environments.

Due to the good behavior of the CCD under irradiation and to the very low dark current generation, the CCD can be operated at $+20\,^{\circ}\mathrm{C}$ without affecting the end-of-life performance of the sensor.

A two-stage *thermo electric cooler* is used to keep a CCD operative temperature of $20\,^{\circ}\mathrm{C}$ for a base-plate temperature up to $60\,^{\circ}\mathrm{C}$.

The proposed sunshade is a small and light off-the-shelf sunshade having the following characteristics:

The sensor exchanges data/commands with the AOCS via a MIL 1553 B bus interface. The star tracker is connected to the dual-redundant bus and acts as a Remote Terminal Unit.

A-STR operations are synchronized by externally provided timing pulses (i.e. a pulse per second).

The *autonomous attitude determination mode* is the main operational mode of the sensor and it is composed of two operative sub-modes, internally managed: the *autonomous acquisition and coarse attitude determination mode* (AA&CAD) and the *autonomous tracking and fine attitude determination mode* (AT&FAD).

The AA&CAD mode is entered after the reception of a command from the AOCS or as a fallback from the AT&FAD mode. This mode allows the acquisition of the initial attitude and the estimation of the angular rate without the need for any a-priori attitude information (the so called *lost in space* solution). In each cycle, the whole FOV image collected from the CCD is processed on-line for pixel data reduction by adaptive thresholding algorithms and run-length encoding based algorithms implemented in the ASIC. The whole FOV is used during the attitude acquisition mode to increase as much as possible the probability of stars matching. This data is then processed off-line for segment clustering, star image and extended target identification (if any) as well as for false events and hot pixel (if any) rejection. Several coherence criteria are implemented to identify a cluster as a candidate star (i.e. clusters of one pixel, clusters with dimensions exceeding a predefined threshold, clusters with saturated pixels or bad pixels, etc.). The clusters associated with the identified star images are processed for star position and magnitude determination. At the end, a list containing the position and the magnitude of the detected stars is fed to the *pattern identification module*. The pattern identification module performs the star pattern identification through matching of star triplets (triads) generated using the detected stars, with the on-board triad database. Once the star pattern has been identified, the stars belonging to the matched triads and the corresponding reference stars, contained in the reference star catalog, are used for attitude determination by using the Q-method algorithm. The attitude initialization is obtained using attitude data from two consecutive cycles: attitude data is compared for rate estimation and for an attitude consistency check taking into account the maximum allowable rate. If the attitude information is consistent, then the AT&FAD mode is entered, otherwise an additional cycle is executed and a message is sent to the AOCS.

The AT&FAD mode is autonomously entered after successful completion of the AA&CAD, receiving as input from the previous mode the coarse attitude and estimated rate. In this mode, only windows commanded around stars selected from the on-board reference star catalog will be read and processed for star position and magnitude calculation. The coordinates of each star are corrected for relativistic aberration by using the S/C velocity information provided by the AOCS. From the tracked stars' data and star reference catalog, the sensor determines the fine attitude by using the Q-method algorithm. The sensor selects autonomously the stars (up to 10) to be tracked in the next cycle, managing the stars moving outside the FOV and entering the FOV, and predicts their position on the CCD reference frame by using reference star catalog data and attitude data. Reference stars are selected to optimize the error on the overall attitude estimation and can be updated at each cycle. In particular, to operate with high quality star spot and reduced optical distortion, the useful FOV is limited to a circle with 8.2° half cone during the fine attitude measurement. Until at least two selected stars are identified the sensor will continue the AT&FAD mode updating the attitude data. When less than two stars are detected, or a large object prevents the tracking of stars, the AT&FAD mode will fall back autonomously to the AA&CAD mode. For each cycle the STR computes the information matrix which is the inverse of the covariance matrix. The three elements of the diagonal of the information matrix are used as attitude measurement quality factor and provided to the AOCS.

9.2.5.1 Star Tracker Errors

The star tracker errors can be divided in the following three categories:

- systematic (bias) errors;
- single star errors (low frequency and random);
- attitude errors (low frequency and random).

Bias Errors

The bias errors category includes alignment uncertainties, biases and all long term errors. These errors don't affect the measurement of the star position and are independent of the apparent star motion rate and pointing direction.

Single Star Low Frequency Errors

These errors include all the low frequency spatial error sources affecting the star position measurement within the FOV and depend on the distribution of stars within the FOV but not on the slew rate.

The low frequency errors class includes the following errors:

- residual error due to the relativistic aberration,
- star catalog errors,
- residual optical distortion error after calibration,
- residual focal length error after calibration,
- error caused by focal length variation with the temperature.

Single Star Random Errors

The random error includes all the spatial error sources varying on pixel scale and the temporal error sources, which affect the position measurement of the single star over a short period. The temporal and high frequency spatial error sources include the following errors:

- centroiding algorithm error,
- star image smearing,
- CCD non-uniformities,
- shot noise (star signal, dark current plus straylight background),
- readout noise,
- arithmetic roundoff error,
- AD converter quantization error.

Attitude Measurement Accuracy Errors

The attitude accuracy is dependent on the measurement accuracy of single star position, number of stars available for attitude measurement and their position in the FOV. The accuracy of the attitude determination typically can be assessed through

Table 9.1 Typical Star tracker error accuracy summary

	Bias error arcsec	Low Freq. error arcsec	Random error arcsec at 0.5 deg/sec
Boresight axis	10	20	70
Pitch/Yaw axis	10	3	7

Montecarlo simulations by using a SW simulator which generates stars pattern within the FOV as would be seen by the sensor.

Table 9.1 shows a typical star tracker accuracy performance for a medium class sensor considering the different error classes. The values reported are worst case or maximum guaranteed error.

9.2.6 Magnetometer

The magnetometers are commonly used for the attitude determination in LEO and MEO. They measure both the direction and the magnitude of the magnetic field. This sensor requires low power in input, it is lightweight and highly reliable. It can operate over a wide range of temperatures and has no moving parts. Magnetometers are not very accurate as attitude sensors since the magnetic field in space is not precisely known and changes also in function of the solar activity.

Magnetic field sensor outputs are processed by the electronic unit generally integrated in a single box with the detector, then the data output is forwarded to the avionic attitude determination software. In order to determine the spacecraft's orientation, the signals that represent the measured magnetic field in **BRF** are compared to the estimated reference magnetic field in **ORF**; this last information is obtained using the position of the satellite provided by the orbital propagator (see Chap. 4) and an Earth magnetic field model like the IGRF (international geomagnetic reference field).

The magnetometer's main function is to provide three analogue signals proportional to the three mutually perpendicular projections of the Earth's magnetic field vector.

In the commercial market we find many suppliers (e.g. IAI, Lusospace, Zarm, Goodrich, SSTL and others) that apply different designs and different technological principles. We will show in the next section a specific product example.

The geo-magnetic field measurement provided by the MGM is used for three main purposes:

- in safe mode, to damp out the high angular rates (e.g. to drive the B-dot control low);
- to support the use of the magnetotorquers (MGT) in order to define the MGT input to obtain a desired torque;
- to determine the attitude of the spacecraft together with the Sun's measurement (i.e. triad attitude determination).

Table 9.2 MGM main technical data: courtesy of Lusospace

Specification description	Value
Dynamic field range	±64000 nT
Noise	$\leq$3 nT/Hz$^{0.5}$ (3σ) in the bandwidth
Offset	$\leq$500 nT(3œ)
Linearity	<±0.5 %(Full scale:±64 000 nT)
Bandwidth (−3 dB)	50 Hz
Power consumption	<1W
Axis alignments	±1°(max.) and measurement accuracy of 0.2° (goal 0.1°)
Accuracy	Measurement accuracy <0.2°
Scale factor	10 v/Gauss
Reliability	>0.9997
Mission life	10 years life in LEO
Storage	10 Years on ground (testing and storage time)
Supply voltage	From ±13.95 V to ± 16.05 V
Output single range	−5.12 V to ±16.05 V
Test connector	Full scale
Output impedance	Typical operational amplifier
Mass	<500 g
Dimension **L** × **W** × **H**	100 × 83 × 75 mm (including mounting feet)
Mounting bolts	4 × M4

Different technologies can be used to design a magnetometer sensor for the space environment. The more used technologies are: the Hall effect, the fluxgate detector and the anisotropic magneto-resistive (AMR) detector (see [3] for details). The magnetometer sensor whose parameters are shown in Table 9.2 is based on the AMR technology.

The magneto-resistance effect was studied by William Thomson in the middle of the nineteenth century but only in this last 20 years after the development of the thin film technology, the anisotropic magneto-resistive (AMR) detector has been utilized in the development of magnetometer sensors.

Thomson found that a magnetic field applied to a ferromagnetic material produce a change in its electric resistivity. The resistivity is a function of the magnetic field magnitude and direction.

An AMR detector may be realized using nickel-iron magnetized thin film strips deposited on a silicon wafer. The detector of each axis is composed by four elements acting as resistors with a resistance sensible to the external magnetic field. The four elements are connected in a shaped Wheatstone bridge configuration as depicted in Fig. 9.12.

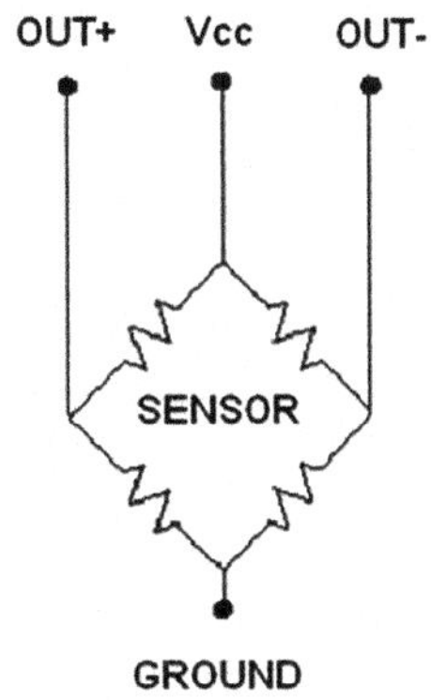

Fig. 9.12 MGM sensor bridge schematic

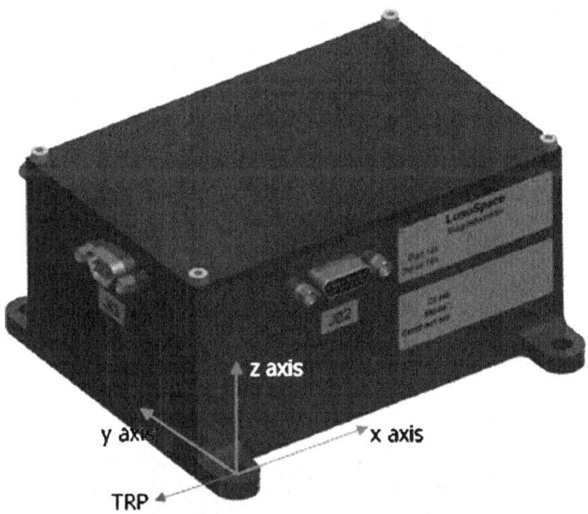

Fig. 9.13 MGM: courtesy of lusospace

Each MGM has three bridges, with orthogonal measurement axes, so that both magnitude and direction of the external magnetic field can be measured.

Special care in the design has to be taken due to a wide range of temperatures in which the MGM typically operates, since the characteristics of the materials and the electronic components may undergo great changes depending on the temperature of operation.

In order to improve the sensor accuracy the magnetometer is provided with a temperature sensor to allow the sensor output calibration by the on board SW. The technical data of the MGM developed by Lusospace is reported in Table 9.2.

The equipment is shown with its mechanical axes in Fig. 9.13.

9.2.7 GNSS Receiver

The GNSS assembly includes the GNSS receiver, the antenna and their interconnection cable. The satellite receives through the GNSS antenna the signals from a constellation of navigation satellites (e.g. GPS or Galileo), here we use the language of the GPS constellation, although for the Galileo constellation we have similar concepts.

Many space qualified receivers are available on the market from different qualified suppliers (e.g. Thales Alenia Space, Airbus Defence and Space, RUAG, SSTL, General Dynamics, JPL and others) with different performance and capabilities and costs.

One of the major capabilities is related to having a single (i.e. L1) or dual frequency (i.e. L1 and L2).

The described commercial GNSS receiver designed by Thales Alenia Space, shown in Fig. 9.14, has dual frequency RF channels for S/C applications able to determine position, velocity and time using the signals generated by the GNSS Constellations GPS and Galileo.

The receiver provides real-time navigation data obtained using all instantaneous measured ranges the so-called SPS solution which does not use any dynamic filtering. Another output is provided with improved accuracy—the NKF solution—using an orbital Kalman filter.

As described in Chap. 4, a precise orbit determination (POD) EKF designed for the specific custom application, implementable in the OBC orbital pre-processing S/W, can be used to additionally improve the real time knowledge of the orbit.

Fig. 9.14 GNSS Unit (without antenna): courtesy of thales alenia space

Table 9.3 GNSSR solution in-orbit accuracy

Typical: SOLUTION PERFORMANCES SPECIFICATION	
SPS Position Error (1σ 1D) in all directions (WGS-84 reference frame)	$\pm$20 m
NKF Position Error (1σ 1D) in all directions (WGS-84 reference frame)	$\pm$8 m
SPS Velocity Error (1σ1D)	$\pm$0.5 m/s
NKF Velocity Error (1σ1D)	$\pm$0.1 m/s
SPS Time Measurement Error (1σ, jitter on the Pulse per Second)	$\pm$0,35 μs
NKF Time Measurement Error (1σ, jitter on the Pulse per Second)	$\pm$0,25 μs

Each receiver is able to accurately measure pseudo-ranges and integrated carrier phases (raw data), to be later processed on ground for scientific or high precision purposes.

The receiver can process both L1 and L2 frequency bands, allowing compensation of ionospheric delays. A special codeless tracking scheme is implemented in order to process the encrypted P(Y) signals transmitted in the GNSS L2 frequency band.

The antenna has been designed to have optimal multipath rejection capability with narrow high-gain radiation pattern; moreover it is designed to optimize the navigation satellites' visibility.

The most common receivers are equipped with a redundant MIL-STD-1553 communication interface (RT configuration) for commanding and acquisition of the measured data and housekeeping telemetry by the OBC.

Table 9.3 reports typical GNSS receiver performance for the raw SPS solution and for the NKF solution. Precise orbit determination POD EKF typical error dynamics have been shown in Figs. 4.3 and 4.4.

9.3 Actuators

Actuators such as reaction/momentum wheels, thrusters, control moment gyros and magneto-torquers, provide a torque to the spacecraft for AOC control. The decision to use one actuator instead of another depends on mission pointing and stability requirements, power requirements, weight restrictions, mission orbital characteristics, and the control system's response time.

The mechanical effect of these actuators and their input command on the satellite's dynamics is provided in Sect. 3.7 for RW and CMGs, and in Chap. 8 for the thrusters. In this section different types of actuators are shown, identifying the mechanism and typical performance behind such units.

9.3.1 Reaction and Momentum Wheels

Reaction and momentum wheels use flywheels whose spin axes are fixed in the **BRF**. These actuators are without any doubt the most used primary actuators for all AOC systems.

The basic difference between reaction wheels (RW) and momentum wheels (MW) is the nominal angular velocity of the flywheel. While reaction wheels typically have zero nominal angular velocity, momentum wheels spin at a predefined nominal angular velocity (i.e. nominal momentum bias) and operate with small variations around this reference velocity. The AOCS changes the wheel velocity to absorb the external disturbance torques or to exchange momentum with the satellite in order to implement a slew maneuver. RWs and MWs cannot increase indefinitely their own angular velocity due to physical and design limits (i.e. *maximum safe wheel speed*). In fact once the *maximum operating speed* is reached, it is necessary to apply an external control torque to the satellite—with RCTs or MGTs—to have a *momentum unloading* to recover the nominal wheel operating velocity (i.e. zero or the predefined angular velocity for MWs).

The AOCS typically requires the usage of an assembly of four or more reaction wheels (RW) to apply a continuous 3D fine attitude control over the orbit during nominal operations, to react to the environmental disturbance torque and to implement an agile steering capability. The wheels will also be in charge of keeping a constant attitude during RCT firing in correspondence with orbit maneuvers, to counteract the unavoidable disturbance torques deriving from the variable position of the center of mass (CoM) with respect to the fixed RCT alignment.

The assembly of reaction wheels must fulfill conflicting requirements such as:

- actuation of fine control torques;
- sufficient angular momentum storage capacity;
- responsive reaction torque to external disturbance;
- high reliability and Resilience to failures.

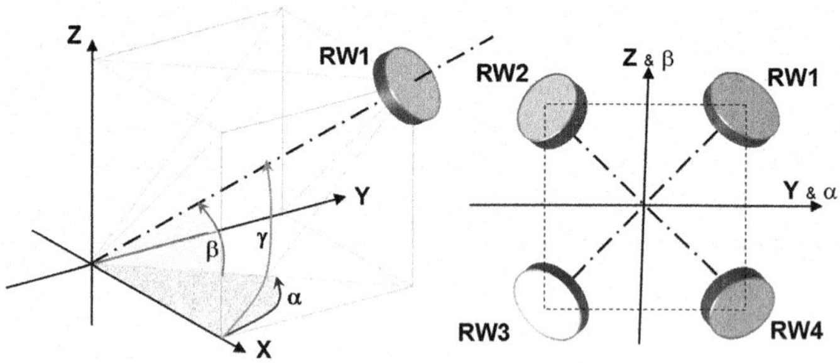

Fig. 9.15 RWs configuration

Table 9.4 RW alignment unit vectors with $|\alpha| = 45°$ and $|\beta| = 35.264°$

			X_{BRF}	Y_{BRF}	Z_{BRF}
RW1	α	β	$\cos\alpha\cos\beta = \frac{1}{\sqrt{3}}$	$\sin\alpha\cos\beta = \frac{1}{\sqrt{3}}$	$\sin\beta = \frac{1}{\sqrt{3}}$
RW2	$-\alpha$	β	$\cos\alpha\cos\beta = \frac{1}{\sqrt{3}}$	$\sin\alpha\cos\beta = -\frac{1}{\sqrt{3}}$	$\sin\beta = \frac{1}{\sqrt{3}}$
RW3	$-\alpha$	$-\beta$	$\cos\alpha\cos\beta = \frac{1}{\sqrt{3}}$	$\sin\alpha\cos\beta = -\frac{1}{\sqrt{3}}$	$\sin\beta = -\frac{1}{\sqrt{3}}$
RW4	α	$-\beta$	$\cos\alpha\cos\beta = \frac{1}{\sqrt{3}}$	$\sin\alpha\cos\beta = \frac{1}{\sqrt{3}}$	$\sin\beta = -\frac{1}{\sqrt{3}}$

A possible solution to these needs is a reaction wheel assembly (RWA) comprising four independent reaction wheels arranged in a typical pyramidal configuration (see Fig. 9.15).

The four RWs can be mounted in a square pyramidal configuration with the pyramid axis along the $+X_{BRF}$ axis. The pyramid's inclination angle is $\alpha = 45°$, where α is the angle between the X_{BRF} axis and the wheel spin axis projection on both the X_{BRF}/Y_{BRF} and the X_{BRF}/Z_{BRF} planes (i.e. all the wheel axis components have the same magnitude: $|u_x| = |u_y| = |u_z| = \frac{1}{\sqrt{3}}$). Calling β the RW elevation angle with respect to the X_{BRF}/Y_{BRF} plane, the RWs unit vectors are given in Table 9.4.

Therefore the β angle is such that: $\sin\beta = \cos\alpha\cos\beta$, thus $\beta = \tan^{-1}(\cos\alpha) = 35.264$. Moreover the angle γ between the wheel spin axis and the X_{BRF} axis is $\gamma = \cos^{-1}(\cos\alpha\cos\beta) = 54.736$, and the pyramid square base is aligned to the Y_{BRF}/Z_{BRF} plane.

This solution provides good torque and momentum capability in any S/C direction either by the simultaneous use of four wheels (nominal configuration) or by the use of only three of them (degraded configuration) in case of a single wheel failure (see Figs. 9.16, 9.17).

In the commercial market we find many suppliers (e.g. Honeywell, Rockwell, Goodrich, IAI, Moog Bradford, SSTL and others) implementing slightly different design solutions. Some new technological approaches like magnetic bearings have not yet developed into real market products. Every product can be found in different wheel and motor sizes in order to cover a wide class of mission specific momentum and torque values.

In the commercial reaction wheel solution shown in Fig. 9.18 the flywheel mass is designed as a spoke wheel, consisting of a hub, five inclined spokes, and a mass rim. The flywheel design has been optimized to give the highest possible ratio of moment of inertia to weight.

The wheel and the motor rotor are mounted to the bearing unit and balanced as an assembly. Correction of static and dynamic unbalance is carried out by adding

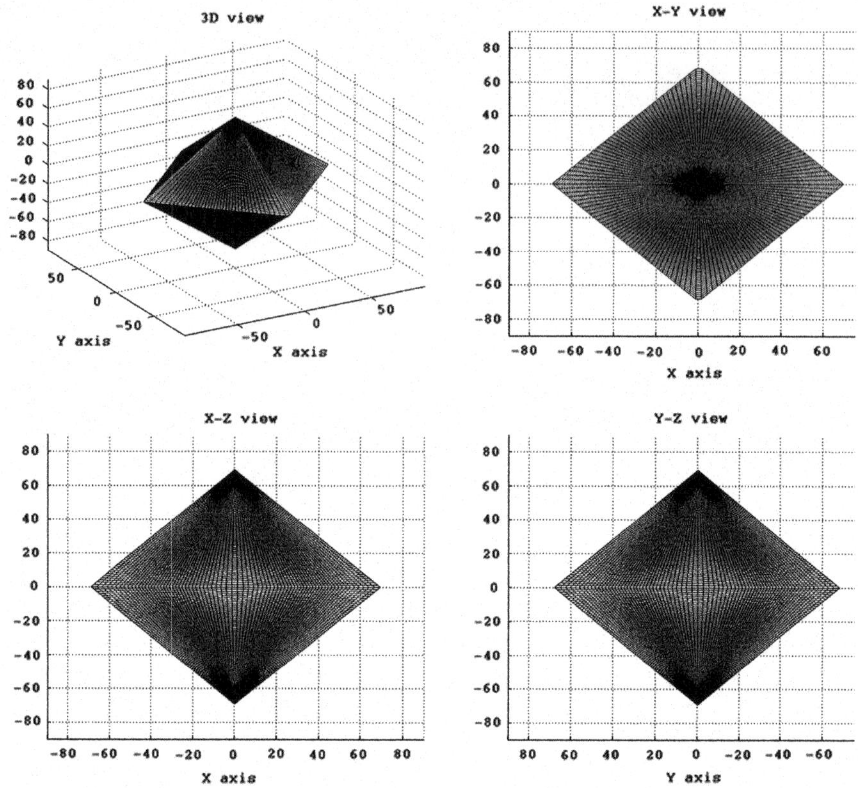

Fig. 9.16 RW angular momentum availability in nominal configuration

balance screws to the inner side of the rim in two separated planes. The bearing unit has to cope with multiple zero crossings. The typical number of zero crossing requested is about 200000 or more zero crossings successfully accumulated during dedicated life tests.

Figure 9.16 shows the envelope of the available angular momentum in the different directions assuming the nominal configuration and that each reaction wheel does not exceed 30 Nms. The envelope may obviously be scaled taking into account different values for the angular momentum available for each reaction wheel.

Each reaction wheel can provide an angular momentum of up to 30 Nms and considering the degraded configuration (three wheels, see Fig. 9.17), in the worst case direction there is an angular momentum of 24.5 Nms, whereas in the coordinate axes directions there is an angular momentum of 34.6 Nms.

When the 4 wheels operate all together, the transfer matrix from the wheels axes to the **BRF** axes is a 3×4 matrix that has a one dimensional kernel, identified by the vector of wheel commanded torques which provides a null torque effect on the satellite. The matrix which transforms the torques from the wheel axes to **BRF** axes is the same that transforms the momenta. Therefore, we have a one dimensional vector of non null wheel momenta which realizes a 0-bias momentum in **BRF**. We can

Table 9.5 Typical RW key performances

Parameter	Value	Unit
Angular momentum	$\pm 2 \ldots \pm 200$	Nms
Torque max levels	$\pm 0.04 \ldots \pm 0.4$	Nm
Friction at max speed	$<0.002 \ldots <0.02$	Nm
Tacho Pulses per revolution	8/18/24/72	
Static Unbalance	<2	gcm
Dynamic Unbalance	<20	gcm^2

bias the wheel momenta along the kernel direction without any visible effect on the satellite dynamics, creating a favorable condition for the wheels life and reliability by minimizing low speed periods and zero crossings.

Table 9.5 shows key performance parameters for different classes of commercial reaction wheels. The large spread of values depend on the different classes of satellites and system performances.

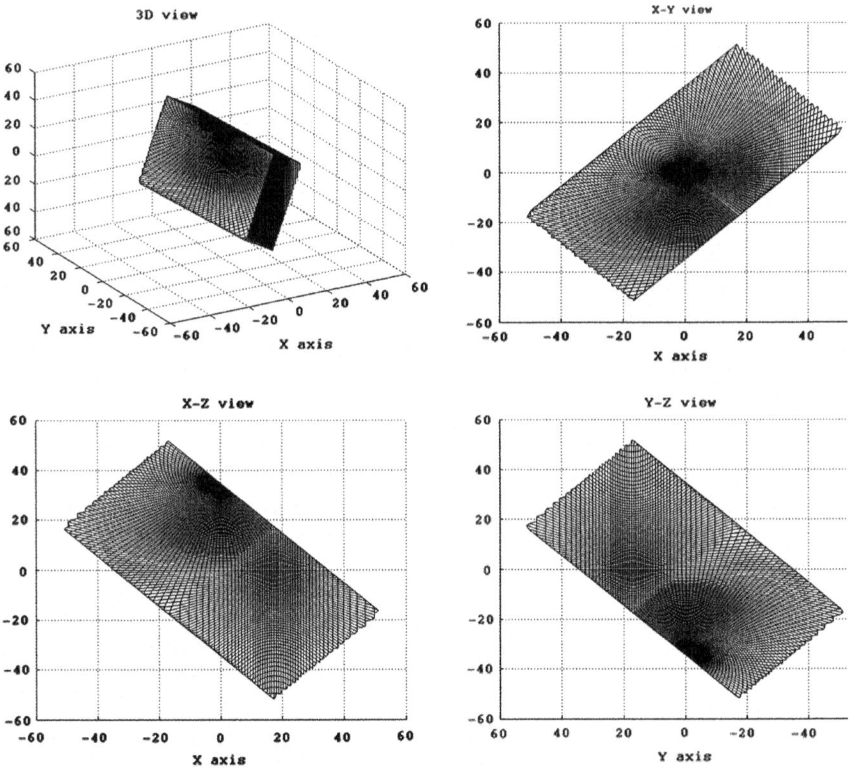

Fig. 9.17 RW angular momentum availability in degraded configuration

Fig. 9.18 Reaction wheel: courtesy of rockwell collins deutschland GmbH

9.3.2 Control Moment Gyros

The control moment gyro (CMG) is a momentum management system device, able to provide, with relatively low power and mass figures, extremely high torques to the satellite if compared with reaction wheels. A CMG contains a spinning rotor (flywheel) at constant or variable speed, the rotor spinning axis can be reoriented using a gimbal. The gyroscopic reaction torque produced by the CMG gimbal rotation is orthogonal both to the rotor spin axis and to the gimbal axis (see Fig. 9.19, Sect. 3.7 and Eqs. 3.56, 3.78).

The most widely used type of CMGs for satellites is the single gimbal CMG (SGCMG); dual gimbal CMG and variable speed CMG (VSCMG) have been used and are subject of research.

A SGCMG is a CMG with a constant speed momentum wheel, gimballed in one axis. The type and number of CMGs that can be used in an AOC system is a trade off result between performance, cost, mechanical and algorithmic complexity.

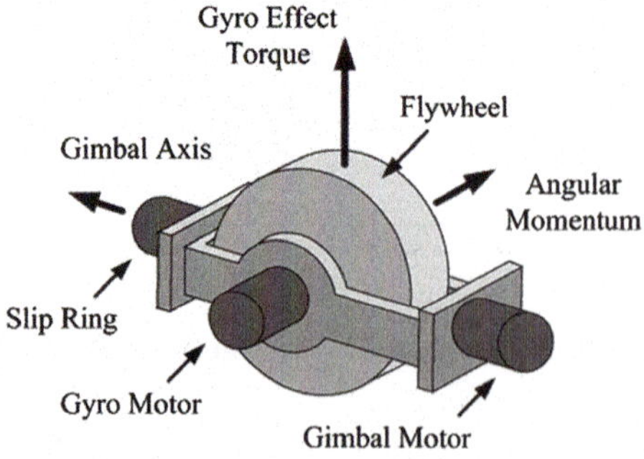

Fig. 9.19 Conceptual CMG unit

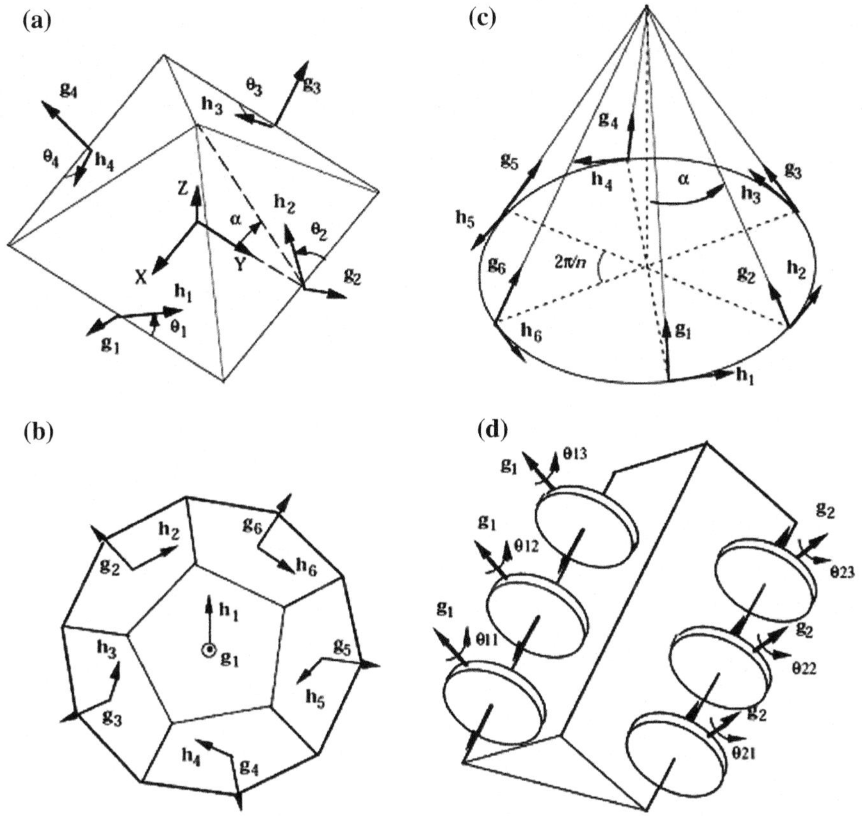

Fig. 9.20 CMG system types. *Credit* Dr. H. Kurokawa [5]

SGCMGs require a minimum of four units to implement a failure tolerant 3-axis control system. A 4 SGCMG configuration map the 4D gimbal space into the 3D momentum space, the Jacobian of this mapping has almost everywhere rank 3 but exist surfaces in the gimbal space where the rank of the Jacobian becomes 2. Such *singular surfaces* can be escaped under certain conditions with specific steering laws (see for example [4]) that exploit the so-called null motion.

SGCMGs have been systematically studied in the past and have been used in few space missions where satellite agility is one of the key requirements.

CMGs and agility change the way in which we operate the satellites, increasing the services and the capability to collect more data in the satellite access area.

In the commercial market we find few suppliers (e.g. Honeywell, Airbus Defence and Space, Thales Alenia Space, L3 and few others) implementing different design solutions. Being this product used mainly for very agile applications it is normally quite costly and available for specific classes of platforms.

Typical SGCMG configurations have certain kinds of symmetries, which can be classified into two types: *Independent* and *Multiple*. In Fig. 9.20 some different types of architectures are illustrated, h is the momentum axis which can be rotated around the gimbal axis g. Independent type CMGs have no pairs with parallel gimbal axis.

- *Independent types*: Two categories of independent type CMGs have been the main subject of study: *symmetric types* and *skew types*. In the symmetric type the gimbal axes are arranged symmetrically according to a regular polyhedron. The three, four and six unit systems are denoted as S(3), S(4) and S(6). Possible configurations of this type are three, four, six and ten unit systems, because only surfaces not parallel to each other are considered. The four unit or S(4) system, shown in Fig. 9.20, is called the symmetric *pyramid type*. In the skew type all individual units are arranged in symmetrically around a certain axis.
- *Multiple types*: In this type some units have identical gimbal directions. These arrays are denoted as $M(m_1, m_2, m_3, \ldots)$ hereafter, where m_i is the number of the units with the same gimbal direction g_i.

It is demonstrated in [5] that any independent type system has internal impassable singular surfaces, while multiple type systems with more than five units have no internal impassable singular surfaces. In order to use a minimum number of units (typically 4 units), we need some method to manage the presence of these impassable singular surfaces. The methods proposed in the literature are (see [5]):

- To implement on board a momentum steering which produces the requested torque with some torque error in order to by- pass the singular surfaces. This method produces attitude transients whose acceptability must be evaluated.
- To define in batch (i.e. at the ground station) before any attitude maneuver an angular momentum path plan using the null motion to configure the unit before each maneuver so to avoid the singular surfaces. This method is particularly complex to realize, the gimbal rotation range results to be unlimited but the method can provide a more optimized use of the CMG.
- To restrict the momentum workspace constraining the momentum to operate in a domain without singularities. This method reduces the CMG capability in favour

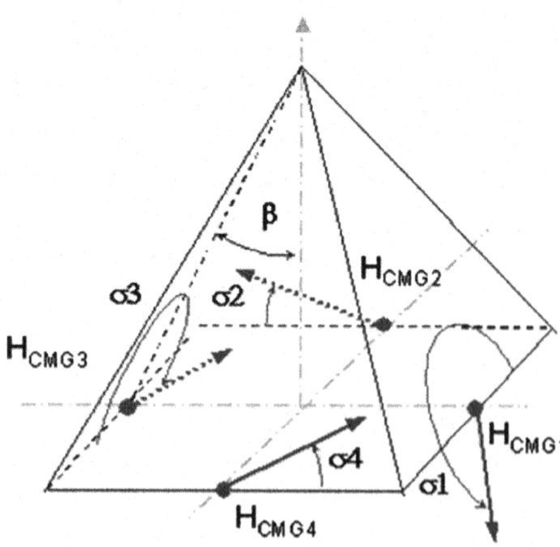

Fig. 9.21 CMGs pyramid configuration

Table 9.6 Typical commercial CMG key performances

Parameter	Value	Unit
Angular momentum	$\pm 2 \ldots \pm 75$	Nms
Torque max levels	$\pm 4 \ldots \pm 75$	Nm
Gimbal rotation	>1 Milion	
Gimbal rotation accuracy	<0.001 ... 3.6	o
Static unbalance	<1	gcm

of simplicity. Momentum and gimbal angles being defined in a one to one relationship, the gimbal rotation range will be limited with possible advantages in the mechanical design of the CMG.

The best design approach is probably to implement a mix of these techniques to be defined for each specific application and CMG configuration.

The *pyramid type* is a classical disposition, where the gimbals of each CMG are placed on a pyramid. The faces of the pyramid represent the plane on which each CMG can give its torque. Depending on the angle β of the pyramid (see Fig. 9.21), this configuration allows for more torque in the axis of the pyramid or in the orthogonal directions.

The following Table 9.6 shows key performance parameters for different classes of control momentum gyro unit. The large spread of values depend on the different classes of satellite or system performances.

9.3.3 Magnetic Torquers

The LEO and MEO missions typically require the use of a magneto-torquer (MGT) assembly, composed by two or three magnetic torque-rods each including, in a single housing, two coils commanded in cold redundancy (i.e. one at time). The MGT, with a proper orientation within the satellite, will be used by the AOC system to provide torques on the three **BRF** axes to implement a magnetic safe mode (i.e. *B-dot* algorithm shown in Chap. 4) or for *momentum management systems* momentum unloading purposes.

In the commercial market we find many suppliers (e.g. Andrews Space, IAI-Tamam, Zarm, Goodrich, SSTL and many others) implementing different size of torque-rods.

The MGT torque is produced by the magnetic dipole moment of the torque-rods interacting with the Earth's magnetic field, therefore it is impossible to produce a torque vector co-linear to the Earth's magnetic field. Moreover, to properly generate torque, the knowledge of the Earth magnetic field vector B is needed, and it has to be provided by the magnetometer (MGM) sensor or through the knowledge of the current attitude and position combined with the availability of an Earth magnetic field model.

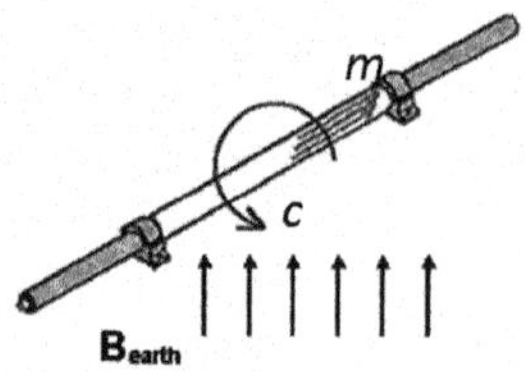

$$c = m \times B$$

Fig. 9.22 MGT principle of operation

Since the control torque c_c is normal to the plane defined by m-MGT dipole momentum- and B (see Fig. 9.22), the component of m parallel to B does not produce any torque, although it requires electrical power from the MGT driver. For this reason in the safe mode control scheme based on a classic B-dot controller, before outputting the physical command to the MGT driver, we need to subtract from m the component parallel to B, in order to always command m normal to the B vector.

When torque-rods are used to unload a certain Δh momentum from the RWs the dipole momentum is computed normal to the B vector with the following law: $m = k\dfrac{\Delta h \times B}{|B|}$.

The AOCS controller needs to know the Earth magnetic field in S/C axes to compute the dipole momentum in order to generate the proper torque.

The MGT actuation interferes with MGM measurements. Therefore when the MGM is working, the MGT is switched off, performing a predefined duty cycle according to the needs of specific AOC modes.

The dipole momentum for wheel unloading is computed as $m = k\dfrac{\Delta h \times B}{|B|}$ and the corresponding control torque $c_c = m \times B = k\dfrac{\Delta h \times B}{|B|} \times B$ is feed-forward compensated on the wheels to minimize the disturbance effects on the attitude.

Torque-rods consist of a solenoid wrapped around a ferromagnetic core. The material chosen for the core is very important. In order to provide a large dipole momentum with low weight, the material must have a high saturation point. In order to ensure efficiency, the material must also have a high permeability. The hysteresis curve is another critical property of the material (see Fig. 9.23).

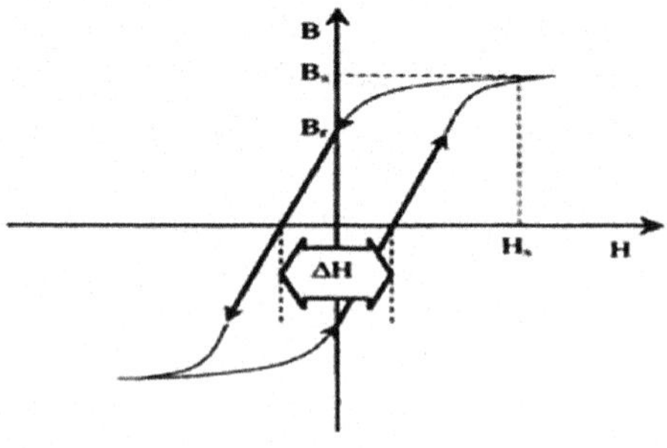

Fig. 9.23 MGT core material hysteresis

Table 9.7 MGT Requirements

Parameter	Performances
Linear dipole range	$\pm 300\,\text{Am}^2$
Linear power consumption	$<7.5\,\text{W}$
Linear supply voltage range	$\pm 12\,\text{V}$
Scale factor stability	$\pm 10\,\%$
Non linear voltage	$\pm 16\,\text{V}$
Non linear dipole range	$\pm 375\,\text{Am}^2$
Residual dipole	$\pm 1.8\,\text{Am}^2 (3\sigma)$
Dipole step response time	$<200\,\text{ms}$
No. of coils	2

The curve illustrates that when H is increased to saturation and decreased back to zero, there is still a substantial B field remaining (and consequently a dipole moment) denoted by B_r.

The design drivers of a torque-rod are an high H-mass ratio and a good linear relationship between the input current and dipole moment.

Since the B field varies with the H field within the material, the inductance is not constant. Therefore, it is enough to be able to calculate the average inductance for switching the direction of the moment. This is easily calculated as the change in magnetic flux divided by the change in current.

The area within the hysteresis loop is equal to the energy loss per time cycle per unit of the volume of the core.

The three torque-rods are typically placed in a three orthogonal configuration (three mutually perpendicular axes). The position of the torque-rods on the S/C is chosen to minimize the effects of mutual interaction among the acting fields and of the stray magnetic fields produced by the torque-rods.

The large static magnetic fields induced by the torque-rods could be critical for sensitive items within the spacecraft or produce the magnetization of other spacecraft metallic parts (i.e. tanks), therefore the sensibility of each satellite item to magnetic field should be verified and the location of such items will be chosen to be compatible with the field levels determined by the rods' design according to the analysis/test data on transverse-axial flux densities provided by the supplier. The mutual interaction among the torque-rods has to be considered as well, in order to avoid two rods being placed in the same corner too close to each other.

In Table 9.7 the main performance parameters of a commercial MGT are summarized.

References

1. M.D. Shuster, S.D. Oh, Three-axis attitude determination from vector observations. J. Guidance Control **4**(1), 70–77 (1981)
2. M.L. Psiaki, Extended quest attitude determination filtering. In *Flight Mechanics Symposium*, NASA Goddard Space Flight Center (1999)
3. P. Ripka, *Magnetic Sensors and Magnetometers* (Artech House, 2001)
4. B. Wie, Singularity analysis and visualization for single-gimbal control moment Gyro systems. J. Guidance, Control Dyn. **27**(2), (2004)
5. H. Kurokawa, *A Geometric Study of Single Gimbal Control Moment Gyros. Singularity problems and Steering laws*, Mechanical Engineering Laboratory, Agency of Industrial Technology and Science, Japan, Report No. 175 (1998)

Appendix A
Complements to Body Kinematics

A.1 Integration in the Body Domain

The kinematic model is a regular application from the reference configuration to the actual configuration of the body, that is: $r_i \rightarrow x_i$.

An example is the transformation Eq. 2.14:

$$x_i = x_i^o + M_{ij}(\theta)(r_j + \phi_j^k(r)\varepsilon_k).$$

The integrals that define the kinetic energy, the potential energy and the mass properties are of the form:

$$\int_V \xi(r)\rho(r)dV$$

where $dV = dx_1dx_2dx_3$ and $\xi(r)$, is a generic function.

We can integrate in $dV_o = dr_1dr_2dr_3$ using the Jacobian of the transformation, provided that the transformation is regular and the Jacobian non singular in the domain of integration:

$$\int_V \xi(r)\rho(r)dV = \int_{V_o} \xi(r)|det\left(\frac{\partial x_i}{\partial r_j}\right)|\rho(r)dV_o$$

The term $|det\left(\frac{\partial x_i}{\partial r_j}\right)|\rho(r)$ is equal to $\rho_o(r)$ due to the invariance of the mass contained in any volume subjected to the given transformation. Therefore in conclusion we have the equivalence:

$$\int_V \xi(r)\rho(r)dV = \int_{V_o} \xi(r)\rho_o(r)dV_o$$

© Springer International Publishing Switzerland 2016

L. Mazzini, *Flexible Spacecraft Dynamics, Control and Guidance*,
Springer Aerospace Technology, DOI 10.1007/978-3-319-25540-8

The regularity of the transformation is guaranteed for small enough deformations, since

$$|det\left(\frac{\partial x_i}{\partial r_j}\right)| = |det[\delta_{ij} + \frac{\partial \phi_{ij}^k}{\partial r_j}\varepsilon_k]| \cong 1 + \Phi_{s,s}^k \varepsilon_k$$

A.2 Some Properties of the Rotation Matrix

We will derive in this Appendix some useful properties of the rotation matrix M and the μ matrix derivatives. We synthesize these properties in the formulas A2.1 and A2.2.

A2.1

A2.1 has various formulations all equivalent:

$$\frac{\partial M_{es}}{\partial \theta_k} = M_{en} e_{nms} \mu_{mk} \Rightarrow \mu_{qk}^{\#} \frac{\partial M_{es}}{\partial \theta_k} = M_{en} e_{nqs} \Rightarrow M_{en} \frac{\partial M_{es}}{\partial \theta_k} = e_{nms} \mu_{mk}$$

The demonstration is simple and starts from the definition of the matrices M and μ and the vector ω.

$$\dot{M}_{ij} = M_{ie} e_{esj} \omega_s = \frac{\partial M_{ij}}{\partial \theta_k} \dot{\theta}_k = M_{ie} e_{esj} \mu_{sk} \dot{\theta}_k \quad \forall \dot{\theta}_k$$

$$\Rightarrow \quad \frac{\partial M_{es}}{\partial \theta_k} = M_{en} e_{nms} \mu_{mk} \quad \Rightarrow \mu_{qk}^{\#} \frac{\partial M_{es}}{\partial \theta_k} = M_{en} e_{nqs}$$

The matrix $\mu^{\#}$ is defined by $\mu^{\#} = \mu^{-1^T}$ (see Eq. 3.11).

A2.2

We want to demonstrate in general the relationship:

$$\mu_{qk}^{\#} \left(\frac{\partial \mu_{sk}}{\partial \theta_e} - \frac{\partial \mu_{se}}{\partial \theta_k}\right) \dot{\theta}_e = e_{qns} \omega_n$$

Before implementing the general demonstration we will see a specific example. Let us introduce the Euler angles ϕ, ϑ, ψ relevant to the three counterclockwise rotations (3, 1, 3) which bring **ORF** into **BRF**. From Sect. 2.3.2.2 we have:

$$\omega_s = \mu_{sk} \dot{\theta}_k$$

where $\theta_1 = \varphi, \theta_2 = \vartheta, \theta_3 = \psi$ and

$$\mu_{sk} = \begin{pmatrix} \sin\vartheta\sin\psi & \cos\psi & 0 \\ \sin\vartheta\cos\psi & -\sin\psi & 0 \\ \cos\vartheta & 0 & 1 \end{pmatrix}$$

the columns of μ_{sk} are, in order, the unit vector of the inertial Z_{ORF} axis, the unit vector of the line of the nodes between the X/Y **ORF** and the X/Y **BRF** planes, the unit vector of the body Z_{BRF} axis all represented in the **BRF** triad (same as ω_s). These unit vectors define a not-orthogonal triad that we call in the following the μ triad.

We then calculate the derivatives of μ_{sk} (here—misusing the indicial notation- for $\mu_{s1}, \mu_{s2}, \mu_{s3}$ we intend the first, second and third column vectors of the matrix μ):

$$\frac{\partial \mu_{s1}}{\partial \theta_1} = 0; \quad \frac{\partial \mu_{s1}}{\partial \theta_2} = \begin{bmatrix} \cos\vartheta\sin\psi \\ \cos\vartheta\cos\psi \\ -\sin\vartheta \end{bmatrix} = \mu_{s1} \times \mu_{s2};$$

$$\frac{\partial \mu_{s1}}{\partial \theta_3} = \begin{bmatrix} \sin\vartheta\cos\psi \\ -\sin\vartheta\sin\psi \\ 0 \end{bmatrix} = \mu_{s1} \times \mu_{s3}$$

$$\frac{\partial \mu_{s2}}{\partial \theta_1} = 0; \quad \frac{\partial \mu_{s2}}{\partial \theta_2} = 0; \quad \frac{\partial \mu_{s2}}{\partial \theta_3} = \begin{bmatrix} -\sin\psi \\ -\cos\psi \\ 0 \end{bmatrix} = \mu_{s2} \times \mu_{s3}$$

$$\frac{\partial \mu_{s3}}{\partial \theta_1} = 0; \quad \frac{\partial \mu_{s3}}{\partial \theta_2} = 0; \quad \frac{\partial \mu_{s3}}{\partial \theta_3} = 0$$

Analyzing these equations it becomes clear that:

$$\frac{\partial \mu_{sk}}{\partial \theta_e} - \frac{\partial \mu_{se}}{\partial \theta_k} = \mu_{sk} \times \mu_{se}$$

This result is equivalent to A2.2, in fact switching back to the correct indicial notation we can write for this local parametrization of $SO(3)$,

$$\mu^{\#}_{qk}\left(\frac{\partial \mu_{sk}}{\partial \theta_e} - \frac{\partial \mu_{se}}{\partial \theta_k}\right)\dot{\theta}_e = \mu^{\#}_{qk} e_{smn} \mu_{mk} \mu_{ne} \dot{\theta}_e = e_{qns}\omega_n$$

Now we produce the general demonstration of A2.2, regardless of the choice of the $SO(3)$ parameters.

The following analytical identities start from one of the A2.1 and can be derived sequentially by derivation, summation or index rotation of the previous:

$$M_{la}\frac{\partial M_{ls}}{\partial\theta_k} = e_{ams}\mu_{mk}$$

$$\frac{\partial M_{la}}{\partial\theta_b}\frac{\partial M_{ls}}{\partial\theta_k} - M_{la}\frac{\partial M_{ls}}{\partial\theta_k\partial\theta_b} = e_{ams}\frac{\partial\mu_{mk}}{\partial\theta_b};$$
$$\frac{\partial M_{la}}{\partial\theta_k}\frac{\partial M_{ls}}{\partial\theta_b} - M_{la}\frac{\partial M_{ls}}{\partial\theta_k\partial\theta_b} = e_{ams}\frac{\partial\mu_{mb}}{\partial\theta_k};$$

$$\frac{\partial M_{la}}{\partial\theta_b}\frac{\partial M_{ls}}{\partial\theta_k} - \frac{\partial M_{la}}{\partial\theta_k}\frac{\partial M_{ls}}{\partial\theta_b} = e_{ams}(\frac{\partial\mu_{mk}}{\partial\theta_b} - \frac{\partial\mu_{mb}}{\partial\theta_k})$$
$$= M_{lp}e_{pia}\mu_{ib}M_{lq}e_{qjs}\mu_{jk} - M_{lp}e_{pia}\mu_{ik}M_{lq}e_{qjs}\mu_{jb}$$
$$e_{ria}e_{rjs}(\mu_{ib}\mu_{jk} - \mu_{ik}\mu_{jb}) = e_{ams}(\frac{\partial\mu_{mk}}{\partial\theta_b} - \frac{\partial\mu_{mb}}{\partial\theta_k})$$

By using the rules for simplifying the scalar and the vector products one can easily derive that:

$$e_{ria}e_{rjs} = \delta_{ij}\delta_{as} - \delta_{is}\delta_{aj}$$

therefore the sequence of equations ends in:

$$\mu_{sk}\mu_{ab} - \mu_{sb}\mu_{ak} = e_{ams}(\frac{\partial\mu_{mk}}{\partial\theta_b} - \frac{\partial\mu_{mb}}{\partial\theta_k})$$

By multiplying for $\dot{\theta}_b$ and $\mu^{\#}_{ck}$ and using the definition of ω we have:

$$\delta_{cs}\omega_a - \delta_{ca}\omega_s = e_{ams}\mu^{\#}_{ck}(\frac{\partial\mu_{mk}}{\partial\theta_b} - \frac{\partial\mu_{mb}}{\partial\theta_k})\dot{\theta}_b$$

If now we multiply for e_{ras} and we sum over the subscripts a and s, considering that:

$$e_{ras}(\delta_{cs}\omega_a - \delta_{ca}\omega_s) = e_{rac}\omega_a - e_{rcs}\omega_s = 2e_{rac}\omega_a$$
$$e_{ras}e_{ams} = -e_{ars}e_{ams} = -\delta_{rm}\delta_{ss} + \delta_{rs}\delta_{ms} = (-3+1)\delta_{rm} = -2\delta_{rm}$$

it follows:

$$e_{rac}\omega_a = -\delta_{rm}\mu^{\#}_{ck}(\frac{\partial\mu_{mk}}{\partial\theta_b} - \frac{\partial\mu_{mb}}{\partial\theta_k})\dot{\theta}_b$$

or

$$\mu^{\#}_{ck}\left(\frac{\partial \mu_{rk}}{\partial \theta_b} - \frac{\partial \mu_{rb}}{\partial \theta_k}\right)\dot{\theta}_b = e_{car}\omega_a$$

This final equation is valid for all parametrizations of $SO(3)$.

Appendix B
Basics of Lagrangian Equations

B.1 The Derivation of the Lagrangian Equations

Let us consider a system of N particles. If the kth particle of mass m_k specified by the position vector r_k is subject to a force f_k, the D'Alambert's principle states for the entire system that

$$\sum_k (m \frac{d^2 r_k}{dt^2} - f_k) \cdot \delta r_k = 0, \ \forall \, \delta r_k, k = 1, \ldots, N$$

We use here $\cdot$ to indicate the scalar product in the 3D space ($f_k{}^T \delta r_k = f_k \cdot \delta r_k$).

We assume that the r_k, $k = 1, \ldots, N$ are determined by a minimal set of generalized coordinates $q_j, j = 1, \ldots, n$, with $n < N$, such that the following function exists:

$$r_k = r_k(q_1, q_2, \ldots, q_n, t), \ k = 1, \ldots, N$$

The velocity vector v_k of the generic particle can be expressed through the derivatives of the generalized coordinates $\dot{q}_i$ by the formula:

$$v_k = \sum_i \frac{\partial r_k}{\partial q_i} \dot{q}_i + \frac{\partial r_k}{\partial t}$$

At the same time the arbitrary virtual displacement r_k can be connected with the virtual displacements δq_k by

$$\delta r_k = \sum_j \frac{\partial r_k}{\partial q_j} \delta q_j.$$

© Springer International Publishing Switzerland 2016
L. Mazzini, *Flexible Spacecraft Dynamics, Control and Guidance*,
Springer Aerospace Technology, DOI 10.1007/978-3-319-25540-8

No variation of time δt is involved here since a virtual displacement by definition considers only displacements of the coordinates. In terms of the generalized coordinates, the virtual work of the of the f_k becomes:

$$\sum_k f_k \cdot \delta r_k = \sum_{k,j} f_k \cdot \frac{\partial r_k}{\partial q_j} \delta q_j = \sum_j Q_j \delta q_j$$

where the Q_j are called the components of the generalized force, defined as:

$$Q_j = \sum_k f_k \cdot \frac{\partial r_k}{\partial q_j}$$

The virtual work performed by the inertial forces is:

$$\sum_{k,j} m_k \ddot{r}_k \cdot \frac{\partial r_k}{\partial q_j} \delta q_j$$

Let us now consider the relation

$$\sum_k m_k \ddot{r}_k \cdot \frac{\partial r_k}{\partial q_j} = \sum_k [\frac{d}{dt}(m_k \dot{r}_k \cdot \frac{\partial r_k}{\partial q_j}) - m_k \dot{r}_k \cdot \frac{d}{dt}(\frac{\partial r_k}{\partial q_j})] \tag{B.1}$$

In the last term of Eq. (B.1) we can interchange the differentiation with respect to t and q_j:

$$\frac{d}{dt}(\frac{\partial r_k}{\partial q_j}) = \frac{\partial \dot{r}_k}{\partial q_j} = \sum_i \frac{\partial^2 r_k}{\partial q_j \partial q_i} \dot{q}_i + \frac{\partial^2 r_k}{\partial q_j \partial t} - \frac{\partial v_k}{\partial q_j}$$

we also have that:

$$\frac{\partial v_k}{\partial \dot{q}_j} = \frac{\partial r_k}{\partial q_j}$$

Substitution of these quantities into Eq. B.1 leads to the result that:

$$\sum_k m_k \ddot{r}_k \cdot \frac{\partial r_k}{\partial q_j} = \sum_k [\frac{d}{dt}(m_k v_k \cdot \frac{\partial v_k}{\partial \dot{q}_j}) - m_k v_k \cdot (\frac{\partial v_k}{\partial q_j})]$$

D'Alembert's principle can now be rewritten as:

$$\sum_j \left\{ \frac{d}{dt}[\frac{\partial}{\partial \dot{q}_j}(\sum_i \frac{1}{2} m_i v_i \cdot v_i)] - \frac{\partial}{\partial q_j}(\sum_i \frac{1}{2} m_i v_i \cdot v_i) - Q_j \right\} \delta q_j = 0$$

Since the total kinetic energy is given by

$$\mathscr{T} = \sum_i \frac{1}{2} m_i (\frac{dr_k}{dt} \cdot \frac{dr_k}{dt}) = \sum_i \frac{1}{2} m_k v_k \cdot v_k$$

D'Alembert's principle becomes:

$$\sum_j \left\{ \left[\frac{d}{dt} \frac{\partial \mathscr{T}}{\partial \dot{q}_j} - \frac{\partial \mathscr{T}}{\partial q_j} \right] - Q_j \right\} \delta q_j = 0 \tag{B.2}$$

Since the virtual displacement δq_j is independent of δq_k the only way for Eq. (B.2) to hold is for the individual coefficients to vanish:

$$\frac{d}{dt} \frac{\partial \mathscr{T}}{\partial \dot{q}_j} - \frac{\partial \mathscr{T}}{\partial q_j} - Q_j = 0$$

Suppose that the forces are derivable from a scalar function $\mathscr{V}$, that we call Potential Energy, as follows:

$$f_i = -\nabla_i \mathscr{V}$$

Then the generalized forces can be written as:

$$Q_j = -\sum_i \nabla_i \mathscr{V} \cdot \frac{\partial r_i}{\partial q_j}$$

which is exactly the partial derivative of a function $\mathscr{V}(r_1, r_2, \ldots, r_N, t)$ with respect to q_j

$$Q_j = -\frac{\partial \mathscr{V}}{\partial q_j}$$

By defining now $\mathscr{L} = \mathscr{T} - \mathscr{V}$ we have:

$$\frac{d}{dt} \frac{\partial \mathscr{L}}{\partial \dot{q}_j} - \frac{\partial \mathscr{L}}{\partial q_j} = 0 \quad j = 1, \ldots, n$$

these equations are called Lagrangian equations.

If the forces are partly derived by a potential energy and partly not, we have:

$$\frac{d}{dt} \frac{\partial \mathscr{L}}{\partial \dot{q}_j} - \frac{\partial \mathscr{L}}{\partial q_j} = Q_j \quad j = 1, \ldots, n$$

where the Q_j are the generalized forces calculated only for the interactions not derivable from a scalar Potential Energy.

B.2 The Calculation of the Mass Properties

The mass properties of the Lagrangian equations will be derived here for the simple case of a mechanical system composed of a Rigid Body and a point mass attached to the rigid body with a spring.

We want to illustrate the basic properties of each symbol.

Let the origin of the kinematic model be a fixed point of the rigid body. The mass properties of this body are named: m_R, p_R, I_R, m_R

A point mass of mass m_e is positioned in r_e with respect to the origin and can move in the direction j of a displacement ε. The mass is attached to the point r_e with a spring of stiffness k.

Using the compact notation, being E_n the $n \times n$ Identity matrix we derive the following mass properties from the definitions:

$$\begin{array}{ll}
m = m_R + m_e; & p = p_R + m_e r_e \\
L = m_e j; & S = m_e(r_e \times j) \\
I = I_R + m_e(r_e^T r_e E_3 - r_e r_e^T); & \\
T = m_e; & J = m_e(2r_e^T j E_3 - r_e j^T - j r_e^T) \\
K = m_e(E_3 - jj^T); & P = 0
\end{array}$$

For the potential energy we get: $\mathcal{V} = 1/2k\varepsilon^2$

Let us introduce a new mode. The point mass is allowed to move in the two directions j_a and j_b with stiffnesses k_a and k_b.

$[a, b]$ denote a matrix whose columns are a and b. We have:

$$\begin{array}{ll}
m = m_R + m_e; & p = p_R + m_e r_e \\
L = m_e[j_a, j_b]; & S = m_e[(r_e \times j_a), (r_e \times j_b)] \\
I = I_R + m_e(r_e^T r_e E_3 - r_e r_e^T); & T = m_e E_2 \\
J^a = m_e(2r_e^T j_a E_3 - r_e j_a^T - j_a r_e^T); & J^b = m_e(2r_e^T j_b E_3 - r_e j_b^T - j_b r_e^T) \\
K^{ab} = m_e(j_a^T j_b E_3 - j_a j_b^T); & P^{ab} = m_e(j_a \times j_b)
\end{array}$$

Here K^{ab} is a 3×3 matrix and P^{ab} is a 3D vector. For the potential energy we get: $\mathcal{V} = 1/2k_a\varepsilon_a^2 + 1/2k_b\varepsilon_b^2$ It is worth noting that in this case the matrix P that accounts for the Coriolis coupling between rotation and deformation is different from zero.

Appendix C
Basics of Hamiltonian Systems

C.1 The Hamiltonian and Hamiltonian Equations

Let's consider a N-dof dynamical system. Let the equations of motion be expressed as $2N$ first-order homogeneous *Lagrange* equations.

We define a new set of $2N$ generalized variables:

- N generalized coordinates q_i,
- N generalized momenta $p_i = \frac{\partial \mathscr{L}(q_j, \dot{q}_j, t)}{\partial \dot{q}_i}$
  where $\mathscr{L}$ is the Lagrangian function.

If a function $H = H(q_i, p_i)$ exists such that the Lagrangian equations are equivalent to:

$$\begin{aligned} \dot{q}_i &= \frac{\partial H}{\partial p_i}, \quad i = 1, \ldots, N \\ \dot{p}_i &= -\frac{\partial H}{\partial q_i}, \quad i = 1, \ldots, N \end{aligned} \tag{C.1}$$

then the system is said to be a *Hamiltonian system*, H is called the *Hamiltonian function* and (q, p) are *canonical variables*.

Equation C.1 are the *Hamilton equations of motion*. The *Hamiltonian function* can be obtained from the Lagrangian by means of the *Legendre transformation*:

$$H(q, p, t) = \dot{q}_i p_i - \mathscr{L}(q, \dot{q}, t) \tag{C.2}$$

where:

$$p_i = \frac{\partial \mathscr{L}(q_j, \dot{q}_j, t)}{\partial \dot{q}_i} \tag{C.3}$$

© Springer International Publishing Switzerland 2016
L. Mazzini, *Flexible Spacecraft Dynamics, Control and Guidance*,
Springer Aerospace Technology, DOI 10.1007/978-3-319-25540-8

It is important to notice that, from a Lagrangian viewpoint, the only independent variables are the q_i's, while the generalized velocities $\dot{q}_i$ are simply their derivatives. In Hamiltonian mechanics, on the contrary, both halves of the set of canonical variables, q_i and p_i, are independent and equally significant.

The motion is therefore represented in a $2N$-dimensional space, which is called *phase space*.

C.2 Some Properties of the Hamiltonian Systems

Proposition

If the Lagrangian $\mathscr{L}$ is expressed as a sum of homogeneous, zero, first or second degree functions of the generalized velocities $\dot{q}_i$,
the generalized coordinates do not depend on time in their definition, all forces can be derived from a conservative potential $\mathscr{V}$;
Then: the Hamiltonian function represents the total energy: $H = \mathscr{T} + \mathscr{V} = \mathscr{E}$

A property of Hamiltonian systems is related to *cyclic variables*. A canonical variable is said to be *cyclic* if the Hamiltonian function does not contain it explicitly.

Property 1 *The conjugate momentum corresponding to a cyclic variable is a first integral of motion.*

Actually, if the q_j coordinate does not appear explicitly in H, the corresponding Hamilton equation for the conjugate momentum gives:

$$\dot{p}_j = -\frac{\partial H}{\partial q_j} = 0 \rightarrow p_j = \text{constant} \tag{C.4}$$

The above mentioned Property 1 shows how effective the Hamiltonian formulation can be for the solution of problems containing some cyclic coordinates, because the actual number of degrees of freedom reduces automatically.

In fact:

$$\begin{cases} \frac{dH}{dt} = \frac{\partial H}{\partial q_i}\dot{q}_i + \frac{\partial H}{\partial p_i}\dot{p}_i + \frac{\partial H}{\partial t} \\ \dot{q}_i = \frac{\partial H}{\partial p_i} \\ \dot{p}_i = -\frac{\partial H}{\partial q_i} \end{cases} \rightarrow \frac{dH}{dt} = \frac{\partial H}{\partial t} = 0 \tag{C.5}$$

C.3 Canonical Transformations

In many cases the solution of the $2N$ Hamilton equations of motion can be extremely simplified by introducing some coordinate transformations in order to obtain a system in which H is a constant and all the generalized coordinates q_i are cyclic variables. Then all conjugate momenta are constant and all generalized coordinates are expressed as linear functions of time:

$$\begin{cases} p_i = \alpha_i \\ \dot{q}_i = \frac{\partial H}{\partial p_i}|_{p=\alpha} = \omega_i \end{cases} \quad i = 1, \ldots, N \tag{C.6}$$

So

$$q_i = \omega_i t + \beta_i \tag{C.7}$$

where β_i are the initial conditions for q_i.

Unlike the Lagrangian formulation, not all coordinate transformations lead to a Hamiltonian system in the new variables. The general form of a transformation is

$$\begin{cases} Q_i = Q_i\,(q, p, t) \\ P_i = P_i\,(q, p, t) \end{cases} \tag{C.8}$$

where Q_i, P_i are the new generalized coordinates and momenta.

Among all the possible transformations of this form we are only interested in the ones for which Q_i, P_i are *canonical variables*, so that a new Hamiltonian function $K = K\,(Q, P, t)$ exists such that the dynamics in the new variables is:

$$\begin{cases} \dot{Q}_i = \frac{\partial K}{\partial P_i} \\ \dot{P}_i = -\frac{\partial K}{\partial Q_i} \end{cases} \quad i = 1, \ldots, N \tag{C.9}$$

A transformation that brings a Hamiltonian system into another Hamiltonian system is a *canonical transformation*; it also has to be invertible.

Canonical transformations are obtained by *generating functions*.

For orbital problems the typical form of a canonical transformation is:

$$\begin{cases} p_i = \frac{\partial F_2}{\partial q_i} \\ Q_i = \frac{\partial F_2}{\partial P_i} \\ K = H + \frac{\partial F_2}{\partial t} \end{cases} \tag{C.10}$$

where $F_2(q_i, P_i)$ is a *generating function* of type 2 (refer to Goldstein 1980 for the other types of generating functions), any regular function depending on the old coordinates q_i and the new momenta P_i is eligible as *generating function* of type 2.

C.4 Hamiltonian Approach to Orbit Motion

The Lagrangian function for the restricted two-body orbit problem can be written in spherical coordinates:

$$\mathscr{L} = \mathscr{T} - \mathscr{V} = \frac{\dot{r}^2 + r^2\dot{\theta}^2 + r^2 cos^2\theta\dot{\phi}^2}{2} + \frac{\mu}{r}$$

The generalized momenta p_i associated to the coordinates r, θ, ϕ are:

$$p_i = \frac{\partial \mathscr{L}}{\partial \dot{q}_i} \quad \rightarrow \quad \begin{cases} p_1 = \dot{r} \\ p_2 = r^2\dot{\theta} \\ p_3 = r^2 cos^2\theta\dot{\phi} \end{cases}$$

The Hamiltonian is found, through the application of a Legendre transformation on the Lagrangian:

$$H(q, p, t) = \dot{q}_i p_i - \mathscr{L}(q, \dot{q}, t)$$

the Hamiltonian turns out to be:

$$H(q, p) = \frac{\dot{r}^2 + r^2\dot{\theta}^2 + r^2 cos\theta^2\dot{\phi}^2}{2} - \frac{\mu}{r} = \mathscr{T} + \mathscr{V} = \mathscr{E}$$

and explicitly using the momenta:

$$H(q, p) = \frac{p_1^2 + \frac{p_2^2}{r^2} + \frac{p_3^2}{r^2 cos\theta^2}}{2} - \frac{\mu}{r}$$

From this Hamiltonian, using the method of the generating function we can construct any new set of canonical variables like the Hill variables.
The formulation of Kepler's problem in Hill variables can be found through the application of a canonical transformation using the *generating function* of type 2 $S(r, \theta, \phi, p, G, H)$ (see also to [1]):

$$S = rp + \int_0^\theta \sqrt{G^2 - \frac{H^2}{cos\theta^2}} d\theta + H\phi$$

from the set $(r, \theta, \phi, p_1, p_2, p_3)$ to a new set in which the momenta are

$$\begin{cases} p = \dot{r} \\ G = r^2\sqrt{\dot{\phi}^2 cos^2\theta + \dot{\theta}^2} \\ H = r^2 cos^2\theta\dot{\phi} \end{cases}$$

The new coordinates and the transformed Hamiltonian are then found using Eq. C.10 (see also to [2]):

$$\begin{cases} \frac{\partial S}{\partial r} = r \\ \frac{\partial S}{\partial G} = g + \nu \\ \frac{\partial S}{\partial H} = h \end{cases}$$

The new form of the Hamiltonian in the Hill coordinates is:

$$F = \mathscr{E} = -\frac{1}{2}\left(-p^2 + 2\frac{\mu}{r} - \frac{G^2}{r^2}\right) \tag{C.11}$$

We can write the Kepler problem in the Hill variables simply using C.1, therefore G, H and h are constants of motion together with the Hamiltonian itself.

The full set of the Hamilton equations for Hill variables is:

$$\begin{cases} \dot{r} = \frac{\partial F}{\partial p} = p \\ \dot{u} = \frac{\partial F}{\partial G} = \frac{G}{r^2} \\ \dot{h} = \frac{\partial F}{\partial H} = 0 \\ \dot{p} = -\frac{\partial F}{\partial r} = \frac{G^2}{r^3} - \frac{\mu}{r^2} \\ \dot{G} = -\frac{\partial F}{\partial u} = 0 \\ \dot{H} = -\frac{\partial F}{\partial h} = 0 \end{cases} \tag{C.12}$$

References

1. H. Goldstein, Classical Mechanics, Zanichelli (1980)
2. K. Aksnes, A second-Order Artificial Satellite Theory Based on an Intermediate Orbit, NASA Technical Report 32-15071 (1970).

Appendix D
The J_2 Perturbation Solution

In this chapter we give an example of the use of the Hamiltonian method to get series expansions of the solution of a perturbation problem. Following a classic paper [1] we present here a *first order* solution of the J_2 zonal problem obtained with the Hamiltonian techniques. We do not calculate all the transformations because our aim is to show the methodology and the properties of the final solution (all the detailed information can be found in [1])

A perturbation problem can be written by defining the Hamiltonian as the sum of the unperturbed term F_0 plus a small perturbation term F_1.

The expression of Earth gravity potential energy in Eq. 3.91 is therefore simplified to

$$U = -\frac{\mu}{r}\left[1 - J_2\left(\frac{R_e}{r}\right)^2 \frac{1}{2}\left(3\sin^2(\mathfrak{L}) - 1\right)\right] \tag{D.1}$$

Defining U_0 the unperturbed energy potential term, then:

$$U = U_0 + R$$

with R the perturbing term.

The Hamiltonian of the J_2 zonal problem is then

$$F = \mathscr{E} = \mathscr{E}_0 + R \quad \text{with} \begin{cases} \mathscr{E}_0 = \frac{v^T v}{2} - \frac{\mu}{|x|} = -\frac{\mu}{2a} \text{ energy of the unperturbed motion} \\ R = \text{perturbing Hamiltonian term} \end{cases} \tag{D.2}$$

The solution is developed in the classical set of *Delaunay variables*, because it is a canonical set where all the generalized coordinates are cyclic with respect to the Hamiltonian of the unperturbed Keplerian motion.

Delaunay variables L, G, H, l, g, h are related to the classical orbit elements set as follows:

© Springer International Publishing Switzerland 2016

L. Mazzini, *Flexible Spacecraft Dynamics, Control and Guidance*,
Springer Aerospace Technology, DOI 10.1007/978-3-319-25540-8

Generalized momenta	Generalized coordinates
$L = \sqrt{\mu a}$	$l =$ mean anomaly
$G = L\sqrt{1-e^2}$	$g = \omega =$ argument of perigee
$H = G\cos I$	$h = \Omega =$ right ascension of the ascending node

where the meaning of variables G and H is the same as for Hill variables, i.e. the modulus and the Z_{ECI} projection of angular momentum, respectively.

Substituting the new variables in the J_2 disturbing potential, leads to:

$$R = -\frac{\mu k_2}{r^3}\left(1 - 3\sin^2 \mathfrak{L}\right) = -\frac{\mu^4 k_2}{L^6}\left[\left(-\frac{1}{2} + \frac{3}{2}\frac{H^2}{G^2}\right)\left(\frac{a}{r}\right)^3 + \left(\frac{3}{2} - \frac{3}{2}\frac{H^2}{G^2}\right)\left(\frac{a}{r}\right)^3 \cos\left(2g + 2f\right)\right] \tag{D.3}$$

where $k_2 = \frac{1}{2}R_e^2 J_2$. The terms not yet expressed in Delaunay variables can be expanded in Fourier series as follows:

$$\begin{cases} \left(\frac{a}{r}\right)^3 = \left(\frac{L}{G}\right)^3 + \sum\limits_{j=1}^{\infty} 2P_j \cos jl = \left(\frac{L}{G}\right)^3 + \sigma_1 \\ \left(\frac{a}{r}\right)^3 \cos\left(2g + 2f\right) = \sum\limits_{j=-\infty}^{+\infty} Q_j \cos\left(2g + jl\right) = \sigma_2 \end{cases} \tag{D.4}$$

where the coefficients P_j e Q_j depend on eccentricity only (see Brouwer 1959 for the details).

The expression of the Hamiltonian in terms of Delaunay variables is then

$$F = -\frac{\mu^2}{2L^2} - \frac{\mu^4 k_2}{L^6}\left\{\left(-\frac{1}{2} + \frac{3}{2}\frac{H^2}{G^2}\right)\left[\left(\frac{L}{G}\right)^3 + \sum_{j=1}^{\infty} 2P_j \cos jl\right] + \left(\frac{3}{2} - \frac{3}{2}\frac{H^2}{G^2}\right)\sum_{j=-\infty}^{+\infty} Q_j \cos\left(2g + jl\right)\right\} \tag{D.5}$$

The equations of motion are then simply given by Hamilton equations:

$$\begin{cases} \frac{dL}{dt} = -\frac{\partial F}{\partial l} \\ \frac{dG}{dt} = -\frac{\partial F}{\partial g} \\ \frac{dH}{dt} = -\frac{\partial F}{\partial h} \end{cases} \quad \begin{cases} \frac{dl}{dt} = \frac{\partial F}{\partial L} \\ \frac{dg}{dt} = \frac{\partial F}{\partial G} \\ \frac{dh}{dt} = \frac{\partial F}{\partial H} \end{cases} \tag{D.6}$$

The Hamiltonian F reveals two main properties of the dynamical system:

1. The angle coordinate h doesn't explicitly appear in F, therefore it is cyclic. This means that the momentum coordinate H is an integral of motion.
2. The Hamiltonian does not depend explicitly on time, so it is an integral of motion itself.
3. The Hamiltonian when k_2 is null becomes cyclic in the angular variables.

The classical solution method of this Hamiltonian system is based on the sequential application of two canonical transformations, with the objective to obtain a system in trivial form, where all the three angle variable are cyclic.

Each transformation is described by a generating function of type 2, depending on the old angles and the new momenta.

Both transformations will be close to the identity transformation apart from factors proportional to k_2 and ${k_2}^2$ which will be used to make the dynamics in the new coordinates cyclic in the angular variables (l, g, h).

We will use the Brouwer notation to express the cyclic variables in the Hamiltonian with a $-$ sign, then, the Hamiltonian of the J_2 problem is $F = F\,(L, G, H, l, g, -)$. We also use the prime apex $'$ for the transformed variables and the star apex $*$ for the transformed Hamiltonian, so the equations of the generic canonical transformation are:

$$\begin{cases} L = \frac{\partial S}{\partial l} \\ G = \frac{\partial S}{\partial g} \\ H = \frac{\partial S}{\partial h} \end{cases} \begin{cases} l' = \frac{\partial S}{\partial L'} \\ g' = \frac{\partial S}{\partial G'} \\ h' = \frac{\partial S}{\partial H'} \end{cases} \tag{D.7}$$

$$\mathrm{F}^*\left(L', G', H', l', g', -\right) = \mathrm{F}\,(L, G, H, l, g, -)$$

and the new equations of motion are

$$\begin{cases} \frac{dL'}{dt} = -\frac{\partial F^*}{\partial l'} \\ \frac{dG'}{dt} = -\frac{\partial F^*}{\partial g'} \\ \frac{dH'}{dt} = -\frac{\partial F^*}{\partial h'} \end{cases} \begin{cases} \frac{dl'}{dt} = \frac{\partial F^*}{\partial L'} \\ \frac{dg'}{dt} = \frac{\partial F^*}{\partial G'} \\ \frac{dh'}{dt} = \frac{\partial F^*}{\partial H'} \end{cases} \tag{D.8}$$

The two canonical transformations to be applied have to eliminate respectively the Hamiltonian's dependency upon mean anomaly l' and the argument of perigee g'. The application of such transformations is equivalent to an *averaging* operation on the characteristic period of the new-cyclic variable.
The elimination of l' from the Hamiltonian is therefore equivalent to an averaging on the *short period*, that is the time interval in which the mean anomaly completes one revolution (i.e. orbit period). The second transformation acts on the *long period*, eliminating all the variations whose period is of the order of magnitude of the time in which the argument of perigee completes one full rotation due to the J_2 effect.

After such manipulation, the double-averaged dynamics will not contain periodic terms anymore: the transformed variables will be constant or linear in time. After solving the trivial equations of motion in double-averaged variables, the inverse canonical transformations will have to be applied to re-introduce the long- and short-period dynamics.

The generating functions of the two transformations, that will be called S and S^* from now on, have to be identified by imposing the new Hamiltonian to be cyclic in the new variables.

A classical method to find the canonical transformations is Von Zeipel's, that consists in writing the generating function and the transformed Hamiltonian as power series of the small perturbation parameter (here k_2), writing the canonical transformation equation while imposing the cyclic variable not to appear in the transformed Hamiltonian, then finally identifying the series coefficients of the expanded generating function by direct comparison of the terms of the same order appearing on the two sides of the transformation equation. The accuracy of the final solution depends on the order of truncation of such series (i.e. order of the perturbation method), that is the level of detail in the characterization of the two canonical transformations.

Considering the first transformation with generating function S, we have

$$\begin{cases} S = S_0 + S_1 + S_2 ... \\ F^* = F_0^* + F_1^* + F_2^* ... \end{cases} \quad \text{with } S_0 = L'l + G'g + H'h \tag{D.9}$$

The subscript 0, 1, 2 in the preceding expressions means that the term is multiplied by the same power of k_2.

To impose the cyclic condition in the new variable l' we use the relationship connecting the new and the original Hamiltonian, using the canonical transformation to write L, G, H, l', g', h':

$$F\left(L, G, H, l, g, -\right) = F^*\left(L', G', H', -, g', -\right) \tag{D.10}$$

$$\Downarrow$$

$$F_0\left(\frac{\partial S}{\partial l}\right) + F_1\left(\frac{\partial S}{\partial l}, \frac{\partial S}{\partial g}, \frac{\partial S}{\partial h}, l, g, -\right) = F_0^* + F_1^*\left(L', G', H', -, \frac{\partial S}{\partial G'}, -\right) + F_2^*\left(L', G', H', -, \frac{\partial S}{\partial G'}, -\right) \tag{D.11}$$

The determination of the generating function is carried out through the direct comparison of the terms of each order on both sides of the last expression, after the Taylor series expansion:

$$\begin{aligned} &F_0\left(L'\right) + \frac{\partial F_0}{\partial L'}\frac{\partial S_1}{\partial l} + \frac{\partial F_0}{\partial L'}\frac{\partial S_2}{\partial l} + \frac{1}{2}\frac{\partial^2 F_0}{\partial L'^2}\left(\frac{\partial S_1}{\partial l}\right)^2 \\ &+ F_1\left(L', G', H', l, g, -\right) + \frac{\partial F_1}{\partial L'}\frac{\partial S_1}{\partial l} + \frac{\partial F_1}{\partial G'}\frac{\partial S_1}{\partial g} \\ &= F_0^* + F_1^*\left(L', G', H', -, g, -\right) + \frac{\partial F_1^*}{\partial g}\frac{\partial S_1}{\partial G'} + F_2^*\left(L', G', H', -, g, -\right) \end{aligned} \tag{D.12}$$

The following equalities (the functions must be considered dependent on (L', G', H', l, g, h)) are then found:

Order 0: $F_0\left(L'\right) = F_0^*\left(L'\right)$
Order 1: $\frac{\partial F_0}{\partial L'}\frac{\partial S_1}{\partial l} + F_1 = F_1^*$
Order 2: $\frac{\partial F_0}{\partial L'}\frac{\partial S_2}{\partial l} + \frac{1}{2}\frac{\partial^2 F_0}{\partial L'^2}\left(\frac{\partial S_1}{\partial l}\right)^2 + \frac{\partial F_1}{\partial L'}\frac{\partial S_1}{\partial l} + \frac{\partial F_1}{\partial G'}\frac{\partial S_1}{\partial g} = F_2^* + \frac{\partial F_1^*}{\partial g}\frac{\partial S_1}{\partial G'}$

In the equation at order 1 it is possible to express F_1 as the sum of one term containing the variable to be eliminated l plus one term that is independent on l. The first term has the meaning of the periodic portion of the first order perturbation potential, while the second one represents the secular part:

$$F_1 = F_{1p} + F_{1s} \tag{D.13}$$

Since F_1^* has to be independent on l, the equation at order 1 can be split into:

$$\begin{cases} \frac{\partial S_1}{\partial l} = -\frac{1}{\frac{\partial F_0}{\partial L'}} F_{1p} \\ F_1^* = F_{1s} \end{cases} \tag{D.14}$$

All the quantities on the right side are known, so S_1 can be found by integration.

With the generating function in closed form, the transformation equations at order 1 can be expressed:

$$\begin{cases} L = L' + \frac{\partial S_1}{\partial l} \\ G = G' + \frac{\partial S_1}{\partial g} \\ H = H' \end{cases} \quad \begin{cases} l = l' - \frac{\partial S_1}{\partial L'} \\ g = g' - \frac{\partial S_1}{\partial G'} \\ h = h' - \frac{\partial S_1}{\partial H'} \end{cases} \tag{D.15}$$

For the complete expressions of the generating function and transformation equations see the work of [1].

Since we consider a *first order* theory, the periodic terms will be dropped in the order 2 equation, without contributing to S_2. Instead the secular term of the order 2 equation is included in the definition of the new Hamiltonian:

$$F_2^* = \left[\frac{1}{2}\frac{\partial^2 F_0}{\partial L'^2}\left(\frac{\partial S_1}{\partial l}\right)^2 + \frac{\partial F_{1p}}{\partial L'}\frac{\partial S_1}{\partial l} + \frac{\partial F_1}{\partial G'}\frac{\partial S_1}{\partial g}\right]\Bigg|_s \tag{D.16}$$

The last step shows the meaning of *first order* theory: the terms representing secular perturbations are developed in series up to k_2^2, while the periodic ones only up to k_2.

The transformed system with Hamiltonian F^* is not yet in trivial form, but it contains only secular and long-period terms. To eliminate the long-period dynamics, the second transformation generating function S^* has to be found:

$$F^* = F_0^* + F_1^* + F_2^* \tag{D.17}$$

with $F_2^* = F_{2s}^* + F_{2p}^*$ where the periodic part contains the dynamics of the average argument of perigee g'.

By writing the new generating function and the new Hamiltonian, and applying the same method as done before:

$$\begin{cases} S^* = S_0^* + S_1^* \\ F^{**} = F_0^{**} + F_1^{**} + F_2^{**} \end{cases} \text{with } S_0^* = L''l' + G''g' + H''h' \tag{D.18}$$

developing the equality:

$$F^*\left(L', G', H', -, g', -\right) = F^{**}\left(L'', G'', H'', -, -, -\right) \tag{D.19}$$

the relations connecting the components of F^{**} and S^* with F^* are found:

Order 0: $F_0^* = F_0^{**}$

Order 1: $F_1^* = F_1^{**}$

Order 2: $\begin{cases} F_{2p}^* + \frac{\partial F_1^*}{\partial G''}\frac{\partial S_1^*}{\partial g'} = 0 \\ F_{2s}^* = F_2^{**} \end{cases}$

The expression of S_1^* is found, by integration of the periodic part of the last equation:

$$\frac{\partial S_1^*}{\partial g'} = -\frac{1}{\frac{\partial F_1^*}{\partial G''}} F_{2p}^* \tag{D.20}$$

therefore we have:

$$\begin{cases} L' = L'' \\ G' = G'' + \frac{\partial S_1^*}{\partial g'} \\ H' = H'' = H \end{cases} \quad \begin{cases} l' = l'' - \frac{\partial S_1^*}{\partial L''} \\ g' = g'' - \frac{\partial S_1^*}{\partial G''} \\ h' = h'' - \frac{\partial S_1^*}{\partial H''} \end{cases} \tag{D.21}$$

The Hamiltonian of the transformed system depends now exclusively on generalized momenta, so we have the trivial solution:

$$\begin{cases} L'' = L' = L_0' \\ G'' = G_0'' \\ H'' = H = H_0 \end{cases} \quad \begin{cases} l'' = l_0'' + n_l t \\ g'' = g_0'' + n_g t \\ h'' = h_0'' + n_h t \end{cases} \tag{D.22}$$

Where n_l, n_g, n_h are the constant values of the derivatives of F^{**} with respect to the three momenta.

$$\begin{aligned} n_l &= \frac{\partial F^{**}}{\partial L''} \\ n_g &= \frac{\partial F^{**}}{\partial G''} \\ n_h &= \frac{\partial F^{**}}{\partial H''} \end{aligned} \tag{D.23}$$

The motion described by such solution contains only secular variations, so the double-primed Delaunay variables are also called *mean* variables, where two kinds of periodicity have been eliminated (long and short-term).

Now that we have a closed form solution in double primed variables, the two inverse canonical transformations have to be applied to reintroduce the periodic components.

After the application of the two inverse transformations, we have the following analytical equations providing the final state at time t as a function of the initial conditions in terms of double-primed coordinates. For convenience, the classical elements a, e, I are used in these equations, instead of the generalized momenta L, G, H, and the sequential substitutions corresponding to the inverse canonical transformations are highlighted.

The parameters contained in the expression have the following definitions:

$$n_0 = \sqrt{\frac{\mu}{a''^3}}; \quad \eta = \sqrt{1 - e''^2}; \quad \theta = \cos I'' \tag{D.24}$$

$$\gamma_2 = \frac{k_2}{a''^2} \quad ; \gamma_2' = \frac{\gamma_2}{\eta^4} \tag{D.25}$$

secular dynamics

$$\begin{aligned} a'' &= a_0'' \\ e'' &= e_0'' \\ I'' &= I_0'' \\ l'' &= n_l t + l_0'' \\ g'' &= n_g t + g_0'' \\ h'' &= n_h t + h_0'' \end{aligned} \tag{D.26}$$

With

$$\begin{aligned} n_l &= n_0 \left\{1 + \tfrac{3}{2}\gamma_2'\eta\left(-1 + 3\theta^2\right) + \tfrac{3}{32}\gamma'^2_2\eta\left[-15 + 16\eta + 25\eta^2 \right.\right. \\ &\qquad \left.\left. + \left(30 - 96\eta - 90\eta^2\right)\theta^2 + \left(105 + 144\eta + 25\eta^2\right)\theta^4\right]\right\} \\ n_g &= n_0 \left\{\tfrac{3}{2}\gamma_2'\left(-1 + 5\theta^2\right) + \tfrac{3}{32}\gamma'^2_2\left[-35 + 24\eta + 25\eta^2 \right.\right. \\ &\qquad \left.\left. + \left(90 - 192\eta - 126\eta^2\right)\theta^2 + \left(385 + 360\eta + 45\eta^2\right)\theta^4\right]\right\} \\ n_h &= n_0 \left\{-3\gamma_2'\theta + \tfrac{3}{8}\gamma'^2_2\left[\left(-5 + 12\eta + 9\eta^2\right)\theta + \left(-35 - 36\eta - 5\eta^2\right)\theta^3\right]\right\} \end{aligned} \tag{D.27}$$

long term dynamics

$$a' = a''$$

$$e' = e'' + \tfrac{1}{8}\gamma_2' e''\eta^2\left(1 - 11\theta^2 - \tfrac{40\theta^4}{1-5\theta^2}\right)\cos 2g''$$

$$I' = I'' - \tfrac{1}{8}\gamma_2' e''^2 \tan I''^{-1}\left(1 - 11\theta^2 - \tfrac{40\theta^4}{1-5\theta^2}\right)\cos 2g''$$

$$l' = l'' + \left\{\tfrac{1}{8}\gamma_2'\eta^3\left[1 - 11\theta^2 - 40\theta^4\left(1-5\theta^2\right)^{-1}\right]\right\}\sin 2g''$$

$$\begin{aligned} g' = g'' + \Big\{-\tfrac{1}{16}\gamma_2'\Big[&2 + e''^2 - 11\left(2+3e''^2\right)\theta^2 - 40\left(2+5e''^2\right)\theta^4\left(1-5\theta^2\right)^{-1} \\ &- 400e''^2\theta^6\left(1-5\theta^2\right)^{-2}\Big]\Big\}\sin 2g'' \end{aligned}$$

$$h' = h'' + \left\{-\tfrac{1}{8}\gamma_2' e''^2\theta\left[11 + 80\theta^2\left(1-5\theta^2\right)^{-1} + 200\theta^4\left(1-5\theta^2\right)^{-2}\right]\right\}\sin 2g'' \tag{D.28}$$

short term dynamics

$$a = a''\left\{1 + \gamma_2\left[\left(-1+3\theta^2\right)\left(\tfrac{a''^3}{r'^3} - \eta^{-3}\right) + 3\left(1-\theta^2\right)\tfrac{a''^3}{r'^3}\cos\left(2g'+2f'\right)\right]\right\}$$

$$\begin{aligned} e = e' + \tfrac{\eta^2}{2e''}\Big\{&\gamma_2\left[\left(-1+3\theta^2\right)\left(\tfrac{a''^3}{r'^3} - \eta^{-3}\right) + 3\left(1-\theta^2\right)\left(\tfrac{a''^3}{r'^3} - \eta^{-4}\right)\cos\left(2g'+2f'\right)\right] \\ &- \gamma_2'\left(1-\theta^2\right)\left[3e''\cos\left(2g'+f'\right) + e''\cos\left(2g'+3f'\right)\right]\Big\} \end{aligned}$$

$$I = I' + \tfrac{1}{2}\gamma_2'\theta\left(1-\theta^2\right)^{\frac{1}{2}}\left[3\cos\left(2g'+2f'\right) + 3e''\cos\left(2g'+f'\right) + e''\cos\left(2g'+3f'\right)\right]$$

$$\begin{aligned} l = l' - \tfrac{\eta^3}{4e''}\gamma_2'\Big\{&2\left(-1+3\theta^2\right)\left(\tfrac{a''^2}{r'^2}\eta^2 + \tfrac{a''}{r'} + 1\right)\sin f' \\ &+ 3\left(1-\theta^2\right)\left[\left(-\tfrac{a''^2}{r'^2}\eta^2 - \tfrac{a''}{r'} + 1\right)\sin\left(2g'+f'\right) + \left(\tfrac{a''^2}{r'^2}\eta^2 + \tfrac{a''}{r'} + \tfrac{1}{3}\right)\sin\left(2g'+3f'\right)\right]\Big\} \end{aligned}$$

$$\begin{aligned} g = g' + \tfrac{\eta^2}{4e''}\gamma_2'\Big\{&2\left(-1+3\theta^2\right)\left(\tfrac{a''^2}{r'^2}\eta^2 + \tfrac{a''}{r'} + 1\right)\sin f' \\ &+ 3\left(1-\theta^2\right)\left[\left(-\tfrac{a''^2}{r'^2}\eta^2 - \tfrac{a''}{r'} + 1\right)\sin\left(2g'+f'\right) + \left(\tfrac{a''^2}{r'^2}\eta^2 + \tfrac{a''}{r'} + \tfrac{1}{3}\right)\sin\left(2g'+3f'\right)\right]\Big\} \\ &+ \tfrac{1}{4}\gamma_2'\Big\{6\left(-1+5\theta^2\right)\left(f' - l' + e''\sin f'\right) \\ &+ \left(3-5\theta^2\right)\left[3\sin\left(2g'+2f'\right) + 3e''\sin\left(2g'+f'\right) + e''\sin\left(2g'+3f'\right)\right]\Big\} \end{aligned}$$

$$h = h' - \tfrac{1}{2}\gamma_2'\theta\left[6\left(f' - l' + e''\sin f'\right) - 3\sin\left(2g'+2f'\right) - 3e''\sin\left(2g'+f'\right) - e''\sin\left(2g'+3f'\right)\right] \tag{D.29}$$

where f' and r' are defined by

$$
\begin{aligned}
&E' - e'' \sin E' = l' \\
&\tan \tfrac{f'}{2} = \left(\tfrac{1+e''}{1-e''}\right)^{\frac{1}{2}} \tan \tfrac{E'}{2} \\
&\tfrac{a''}{r'} = \tfrac{1+e'' \cos f'}{1-e''^2}
\end{aligned}
$$

Reference

1. D. Brouwer, Solution of The Problem of Artificial Satellite Theory without Drag, The Astronomical Journal 64, No.1274 (1959) 378-396

Appendix E
Complements to Optimal Orbital Transfer

The state evolution equations can be integrated for any impulsive thrust ($w = const.$). The formulas derive from the evolution of the Hill parameters during a maneuver. Being the initial state X_0, the speed change (V_r, V_t, V_n) the final state can be found with the following formulas:

$$
\begin{aligned}
u &= w + X_6 \\
Y_1 &= Y_{1,0} \\
1/X_3 &= (1/X_{3,0} + 2V_t/(Y_1\sqrt{X_{3,0}}) + (V_t^2 + V_n^2)/Y_1^2) \\
X_4 &= X_{4,0}\sqrt{X_3/X_{3,0}} + X_{4,0}\sqrt{X_3}\,V_t/Y_1 - \sqrt{1 - X_{4,0}^2}\sqrt{X_3}\cos u_0 V_n/Y_1 \\
\sin u &= \sin u_0\sqrt{1 - X_{4,0}^2}/\sqrt{1 - X_4^2} \\
X_6 &= u - w \\
X_5 &= X_{5,0} + \arcsin\left(X_{4,0}\sin u_0/\sqrt{\cos u_0{}^2 + X_{4,0}^2 \sin u_0{}^2}\right) \\
&\quad - \arcsin\left(X_4 \sin u/\sqrt{\cos u^2 + X_4^2 \sin u^2}\right) \\
Y_2 &= \sqrt{X_3/X_{3,0}}\,Y_{2,0} - \sqrt{X_3}\,V_r \\
X_1 &= (Y_1 - X_3)\cos u - Y_2 \sin u \\
X_2 &= (Y_1 - X_3)\sin u + Y_2 \cos u
\end{aligned}
\tag{E.1}
$$

The co-state can be computed during the impulsive firing considering that the vector $\mathfrak{l}(V)$ satisfies a very simple integrable equation:

$$
\begin{aligned}
\frac{d\mathfrak{l}_1}{dV} &= 0 \\
\frac{d\mathfrak{l}_2}{dV} &= \mathfrak{l}_3^2\frac{\sqrt{X_3}}{Y_1} \\
\frac{d\mathfrak{l}_3}{dV} &= -\mathfrak{l}_3\mathfrak{l}_2\frac{\sqrt{X_3}}{Y_1}
\end{aligned}
\tag{E.2}
$$

© Springer International Publishing Switzerland 2016
L. Mazzini, *Flexible Spacecraft Dynamics, Control and Guidance*,
Springer Aerospace Technology, DOI 10.1007/978-3-319-25540-8

This equation represents a rotation of the firing vector around the direction of the Earth. During the firing X_3 and X_4 vary and the orbital frame rotates, being the impulsive firing inertially fixed, the $\mathfrak{l}(V)$ unit vector must rotate. Because Y_1 remains fixed during the firing and the variation of X_3 is known from the previous equations:

$$\frac{1}{X_3} = \frac{1}{X_{3,0}} + 2\frac{\mathfrak{l}_2 V}{\sqrt{X_3 Y_1}} + (\mathfrak{l}_2^2 + \mathfrak{l}_3^2)\frac{V^2}{Y_1^2} \tag{E.3}$$

the solution for the firing vector is found to be:

$$\begin{aligned}
\mathfrak{l}_1 &= \mathfrak{l}_{1,0} \\
\mathfrak{l}_2 &= \mathfrak{l}_{2,0}\sqrt{1 + \frac{\mathfrak{l}_{3,0}^2}{\mathfrak{l}_{2,0}^2}\left(1 - \frac{X_3}{X_{3,0}}\right)} \\
\mathfrak{l}_3 &= \mathfrak{l}_{3,0}\frac{\sqrt{X_3}}{\sqrt{X_{3,0}}}
\end{aligned}$$

The co-state $L(V)$ at the beginning of the firing arc is necessary to fix the initial conditions of $l(V)$ with $\mathfrak{l} = B(X, w^*)^T L$, then $\mathfrak{l}(V)$ is determined from the above equations.

Also the evolution of $L(V)$ during the firing can be determined noting that L_5 and L_6 are constant and using the following linear algebraic system:

$$\begin{aligned}
\mathfrak{l} &= B(X, w^*)^T L \\
0 &= \mathfrak{l}^T \frac{\partial B(X, w^*)^T}{\partial w} L
\end{aligned} \tag{E.4}$$

The second equation is derived by:

$$\frac{\partial \widetilde{H}(X, L, w^*)}{\partial w} = 0 \tag{E.5}$$

By direct calculation the linear algebraic system has rank 4 apart in singular points where:

$$\mathfrak{l}_1 \cos(u)\sqrt{1 - {X_4}^2} Y_1 + \mathfrak{l}_2 \cos(u)\sqrt{1 - {X_4}^2} Y_2 + \mathfrak{l}_3 Y_2 X_4 = 0 \tag{E.6}$$

Appendix F
Elements of Plasma Physics

F.1 Maxwell's Equations

Maxwell's equations describing electromagnetic fields are:

$$\nabla \cdot E = \frac{\rho_q}{\varepsilon_0} \tag{F.1}$$

$$\nabla \cdot B = 0 \tag{F.2}$$

$$\nabla \times E = -\frac{\partial B}{\partial t} \tag{F.3}$$

$$\nabla \times B = \mu_0 J + \mu_0 \varepsilon_0 \frac{\partial E}{\partial t}, \tag{F.4}$$

where E and B are the electric and magnetic fields, ρ_q and J (the charge density and the current density, respectively) are the sources of the fields, μ_0 is the magnetic permettivity in vacuum and ε_0 is the dielectric constant in vacuum.

F.2 Plasma Oscillations

Consider a distribution of ions and electrons with rigidly-fixed mutual distances in a quasi-neutral plasma (i.e. with $n_e \simeq n_i$, being n_e and n_i the density of electrons and ions, respectively). Electrons, being much lighter than ions, have a much faster response to local short-time fluctuations than ions. In response to field fluctuations, suppose they are subjected to a small displacement, ξ. An electric field will set on, which tends to bring back the electrons on ions. When electrons transit at the equilibrium position, their inertia will let them move by the amount ξ, until an

© Springer International Publishing Switzerland 2016

L. Mazzini, *Flexible Spacecraft Dynamics, Control and Guidance*,
Springer Aerospace Technology, DOI 10.1007/978-3-319-25540-8

electric force opposite the previous one will bring them back on ions. The resulting oscillations are known as *Langmuir waves*. Solving the 1-D Poisson equation F.1

$$\frac{\partial E_x}{\partial x} = -\frac{n_i e}{\varepsilon_0}, \tag{F.5}$$

being E_x the 1-D electric field, gives $E_x(\xi) = -\frac{n_i e}{\varepsilon_0}\xi$. Thus, the equation of motion is

$$n_i m_e \frac{d^2\xi}{dt^2} = -\frac{n_i^2 e^2}{\varepsilon_0}\xi. \tag{F.6}$$

By substituting the solution $\xi \propto exp(-i\omega t)$ into Eq. F.6 we find

$$-\omega^2 + \frac{n_i e^2}{m_e \varepsilon_0} = 0 \Rightarrow \omega = \sqrt{\frac{n_i e^2}{m_e \varepsilon_0}} \equiv \omega_e. \tag{F.7}$$

The frequency ω_e is the *plasma oscillation electron frequency*. In the same way we can define the *plasma oscillation ion frequency* as

$$\omega_i = \sqrt{\frac{n_i e^2 Z^2}{m_i \varepsilon_0}}, \tag{F.8}$$

where we have generalized to the case of ion charge *Ze*. The *plasma oscillation total frequency* is $\omega_p^2 = \omega_e^2 + \omega_i^2$. In most cases $\omega_e \gg \omega_i$ and then $\omega_p^2 \simeq \omega_e^2$.

F.3 Charged Particles in Electromagnetic Fields

The dynamics of a particle with mass m and charge q in an electromagnetic field is described by the equation of motion

$$m\frac{d^2r}{dt^2} = q\left(E + \frac{dr}{dt} \times B\right), \tag{F.9}$$

where $F = q(E + v \times B)$ is the Lorentz force, which can be written as $F = q(E + v_\perp \times B)$, where $v_\perp$ is the component of the vector velocity perpendicular to the magnetic field. The equation of motion can be analytically solved only in a few simple conditions.

Consider the case of no electric field and a magnetic field constant in direction, for example, $\hat{z}$, such that $B = B_0\hat{z}$. The equation of motion in the direction parallel to the magnetic field (being the velocity component in this direction $v_{||}$), gives $m\frac{dv_{||}}{dt} = 0$, which states that the motion is unperturbed along $\hat{z}$. In direction perpendicular to the magnetic field

$$m\frac{dv_\perp}{dt} = qB_0 v_\perp \times \hat{z}. \tag{F.10}$$

As the magnetic force is always perpendicular to the particle's velocity, the total kinetic energy of the particle must be constant. Moreover, as $v_{||} = constant$, it must be also $|v_\perp| = constant$. The particle moves following a uniform circular motion around the magnetic field lines. The solution of the equation of motion gives $v_x = |v_\perp| sin(\Omega t)$, $v_y = |v_\perp| cos(\Omega t)$, $v_z = v_{||} = constant$, where we introduced the *gyrofrequency* $\Omega = |q|B_0/m$, and the initial condition $v_x(t=0) = 0$. By integrating such relations we find the resulting helicoidal motion wrapped around the magnetic field lines, which corresponds to a circular motion orthogonal to the magnetic field, superposed to a uniform motion parallel to the field. We define the *gyroradius* as the position vector of the particle relative to the center of gyration around the magnetic field lines:

$$r_g = -\frac{m}{qB^2} v_\perp \times B. \tag{F.11}$$

Consider now the case of constant external electric and magnetic fields. Suppose the electric field has both parallel and perpendicular components to the magnetic field, namely $E = E_{0,x}\hat{x} + E_{0,z}\hat{z}$. The solution of the equations of motion gives $v_x = |v_\perp| sin(\Omega t)$, $v_y = |v_\perp| cos(\Omega t) - E_{0,x}/B_0$, and $v_z = qE_{0,z}t/m + v_{0,z}$. The difference from the case $E = 0$ is in the term $E_{0,x}/B_0$ in v_y, which represents a $E \times B$ drift. The resulting motion is helicoidal with a superposed drift perpendicular to both the electric and magnetic fields. The vectorial expression of the drift is given by

$$v_E = \frac{E_\perp \times B}{B_0^2}. \tag{F.12}$$

The drift motion introduced by the presence of both electric and magnetic fields is sometimes termed *Hall effect*. The drift with velocity v_E is sometimes termed *Hall current*, despite it not being a current, in a rigorously physical sense.
Consider the case of a magnetic field gradient along the field direction itself. Let us use a cylindrical geometry with (r, θ, z) the scalar coordinates and the axis $\hat{z}$ directed along the magnetic field. Suppose the poloidal component of the field to be null. Maxwell's equation F.2 can be written as

$$\frac{1}{r}\frac{\partial}{\partial r}(rB_r) + \frac{\partial B_z}{\partial z} = 0. \tag{F.13}$$

By the integration of F.13 we obtain, for $\nabla_{||} B_z$ slowly changing with r,

$$rB_r = -\frac{1}{2}r^2\frac{\partial B_z}{\partial z} \Rightarrow B_r = \frac{-r}{2}\frac{\partial B_z}{\partial z}. \tag{F.14}$$

The Lorentz force along the magnetic field is

$$F_z \equiv (qv \times B)_z = [q(v_r; v_\theta, v_z) \times (B_r; 0; B_z)]_z = -qv_\theta B_r = \frac{qv_\theta r}{2}\frac{\partial B_z}{\partial z}, \quad \text{(F.15)}$$

where we have used the result in F.14. The gyration motion of charged particles may be described by the velocity $v_\theta = -sign(q)|v_\perp|$. By substituting $r \rightarrow r_g$, $r_g = |v_\perp|/\Omega$, $\Omega = qB/m$ and $q = |q|sign(q)$, we find

$$F_z = -\frac{1}{2}\frac{m|v_\perp|^2}{B}\frac{\partial B_z}{\partial z}. \quad \text{(F.16)}$$

Thus, a force opposite to the field gradient arises. Consider a current loop made of a particle spiraling around a magnetic field, being A the area of the loop, and I the resulting current. A magnetic moment $\mu = IA$ is set. The current can be written as $I = q/T$, T being the gyration period $T = 2\pi/\Omega$. The area of the loop is $A = \pi r_g^2 = \pi(|v_\perp|/\Omega)^2$, from it follows that $\mu = m|v_\perp|^2/(2B)$, and the force due to the field gradient is $F_{||} = -\mu\nabla_{||}B$. By multiplying the equation of motion parallel to the field lines times $v_{||}$ we find

$$mv_{||}\frac{dv_{||}}{dt} = -\mu\frac{dB}{dt} \Rightarrow \frac{d}{dt}\left(\frac{1}{2}mv_{||}^2\right) = -\mu\frac{dB}{dt} \Rightarrow \frac{d}{dt}\left(\frac{1}{2}mv_{||}^2\right) + \mu\frac{dB}{dt} = 0. \quad \text{(F.17)}$$

From the conservation of kinetic energy and the expression found for μ we have

$$\frac{d}{dt}\left(\frac{1}{2}mv_{||}^2 + \mu B\right) = 0 \Rightarrow \frac{d}{dt}\left(\frac{1}{2}mv_{||}^2\right) + B\frac{d\mu}{dt} + \mu\frac{dB}{dt} = 0. \quad \text{(F.18)}$$

From F.17 and F.18 it must be that

$$\frac{d\mu}{dt} = 0 \Rightarrow \mu \equiv \frac{m|v_\perp|^2}{2B} = cost. \quad \text{(F.19)}$$

The important consequence of F.19 is that when a particle is moving toward an increasing magnetic strength, $|v_\perp|$ must also increase in order to let μ constant. An increase of $|v_\perp|$ corresponds to a decrease of $v_{||}$, due to the conservation of kinetic energy. If the magnetic field is strong enough, $v_{||}$ can reach zero, and as the gradient force continues to act, the particle comes to a halt and starts to move in the opposite direction. This effect induced by a magnetic field gradient is known as *mirroring* or *magnetic mirror*.

F.4 Debye Shielding and Quasi-Neutrality

A fundamental property of the plasma is the capability of shielding the electric potentials. Suppose to introduce an electric field in a plasma, $E = -\nabla\phi$, by introducing two spheres with charge $\pm Q$. Immediately, a cloud of ions will form around the negative sphere, and a cloud of electrons will form around the positive sphere.

In a hot plasma ($T \neq 0$) the particles of the clouds would have a thermal energy $\sim kT$; and the clouds would be thick due to the motion of particles. In response to thermal fluctuations, the particles at the boundaries of the clouds may have, at a given time, thermal energy large enough to escape the electric potential well.

Statistically speaking, the same number of plasma particles in the vicinity of the boundaries of the clouds will be captured by the cloud itself. We can define the boundary of the clouds as the distance from the charges Q at which $q\phi \simeq kT$. Thus, the potentials up to $\phi \simeq kT/q$ can penetrate the plasma and give rise to finite electric fields.

Consider a test particle with charge $q = +e$ in the center of a frame of reference. In the absence of plasma the potential would be that of Coulomb

$$\phi(r) = \frac{1}{4\pi\varepsilon_0}\frac{e}{r} \equiv \phi_C, \tag{F.20}$$

where r is the radial coordinate. In case of a plasma the potential may be computed from Poisson's equation F.1, with $E = -\nabla\phi$. It reads $\nabla^2\phi(r) = -\rho_q/\varepsilon_0$, being the charge density $\rho_q = \rho_i + \rho_e + e\delta^3(r)$ (the subscripts e and i stem for electrons and ions, respectively), and $\delta^3(r)$ is the δ-Dirac function. Under the assumptions that (i) electrons and ions reach the thermal equilibrium at temperatures T_e and T_i, respectively, with $T_e \neq T_i$; (ii) $e\phi \ll kT_{e,i}$ (weakly coupled plasma); and defining the *Debye length* as

$$\lambda_{D_{e,i}} \equiv \sqrt{\frac{\varepsilon_0 kT_{e,i}}{n_{e,i}e^2}}, \tag{F.21}$$

and the *total Debye length* λ_D with $\frac{1}{\lambda_D^2} = \frac{1}{\lambda_{D_e^2}} + \frac{1}{\lambda_{D_i^2}}$, we find

$$\phi(r) = \frac{1}{4\pi\varepsilon_0}\frac{q}{r}e^{-\frac{r}{\lambda_D}}. \tag{F.22}$$

This means that the electric potential in a plasma decays much faster than that in the vacuum. The Debye shielding introduces a condition for the plasma's quasi-neutrality. In fact, if the system has a characteristic size $L \gg \lambda_D$, and if the plasma is subjected to low-frequency motions (so that the particles have the time to effectively shield each other) then $n_e \simeq n_i \simeq n$.

Index

© Springer International Publishing Switzerland 2016
L. Mazzini, *Flexible Spacecraft Dynamics, Control and Guidance*,
Springer Aerospace Technology, DOI 10.1007/978-3-319-25540-8

CPSIA information can be obtained
at www.ICGtesting.com
Printed in the USA
LVOW01*1720091115
461730LV00001B/5/P

9 783319 255385